George Beadle
An Uncommon Farmer

The Emergence of Genetics in the 20th Century

George Beadle
An Uncommon Farmer

The Emergence of Genetics in the 20th Century

Paul Berg

Stanford University School of Medicine

Maxine Singer

Carnegie Institution of Washington

COLD SPRING HARBOR LABORATORY PRESS
Cold Spring Harbor, New York

George Beadle: An Uncommon Farmer
The Emergence of Genetics in the 20th Century

Printed in the United States of America

Publisher	John Inglis
Commissioning Editor	John Inglis
Developmental Editor	Judy Cuddihy
Project Coordinator	Mary Cozza
Permissions Coordinator	Nora Rice
Production Manager	Denise Weiss
Production Editor	Patricia Barker
Desktop Editor	Susan Schaefer
Copy Editor	Rena Steuer
Cover Designer	Denise Weiss and Michael Albano

Front Cover: George W. Beadle in lab coat. (Photo by Richard Hartt, Pasadena. Courtesy of the California Institute of Technology Archives.)

Title Page: George W. Beadle bagging corn plants. (Courtesy of the California Institute of Technology Archives.)

Back Cover: Beadle's hand-drawn description of experimental design for determining whether the loss of *Neurospora*'s ability to make pantothenic acid is due to a single mutation. The result shows that it is. (Courtesy of the California Institute of Technology Archives.)

Library of Congress Cataloging-in-Publication Data

Berg, Paul.
George Beadle An uncommon farmer : The emergence of genetics in the 20th century / Paul Berg and Maxine Singer.
p. cm.
Includes bibliographical references and index.
ISBN 0-87969-688-5 (alk. paper)
1. Beadle, George Wells, 1903- 2. Geneticists--United States--Biography. 3. Molecular biologists--United States-- Biography. I. Singer, Maxine. II. Title.

QH429.2.B37B37 2003
576.5′092--dc21
[B] 2003051635

10 9 8 7 6 5 4 3 2

All Cold Spring Harbor Laboratory Press publications may be ordered directly from Cold Spring Harbor Laboratory Press, 500 Sunnyside Blvd., Woodbury, NY 11797-2924. Phone: 1-800-843-4388 in Continental U.S. and Canada. All other locations: (516) 422-4100. FAX: (516) 422-4097. E-mail: cshpress@cshl.edu. For a complete catalog of all Cold Spring Harbor Laboratory Press publications, visit our World Wide Web Site http://www.cshlpress.com./

Contents

Preface

Isaac Newton's famous phrase reminds us that major advances in science are made "on the shoulders of giants." Too often, however, we are chagrined to realize that the "giants" in our own field are unknown to many colleagues and students. True, they sometimes know those whom James D. Watson has called the "three Ms of genetics"—Mendel, Morgan, and McClintock—and they surely have heard of Watson himself, as well as Francis Crick, the codiscoverer of the DNA double helix. But ask who was responsible for even Nobel Prize-winning advances such as the discovery that radiation causes mutations or that bacteria have genes, or the establishment of the genetic code, and they are likely to be flummoxed. These and other concepts are taken for granted in everyday teaching and research, although they are less than a hundred years old. The ignorance is perhaps understandable because of the fundamental nature of the concepts and because there is so much biology to learn today. Yet, when presented with the stories of the discoveries and the scientists who made them, people are fascinated and deepen their understanding of the science. Our goal in writing this book was to tell such stories about the giants who, in the first part of the 20th century, established the foundations for the now flourishing science of molecular genetics.

We chose the life story of one of these pioneers, George W. Beadle, as our centerpiece because his life spanned much of the period during which genetics changed from an abstract to a molecular science. Beadle also contributed greatly to that transition. He studied with the leading classical geneticists of the early 20th century and absorbed their lore and the innovative styles with which they did research and taught students. Their life stories, personalities, and accomplishments were to Beadle and his fellow students lodestars to follow in achieving their own ambitions.

Although no one knew the physical nature of genes at the time, Beadle nevertheless set out to discover how genes determine an organism's traits. By age 40, he had demonstrated, to the skepticism of many geneticists and biochemists, that genes are responsible for the production of proteins. In the twelve years following his breakthrough research, genes were proven to be DNA, and their molecular structure was embodied in the double helix, ushering in the era of molecular genetics, a subject whose experimental limits are still being explored. Beadle's achievements as a scientist and subsequently as a brilliant administrator and government advisor landed him the presidency of the University of Chicago. But all of those successes never diminished his sense of himself as a plainspoken farmer who adhered to the old-fashioned values he learned during his boyhood on the family farm in Wahoo, Nebraska.

We owe a special debt to Norman Horowitz, who encouraged us to pursue this project when we were exploring its scope and feasibility. His intimate and long association with "Beets," as he invariably referred to Beadle, provided a rich source of memories of the events leading to and during the discovery of the gene–protein relationship. He also helped bridge the cultures of the Morgan and Beadle eras at Cal Tech.

Many individuals and institutions helped us to plumb the archival sources for the papers, letters, photos, and documents that revealed Beadle in all his dimensions. David Beadle and the late Ruth Beadle gave very generously of their time and recollections about their family, and Ruth Beadle provided materials from her personal archives. We especially thank Raymond Screws, director of the Saunders County Historical Society, and several citizens of Wahoo, Nebraska who knew the young Beadle and his family for providing a glimpse of an early 20th-century Nebraska farm community. The very hospitable staff of the Cal Tech Archives provided access to the Beadle, DuBridge, Delbrück, and other collections, truly treasure troves of Beadle's stewardship of the biology division. Harry Gray at Cal Tech and Phillip Griffiths of the Institute for Advanced Study generously made their facilities available to expedite our writing. P.B. is also grateful to the Center for Advanced Study in the Behavioral Sciences for providing the seclusion that made part of the writing easier. The Special Collections Division of the Regenstein Library at the University of Chicago assembled Beadle's speeches and presidential papers, as well as the Board of Trustees' meeting minutes, without which we could not have understood the successes and failures of his presidency. We also obtained valuable help from the archives of the American Philosophical Society, Rockefeller Foundation and University, Columbia University Oral History Repository, Cornell University, New York Botanical Gardens, American Association for the Advancement of Science, Utah State Historical Society, Cold Spring Harbor Laboratory, and National Research Council, as well as from the libraries and archives at the universities of Nebraska, Stanford, Indiana, Iowa State, at the University of California, San Diego, and at the University of California, Santa Barbara. To all those individuals at these and other institutions whom we may have inadvertently omitted, we express our appreciation and gratitude for their help. The authors will present the letters and papers contributed by Ruth Beadle and others that bear on Beadle's life and activities to the Cal Tech Archive.

Particular individuals whose advice and help enriched our story include Steven Aftergood, Jonathan Aronie (who aided us through the intricacies of the Freedom of Information Act, thus assuring our access to FBI files), Irène Barluet, Redmond Barnett, David Beadle, Ruth Beadle, Lawrence Bogorad, Edward Brody, Simone Brutlag, Richard Burian, Ken Carlson, Gerhard Casper, Anna Chodos, Ed Coe, Harriet Creighton, Amy Crumpton, John Doebley, Renato Dulbecco, Don Duvick, Elaine Engst, Ann Ephrussi, Shelley Erwin, Earl Evans, Norman Giles, Janice Goldblum, Eugene Goldwasser, Judy Goodstein, Hanna Gray, Anthony Griffiths, Peter Geiduschek, Chauncey Harris, Robert Haselkorn, Francis Haskins, David Hogness, Judy Hogness, Norman Horowitz, Hugh H. Iltis, François Jacob, D. Gale Johnson, Becky S. Jordan, A. Dale Kaiser, Lee B. Kass, the late Lily Kay, Wayne Keim, Robert Kohler, Lloyd Kozloff, Lloyd Law, Esther Lederberg, Joshua Lederberg, Kate Levi, Ed Lewis, Pam Lewis, Richard Lewontin, Bonnie Ludt, Victor McElheny, Robert Metzenberg, Rodney W. Moore, Rosalind Morris, Ken Nealson, Charles O'Connell, Ray Owen, David Perkins, Sydney Raffel, David Regnery, Stuart Rice, Ned Rosenheim, Janet Rowley, David Savada, Thomas Shenk, Robert Sinsheimer, Carolyn Slayman, Piotr Slonimski, Chris Somerville, Jozetta Srb, Bernard Strauss, Henry Sturtevant, Ann Sturtevant, Hewson Swift, Marguerite Vogt, Elga Wasserman, Don Weeks, Mary Weiss, Eli Wollman, Charles Yanofsky, Robert Yuritz, and Harriet Zuckerman.

We are grateful to Robert Glaser for his help in obtaining a generous grant from the Markey Trust. Sally Gregory Kohlstedt and several anonymous individuals read an early version of the manuscript and provided valuable suggestions and assistance. We value Dot Potter's editorial assistance throughout the project. We are especially grateful to our Cold Spring Harbor Laboratory Press editor, Judy Cuddihy, for her professionalism and enthusiasm in helping us to improve the manuscript, and to Mary Cozza who, as the Press's project coordinator, guided us through the publication process. Finally, we recognize that the years we have spent on the research and writing of this book were not nearly as interesting or as much fun for our spouses, Millie Berg and Dan Singer, as they were for us; we thank them for their willingness to listen to our frustrations and for their tolerance and support.

Paul Berg
Maxine Singer

Introduction

The word "gene" did not exist at the close of the 19th century. Nearly 40 years earlier, Gregor Mendel had conceived of "factors" that influence the inheritance of traits, but the significance of his elegant experiments with garden peas was largely overlooked until their rediscovery in 1900. In the following half-century, the factors were widely recognized and named genes, and the science of genetics emerged. Although initially rooted in Europe, the successful effort to deepen understanding of the fundamental role of genes in biology was largely an American enterprise led by plant breeders with practical matters in mind and by academic biologists seeking to understand developmental processes. During this time, genes remained abstract entities defined largely by the effect they produced when altered; that is, when they were mutated and did not function properly. Even Mendel's famous wrinkled peas signaled a mutation that kept them from becoming normally plump. Geneticists concentrated on identifying mutations and studying how the mutant genes were transmitted from one generation to the next. They learned that genes affect every conceivable aspect of the development, growth, reproductive capacity, color, and shape of animals and plants. Most important, they learned that genes are located on the structures called chromosomes that are seen within the nucleus of cells. The precisely choreographed "dance" of chromosomes observed during cell division and the formation of eggs and sperm provided a physical basis for Mendel's rules of inheritance. An important extension of these conceptual advances was the recognition that individual genes are located at particular sites on chromosomes. These observations, made over the first two decades of the 20th century, established the chromosomal theory of inheritance and laid the ground for all the advances that followed.

But still, no one knew what genes were made of, and the mechanism by which they exerted their manifold effects on an organism's characteristics and behavior remained a black box. Although there were many speculations, there were but few clues about what genes actually were and what they did. Then, in 1941, George Beadle and his colleagues made a discovery that lifted the lid on one part of this black box. They learned that what a gene does is direct the production of an enzyme. This landmark one gene–one enzyme formulation transformed genetics, for it revealed that the properties of animals and plants are determined by proteins, and that each protein, including each enzyme, is determined by a single gene. An important consequence was that genes had to be thought of as molecular entities, not hypothetical packets of instructions, thereby establishing the significance of the subsequent discoveries that genes are segments of DNA molecules and that DNA itself is a double helix.

The experiments leading to the conclusion that each gene directs the synthesis of one protein, or, as biologists now say, "encodes" one polypeptide, were carried out with a common bread mold, *Neurospora*. Earlier, Beadle was frustrated when he had tried to decipher what genes do by studying several that influence the eye color of flies. These fly experiments demonstrated that sequential chemical reactions leading to the formation of the eye pigments were blocked by mutations. In these experiments, Beadle and Boris Ephrussi used mutant genes that had been identified in the classical way; that is, without any knowledge of the chemical pathways for making the eye pigments. Although Beadle surmised that the mutations were affecting the protein catalysts (the enzymes) responsible for the chemical transformations in the pathway, tracking down the chemistry proved tedious and daunting. He then realized that he might succeed if he turned the question around: Could he find a mutation associated with a defined chemical deficiency? His plan was to create mutations that resulted in detectable nutritional requirements. In this way, he could relate a particular mutation to a defined function. Flies were not practical for such experiments, but the mold *Neurospora* was tractable; its genetic behavior was known, as were its simple nutritional requirements. Soon Beadle and his colleague Edward Tatum accumulated large numbers of mutants, each one blocked in a specific metabolic reaction. These resulted in a dependency on a nutrient that could, in normal *Neurospora*, be made by the organism. Beadle correctly reasoned that each mutation affected a single trait; in this case, the property of a single enzyme.

Beadle's success relied greatly on his astute choice of a suitable organism for the experiments. This set him apart from other geneticists who were typically wedded to a single organism, sometimes for their entire scientific lives. He was not intimidated by having to learn the techniques and extensive background knowledge required when changing organisms. In the course of 14 years he switched from corn to flies to molds as the object of his research. Other biologists eventually took note of Beadle's pragmatic approach and to this day follow his lead in finding the right tool—the right organism—to solve a particular biological question.

In Beadle's mind, these experiments underscored the importance to future advances in biology of merging what had been quite separate sciences, genetics and biochemistry. He wrote and talked incessantly about the necessity to combine these sciences. Putting his belief in the merger to work, he built a model department as chairman of the Division of Biology at Cal Tech. Major discoveries in the following two decades revealed that genes were made exclusively of DNA, not of DNA and/or protein as Beadle and many others thought. Soon after, the double-helical structure of DNA, the biochemical mechanisms for duplicating DNA, and the molecular form of the genetic code were discovered. The code explained precisely how a gene instructs a cell to make a particular protein and provided chemical and mechanistic proof of the one gene–one protein concept. As the 20th century closed, scientists determined the sequence of all the DNA in the human genome. The great importance of this singular achievement emerges from being able to read the code and thus predict the panoply of human proteins. It was Beadle who started us on the road to being able to "read" the genome.

Besides leading to the one gene–one protein concept, Beadle's experiments with *Neurospora* introduced a far-reaching experimental approach for the analysis of biological phenomena. In essence, that strategy is to identify mutants that are affected in one or another of the steps in a particular process. The presumption is that every process is the consequence of many consecutive reactions, each catalyzed, facilitated, or structurally dependent on one or more proteins. For example, the formation of all the cellular constituents and structures, and the regulation of these processes, are dependent on the action of multiple proteins and, therefore, specific genes. Following Beadle's success with *Neurospora*, mutational analysis became the predominant experimental approach to analyzing such complex processes as memory, learning, vision, smell, and even the way in which the shape of an organism is patterned during embryonic and fetal development. Generally, many genetically different mutants affecting the same process are isolated, and then biochemistry and cell biology take over to determine the order in which the normal gene products act. Once a gene has been identified as being associated with a particular process, the effort focuses on identifying the affected gene and protein and determining the protein's function. Examples of the success of applying this approach include the ability to determine the formation of eyes, the construction of muscles, the nature of the circadian clock, and the genetic basis of sexual dimorphism.

Beadle's story reminds us of the fictional characters in Horatio Alger's books as well as real-life American heroes such as Abraham Lincoln, Andrew Carnegie, and Willa Cather, who all succeeded after unfavorable beginnings. When Beadle, the first in his family to go beyond high school, left his father's farm in rural Nebraska to take advantage of the free education at the state university, he expected that his future would be back at the farm. But the faculty recognized his intelligence and drive. His professors engaged him in research that took advantage of the practical knowledge and skills he had learned from his father. They also taught him that the larger world held unimagined opportunities. His intellect and spirit were captured and the farm faded from his future, if not from his sense of self. He went on to great universities—Cornell as a graduate student and Cal Tech as a postdoctoral fellow—to which he brought the hard-working habits learned at the small farm on the harsh Great Plains. He became a member of one of the most productive and exciting scientific communities of his era. Smart and competitive, he moved that community forward with his personal research and his stewardship of two great institutions—the Division of Biology at Cal Tech and then the University of Chicago. When, after World War II, knowledge of genetics became essential to the assessment of radiation hazards from nuclear bomb stockpiling and testing, Beadle worked with his peers to establish wise recommendations for the nation. The times were troubling not only because of the politics surrounding nuclear weapons, but also because of the suspicions and allegations of disloyalty plaguing the scientific community at the start of the Cold War. By publicly defending colleagues against such allegations, Beadle revealed his sterling personal qualities: common sense and an unshakable commitment to fairness and justice. The responsibility Beadle and his colleagues assumed then set the precedent for the continuing involvement of biologists in establishing public policy.

Beadle never lost the independent spirit of the farmer, not to mention the midwestern twang in his voice. How many young scientists would resign a faculty appointment at Harvard after a year because they found the culture inhospitable? How many distinguished scientists take a day out of important policy meetings in Washington to make a round trip home to California to tend their gardens? How many university presidents and Nobel laureates would, upon retirement, return with joy to the tedious field work of corn genetics? There was no hubris in Beadle's personality. There was, instead, a strong sense that his intelligence and eminence should serve the advancement of science, his university, and justice for all.

In our research, we tried and failed to uncover Beadle's flaws, to find people who disliked him, or at least did not admire him. We also sought unsuccessfully to find clues to Beadle's inner thoughts. Interviews with his two sons and the writings of his second wife provide only glimpses of his emotional being. His own writings and speeches, and his letters to colleagues, as well as our interviews with them, failed to provide any greater insight into Beadle's core beliefs. His early experience on the farm, education, approach to research, and pragmatic, honest leadership, rather than his own words, must speak for him.

Beadle's story is, however, more than the story of a single individual. It is also a central element in an amazing century of biology. The century began with a small number of scientists and an abstract approach to studying biological inheritance. It ended with the sequence of the human genome, a vast community of academic biologists, and a thriving biotechnology industry. Beadle's teachers were the great geneticists of the first half of the 20th century and he in turn taught many of those who became the great geneticists of the century's second half. He was a human and intellectual bridge between the world of abstract genetics and the molecular genetics of today.

CHAPTER 1

An Uncommon Farmer[1]

> We were talking about what it is like to spend one's childhood in little towns like these, buried in wheat and corn, under stimulating extremes of climate: burning summers when the world lies green and billowy beneath a brilliant sky, when one is fairly stifled in vegetation, in the colour and smell of strong weeds and heavy harvests; blustery winters with little snow, when the whole country is stripped bare and grey as sheet-iron. We agreed that no one who had not grown up in a little prairie town could know anything about it.[2]

Wahoo, the county seat of Saunders county in the eastern part of Nebraska, was such a town. George Wells Beadle was born and raised on a 40-acre farm a mile and a half south of Wahoo. The state's land was acquired by the U.S. in the Louisiana Purchase of 1803, organized as a territory in 1854, and named Nebraska after the Otoe Indian word meaning flat water. Its grassy flat plains were interrupted only occasionally by forests that covered at most 3 percent of the land. Soon after congress passed the Homestead Act in 1862, settlers began arriving from the east, lured by the extraordinarily fertile and free farm land available in the eastern part of the territory. Railroads followed the settlers, making areas even distant from the Missouri and Platte Rivers accessible to travel and the transportation of farm products. Then, in 1867, Nebraska became the 37th state in the Union.

Farming ran deep in George Beadle's family. His mother Hattie's father, Alexander Albro, was born in 1824 to farming parents, Mr. and Mrs. Samuel Albro. When the family joined the westward migration in 1833, they stopped briefly in Medina County, Ohio, and Peoria (then Fort Clark), Illinois, and by 1836 were settled in Lynn Township, Knox County, Illinois. The Albro's eight children were too important to the success of the family farm to be allowed much time in school and so Alexander had only three months of formal schooling. When he was 17, his father died and he apprenticed in the wagon-maker's trade. This steady, basic business employed him for a decade, by which time he had been married, widowed, and married again. His second wife, Emily Spalding, was born in New York City in 1827 to another pioneering family. Alexander and Emily helped found the town of Galva, Illinois, and by 1855 he was successfully engaged in a variety of commercial enterprises, including a contract to build ten miles of the Rock Island and Peoria Railway outside of town. Service for 2 years with the 112th Illinois Volunteer Infantry Regiment of the Union Army in Kentucky bestowed on him the lifelong title of captain, the rank he held in 1863 when he returned home. Captain Albro was a highly respected and active member of the Republican party, an incorporator and director of the First National Bank of Galva, and, from 1868 to 1872, Deputy United States Revenue Assessor for Henry County. He accumulated property, at least some of which he traded later for land outside Omaha, Nebraska, where he and Emily spent most of their last years. Galva, however, remained their home and they returned there to celebrate their golden wedding anniversary with friends in 1896.[4]

Beadle's gentle mother, Hattie Albro, married Chauncey Elmer Beadle, the handsome farmhand working at her sister and brother-in-law's place, in Elkhorn, Nebraska, in 1892. Hattie's sister, Esther Albro Babbitt, thought a hired hand an inappropriate match, but their father, a self-made success, must have seen promise in Chauncey. The wedding was a simple ceremony at the Albro's home in Galva, Illinois.[3] Being comfortably well off, Alexander Albro purchased the house and land near Wahoo for Hattie, her husband, and their year-old son, Alexander, for $3600.[5] For Chauncey Beadle, a fiercely independent person who had no resources to his name, his father-in-law's generosity meant he could provide for his family and realize his personal ambition.

Chauncey Elmer Beadle, known to all as CE, was also from a farming family that had followed the frontier, albeit with less entrepreneurial success. He had left his parents' farm in Kendallville, Indiana, in 1886 at age 19 and gone west by train to make his own way. CE could trace his ancestry to Samuel and Susanna Beadle who lived near Boston as early as 1656. For the next century, Beadles lived in various parts of New England, and by the late 18th century were recorded, usually as farmers, in New York and Pennsylvania.[6] Colman Beadle, CE's father, was a Quaker born in 1828 in New York State. He migrated west as a young man and married Adelaide Inman in 1852 in Ohio. After five years, they moved to Kendallville where Colman first worked as a peddler. Soon after, he obtained the farm where CE and his seven siblings were born and raised. He died at the farm a year after CE and Hattie were married.

Many years later, CE said, "A farmer has to have determination to survive."[7] He had more than determination. Captain Albro must have been pleased with the energy and intelligence CE brought to the farm and the good life his daughter enjoyed, because in 1898 he deeded title to the farm to Hattie. Sickly and often in pain, Hattie was a constant worry to her parents while she grew up. She had been slow to mature, and at 17, her parents took her to live in Hot Springs, Arkansas for several months in the hope that the fashionable baths would help. Her 6-year-old cousin Cora, who was more like a sister and as close to Hattie as anyone, went along. Although Hattie was intelligent and well read, she never graduated from high school because of her ill health. Her frailties were "borne without complaint and without any lapsing into sadness or any resentment over the fact that she was not just as other young girls were," and she was known for her extraordinary kindness. After she married at age 28, she surprised her family by her robust activities as a farmer's wife and mother with an eight-room house to care for all on her own. "Cooking meals, baking bread and pies and cakes, washing and ironing under the hard conditions of those days, canning fruit, sewing (she was a skillful seamstress), keeping her house in perfect order—it's a marvel that she could do all that, with her little hands and frail body." She welcomed visitors, and Cora, who lived in Wahoo for a while earning her living as a piano teacher, was a frequent guest in the happy Beadle house.[8]

Chauncey's success as a farmer was documented in a United States Department of Agriculture report.[9] By 1907, the Victorian house with its well-kept lawn and surrounding elm, pine, fruit trees, and shrubs was itself estimated to be worth $2000, and the 40 acres, $8000. The report describes the Beadle farm in detail to illustrate how a man could make as good a living on 40 acres as he could in the city, and without working more hours per day. Special plaudits are given to CE's management of his potato crop, particularly his diligent control of the yellow, black-striped Colorado potato bugs through hand picking by hired neighborhood boys, who were paid about five cents per 100 beetles. Other cash earners were chickens, strawberries, and blackberries. For the family especially, there were cows, vegetables, eggs, grapes, rhubarb, and bees, as well as roses, honeysuckle, peonies, lilacs, and other flowers. By depending on his children for substantial chores, CE accomplished all this with only an occasional hired helper.[10]

George Wells Beadle, Hattie and CE's second child, was born on October 22, 1903 when his brother, Alexander, was 8 years old. CE placed the announcement of George's birth in Saunders County's populist newspaper, the *New Era*.[11] The year that George was born, Hattie turned the title of the farm over to CE.[12] Three years later a daughter, Ruth, arrived. By then, CE's success permitted a comfortable life in the house, and Hattie had some time for her interests in literature, music, and art. A piano graced their parlor,[13] and Hattie's paintings hung on the wall of the comfortable looking, well-furnished room. When the Golden Rod Phone Company was organized in 1902, the Beadle family subscribed to telephone service. Before long they had two telephones, each serviced by one of the several telephone companies then operating in Wahoo; that arrangement allowed them to talk with more of their neighbors than if they subscribed to only one company.[14]

Notwithstanding this picture of modest, comfortable success, Nebraska was still a frontier area, and living conditions were not easy. Some families still lived in sod houses. Although written about communities further west in the state, Willa Cather's novels describe life on early Nebraska farms.[15] "Blustery winters" were brutal, especially when it meant trudging two miles to school as George did. Spring and fall meant a great deal of work. "Burning summers" were blazing hot. Wahoo, like the communities Cather described, was a diverse community. The earliest settlers had begun to arrive there from the east in 1865. Some of them were massacred by Pawnee (or perhaps Otoe) Indians who lived in a large permanent camp on the site. Wahoo was established as a town in 1870 and, as the Indians had no doubt feared, their camp had to move. Wagons brought people and goods from the east along Wahoo's mud track streets. There are continuing disputes about how the town got its name, but all explanations agree in one way or another that it derives from the Native Americans' name for the area or the creek that runs nearby.[16]

By 1873, Wahoo had been voted the county seat of Saunders County. In the next year, a two-story courthouse was built, and, in 1875, the *Wahoo Wasp*, a weekly newspaper, was founded. The first high school was built 10 years later. The staunch 1904 courthouse still stands as a monument to the ambitions of the inhabitants. There were good reasons for optimism, including the productivity of the land and the diligent hard work of the people. Those who settled early were prospering as the richness of the soil began turning out very high yields of grain. The confluence by then of three railroad lines, the Union Pacific, the Northwestern, and the Burlington, promised continuing prosperity. The townspeople built good small businesses by servicing the surrounding farms. In 1897, a year after the Beadles settled just outside of town, William Jennings Bryan opened his first presidential campaign in Wahoo.[17] Today, Wahoo remains a market center for the surrounding farms, its population of 3400 just about what it was when George Beadle was born. The railroads no longer run and a single small highway runs through the town connecting it to Interstate 80, 21 miles to the south. One station house and some track preserved behind the Saunders County Historical Society Museum give a nostalgic sense of what George Beadle passed daily on his walk to and from school.

Land itself, bought or obtained for little or nothing before the end of the 19th century, created fortunes by the early decades of the 20th century. CE Beadle, however, did not catch on to the value of real estate until well after

George had grown up and left the farm. During George's youth, he concentrated on making his 40 acres as productive as possible. The family was, and remained, financially well off, if not rich. But its general well-being changed when Hattie Albro, who never really recovered her strength after Ruth was born, died in the spring of 1908. It was she who gave the children religious and cultural foundations. Knowing that the children would lose their devoted "Momma," she suffered over leaving them without her. "You will remember me for quite a while, but not always I am afraid, because you are too young," she wrote poignantly before she died to her "Dear little George." She was especially concerned that CE, who had not joined the Baptist church as she had, would not see to the spiritual needs of the children, "for I don't think that a fathers love is quite the same as a mothers. I want you—oh so much—to be a dear good little boy and grow up to be a Christian man and help the rest of the folks to be Christians too, so that we can all meet in Heaven and be a happy family there."[18] Hattie's concern was well founded. Neither CE nor the children attended church regularly after her death. Late in life George said of his mother that "although my recollections are few, I know her influence in those important formative years was very great indeed."[19] His sister, Ruth, was, perhaps, more honest when she wrote "There has always been a subliminal feeling of deprivation."[20]

Suddenly, CE had responsibilities he had not bargained for, especially the care of three children then 13, 5, and 2 years old. He hired a series of local girls as housekeepers to care for the children and the house. Eventually, his own widowed mother came to live with the family and help out, but the children did not like the "querulous old lady" very much.[21] The general impression in Wahoo was that the children had a difficult time.[22] According to neighbors' recollections, CE demanded a lot in the way of chores. Ruth Beadle recalls him as "a stern father" but one who was their one steady security. On occasion, he resorted to a licking with a switch that the children themselves had to break from the elm tree near the back porch.

When Alexander wanted to attend the university after finishing high school, CE would not allow it, and the young man left the farm for a while to try to succeed on his own. Soon after his return, the harsh reality of the plains visited again. Just five years after Hattie's death, Alexander, then 18, died from injuries suffered when he was kicked by a horse. Talk in Wahoo, remembered to the present, was that CE bore some responsibility for Alexander's death because he did not seek prompt medical attention for his son. Surgery finally took place on the dining room table a week after the accident, and four days later on April 15, 1913, Alexander died at home. Ruth Beadle recalls her gruff father saying that he had told Alexander never to walk behind a horse.[23]

The story of Alexander's death confirmed the townspeople's view that CE was unsympathetic, perhaps even abusive; one story has him saying to an injured farmhand who lost an arm in an accident "you shouldn't have done that."[24] With Alexander's death, the two younger children had lost another important anchor. Their father, the only close relative left, was not openly affectionate and, although he made sure they were well fed and clothed, they largely had to take care of themselves as well as work around the farm.

Just two years after they lost Alexander, Grandmother Beadle herself died and was buried close to her grandson and daughter-in-law in Sunrise Cemetery in Wahoo. Ruth, then 8, recalls that she and George did not miss their grandmother, although on Memorial Day each year they took bouquets of peonies to place on all three graves, each marked by a polished, red granite stone. In the same year, the best-loved and remembered hired housekeeper, Lulu Stapleford, arrived. She was in her early twenties and became a buffer between the children and CE and "the nearest thing to a surrogate mother" that they had. Later, she married Roy Miller of Wahoo and remained close to the Beadle children.[25,26]

During these years, while Ruth worked indoors, George was up before dawn to complete his chores in the fields and with the farm animals. Together, the brother and sister cared for the dogs and many cats, watching for new litters and arguing as to which of them would take charge of the newborns and have the fun of naming them.[27] Sometimes, when their father was absent, they sold farm produce for him. Besides their duties at home, the two children walked almost two miles each way to the small one-room schoolhouse. In winter that often meant shoveling the snow from the path as they went. Ruth depended on George and he took care of her. The closeness that developed in childhood remained with them for the rest of their lives. Ruth had started school a year behind her age group because the long walk was considered too much for her small legs. A few years later her request to be moved up with the children her own age was denied; the teacher thought that she would have a problem doing the school work as well as her considerable chores at home.

As George grew older, school, the local 4H Club, Boy Scouts, and chores on the farm occupied his days. Annual fairs, with their competitions for livestock and farm produce, especially corn, were big events. Nebraska was part of the Corn Belt, and all around Wahoo corn was king and at the center of community life.[28] The U.S. Department of Agriculture supported children's corn clubs. Corn-husking contests were often the high point of the annual corn fairs, where prizes were also given for the longest ears, the tallest stalk, and the most interesting "freak" ear. In cities larger and wealthier than Wahoo, huge, ornate corn palaces were built, their external and internal walls completely covered with decorative patterns of multicolored cobs, stalks, and husks.[29] For two weeks in early fall, they were the scene of corn shows, events that were as popular as tulip shows in 17th-century Holland, with as many as 350,000 people visiting the corn palace in Sioux City in 1888.[30] In a region devoid of mountains, the sea, and the amenities of large eastern cities, they were "the midwest's answer to the Eastern hauteur in the cultural line."[31]

George, by now known as "Beets," attended the corn festivals along with other farm boys. Bee tending earned him some pocket money, as did other farm chores such as picking strawberries and asparagus. He picked enough strawberries and asparagus for CE to send to the market in Omaha that he couldn't stand either of them for the rest of his life.[32] Ruth envied him because he had a chance to learn carpentry and blacksmithing in his father's well-equipped workshop while she, a girl, was not welcome there. And he joined the Home Guard with others too young or too old to be in the armed forces during World War I.

Although a minister from Lincoln officiated at Alexander's funeral at the Baptist church,[33] there is no record that the Beadle family belonged to or regularly attended any of Wahoo's many churches after Hattie died.[34] The many churches reflected a diverse community. When George was born, nearly half of Nebraska's inhabitants had been born abroad.[35] We know from Willa Cather's novels that the Anglo-Saxon, Czech, and Swedish communities tended to remain cohesive and separate, but the books also show the growing interdependence among the three, especially in the towns. Public schools such as the single high school in Wahoo brought the young people together. A photograph of an August 1900 swimming party shows a large group of wet young men and women in typical bathing costumes of the time, their diverse origins revealed by their names.[36]

From his earliest days in Wahoo, CE began to develop a reputation for independence and feistiness. As time went on, he became more and more of a loner and at odds with the Wahoo community, mainly because his strong sense of justice drove him to unpopular actions. Besides the products of his own farm, he began importing potatoes and apples from as far away as Washington State. He traveled a great deal seeking produce and arranging its purchase and shipment. Carloads arrived at the railroad station and he sold the goods at favorable prices. The Wahoo shopkeepers did not like him underselling them and fought back by ensuring passage of a city ordinance requiring CE to pay $25 a day for a license to deliver purchases, even those ordered by telephone. CE took a large ad in the *Wahoo Wasp* for March 10, 1921, announcing that he would no longer deliver, except as a wholesale supplier to grocery stores, which remained permissible under the ordinance. Not easily thwarted, he invited "everyone to come out to my place and get apples and potatoes at wholesale prices." An advertisement in the *Wahoo Wasp* in October 1921 announced the arrival of carloads of apples and potatoes. Early the next spring, additional ads appeared and strawberry plants and oyster shells (for calcium) were added to the list of what was available. Ken Carlson of Wahoo still remembers going out to the farm with his father to buy potatoes out of the "cave."

Anger among Wahoo merchants grew to the point that the Ku Klux Klan burned a cross in a pasture across from the Beadle farm. This was not the Klan's ordinary business in Wahoo. The Nebraska Klan's ire was directed mainly at the Catholic community which, they feared, was "taking over."[37] Later, in 1936, after CE left the farm near Wahoo for another one in an even more remote area, he still appeared Saturday evenings on Wahoo street corners declaiming about the injustices of the world. And he didn't just talk: Over the years, CE instigated several campaigns and lawsuits to eliminate public nuisances, including a dangerous railroad crossing and the town sheriff's terrifying pack of coon hounds.

With a father who behaved like that, it's no wonder that George is remembered as a quiet boy. Laura Motes, who was one year ahead of him in school, tutored a group from his math class. She recalls that he was not withdrawn but never asked for help. He was on his high school debate team and specialized in pole vaulting on the track team. He was certainly seen as "country" by those who lived inside the town limits, and no doubt his father's eccentric behavior affected his own standing. At least one high school classmate, Marie Carney, refused his invitation to the senior banquet in spite of his being a handsome young man because she was sure he would arrive with manure on his shoes.[38]

CE assumed that his son would join him at the farm after he finished high school and George, too, "expected to take over the family farm,"[39] especially since Alexander had not been allowed to go to the university. Then, in 1920, George met Bess McDonald, a chemistry and physics teacher at Wahoo High School who had just graduated from the University of Nebraska. She was about 10 years older than her student and was described by all who knew her as always neat and nice looking, with brown wavy hair. Eventually McDonald, whose mother was also a schoolteacher, married and settled down in Wahoo. But as a single woman, educated and sophisticated, she became George's mentor and friend. He spent afternoons after school at her home. Perhaps he even listened as she played viola in Howard Hanson's string quartet. She introduced the quiet farm boy to books and to the opportunities of the outside world. Most importantly, she talked enthusiastically about the University of Nebraska in Lincoln, less than 25 miles south of Wahoo. From her he learned that money was not a barrier to university attendance because there were no tuition fees. While McDonald succeeded in convincing him to attend the university, he showed no interest in her strong religious beliefs. Neither then nor at any other time in his life was he a churchgoer or believer. Beadle remembered McDonald all his life and often mentioned his debt to her. Some speculated that he had a teenage crush on her, some that he truly fell in love. Whatever, it was his first close relationship with any woman, and she surely provided for him part of what the death of his mother had taken away.

Having succeeded without much formal education, CE had little reason to send his remaining son, who should inherit the farm, off to the university. Beadle later recalled that although "My father thought it was silly to go to college to be a farmer" he finally "allowed it."[40] George entered the university's College of Agriculture at Lincoln in the fall of 1922.[41]

The landscape along the way from Wahoo to Lincoln is flat, broken only by a few, gentle, almost imperceptible, rises. In early summer the cornfields stretch away on both sides of the road as far as the eye can see. The trip could be made by train or in George's Model T Ford automobile. Either CE or his legacy from Grandmother Albro must have provided the living expenses and approximately $500 for the cost of the auto. Gruff as he was, CE was not stingy. Ruth was always able to buy the dresses that caught her fancy in the Wahoo shops.

It was either by chance, or perhaps something about the schooling, but in two decades Wahoo produced, besides George Beadle, four other outstanding men, each in a different field. Sam Crawford started playing baseball for a local Wahoo promoter as a teenager in the 1890s. Known as Wahoo Sam, he was a mainstay of the Detroit Tigers for years and was elected to the Baseball Hall of Fame in 1957. Eleven years younger than Crawford, Clarence W. Anderson was born in Wahoo in 1891 destined to be an outstanding and popular painter of horses. Even more lasting, his book *Billy and Blaze* remains a favorite of American children to this day. Howard Hanson earned major distinction as a composer and director of the Eastman School of Music. Most famous of all was Darryl F. Zanuck, one of the moguls of the early movie industry. Zanuck only remained in Wahoo for 7 years after his birth in 1902, but the town claims him as one of its greats. His name is proudly displayed along with the other four on a billboard at the entrance to town, and his memorabilia are prominently displayed at the County Historical Museum in Wahoo. Even today, Wahoo gets some measure of attention as the site of the fictional office of television talk show host David Letterman.[42]

In later recollections, George sometimes said publicly that Wahoo was "a pretty good place" where "life wasn't dull" and that he had intended to return to the farm as an educated farmer. He even expressed sadness that "so few youngsters are now privileged to live" on a farm.[43] However, he told his family and others that he had had no intention of returning, that he liked neither the farm nor Wahoo.[44] Aside from his father and Ruth, all his ties to Wahoo vanished when he went to the university. A few years later, Ruth too left Wahoo. CE stayed behind, increasingly independent, solitary, and cantankerous in a Wahoo that continued to service the surrounding farms but faced diminished importance after the railroads closed down.

Although George's attachment to the farm and Wahoo disappeared, the legacy of his upbringing remained with him for the rest of his life. His father had given him "plenty of hard work, responsibility, and freedom,"[45] and all of these were at the core of his adult life. The skills George learned in the field and workshop would always be put to use, no matter how menial the job. All his life, he readily assumed responsibility for himself and others and brought to complex issues the strong sense of justice he had imbibed from CE. Perhaps, too, CE's nature and his mother's absence account for George's inclination as an adult to remain a private person who rarely shared his emotional reactions or affections in any obvious way.

CHAPTER 2

Agricultural College at Lincoln

Most of Nebraska's children lived on farms in sparsely populated areas and, even as late as 1967, attended one-room schoolhouses where a single teacher taught all grades. Yet education was important even to the earliest Nebraska settlers, and public education was free in the territory starting in 1855. The University of Nebraska opened in Lincoln in 1871 only 4 years after the territory became a state and 20 years before school education was made compulsory. When Beadle arrived there in 1922, Lincoln was a lively state capital. Building construction was everywhere, commerce was increasing, and amenities such as parks and golf courses were being enhanced. Plans were afoot for an airport, and that spring Charles Lindbergh learned to fly at the local flying school. The elegant Cornhusker Hotel was built in 1926. Hopes were high that development would attract new people because the city's population had been stagnant at about 55,000 for 30 years. "A New Skyline Every Morning" was the chamber of commerce's slogan. The symbol of the city's aspirations was the groundbreaking in 1922 for a new state house. Once built, the severe white "tower on the plains" was visible for great distances. The university's main campus, a collection of handsome stone and brick buildings around a large, green, central quadrangle, was expected to enhance the city's image.[1]

Shortly after Beadle arrived, however, the free tuition Bess McDonald had promised became a thing of the past. Serious economic problems plagued the state after World War I when farm prices fell precipitously. The price of corn itself decreased by more than 65% between 1920 and 1921. Successive governors and the state legislature targeted the neighboring university for budget cutting. Even Charles W. Bryan, mayor of Lincoln from 1915 to 1917, who might have been expected to understand the university's importance to the city's economy, proved unsympathetic when he became governor in the early 1920s. The university's financial situation could have been improved by increasing tuition rates that were, in Beadle's freshman year, only $2 per credit hour. But the regents were uninterested in the academic development of the institution that was taking shape within a short walk from the state house and declined to approve a tuition increase.[2]

It was not that the university itself was in disfavor. Rather, the regents were more interested in football than the academic program. The main campus at Lincoln was competing with the Omaha campus to be the site for construction of a new stadium, and a campaign was launched to raise the three-quarters of a million dollars needed for the stadium and a new gym in Lincoln while academic programs suffered. The Lincoln campus won out, and in the spring of Beadle's freshman year, ground was broken for the new sports facilities. Those who had dedicated their professional lives to building a great university saw their dreams evaporate as sports and the social aspects of college life blossomed. Fraternities and sororities were of more concern to students than their studies. H.L. Mencken wrote in 1929 that "Going to college has come to a sort of social necessity. It almost ranks with having a bathroom and keeping a car."[3] And to make matters worse in the eyes of serious faculty scholars, there was in Nebraska as elsewhere in the nation a growing interest in professional rather than general education. Willa Cather, who had graduated from the university, wrote in 1923, "The classics, the humanities, are having their dark hour. They are in eclipse. Studies that develop taste and enrich personality are not encouraged."[4] The College of Business Administration and the Teachers College came in for her special, scathing criticism. Her view of the College of Agriculture is unrecorded. Known to all as the "Cow College," it was likely considered marginal at best by the central campus faculty, although in its field it had an outstanding reputation.

The College of Agriculture was (and is) two miles northeast of the main university campus. General economic hardships and the particular plight of the corn farmers probably accounted for the falloff in the number of students in the College of Agriculture from 470 in 1920 to 120 in 1926. Another factor may have been the recent introduction of 4-year high schools. Certainly the decrease in students was not caused by a lack of quality in the College of Agriculture.[5] The college was outstanding and, for Beadle, the right place at the right time for a solid, even inspiring, education in agronomy, plant biology, and genetics. Several other students of the college in that era also went on to become distinguished plant scientists. When a national survey was made of the university's departments in 1934, six were deemed worthy of offering a Ph.D. degree, and three of them were plant science departments.[6]

The College of Agriculture had been founded in 1909 through the efforts of Charles Edwin Bessey, who also established its tradition of excellence. By then Bessey had been in Lincoln for a quarter of a century and was a powerful professor and a widely admired botanist and ecologist interested in the grasses of the midwestern plains. Unlike the many 19th-century American scientists who went to Europe for training, he was educated entirely within the United States. At Lincoln he had first created what became a leading department of botany that emphasized the new science of ecology. One of Bessey's most forward-looking ideas was that botany should break from a focus on plant classification and instead emphasize rigorous experimental work, including microscopy. He also believed in engaging students outside of the classroom and for this purpose organized the "Botanical Seminar," in which students engaged the latest research and learned how to talk about science. A man of great energy, he also helped to draft the Hatch Act of 1887, which provided for federal establishment and support of agricultural experiment stations all over the country. Through their research, these stations made and continue to make critical contributions to the productivity of American agriculture, not least in Nebraska.[7]

One of Bessey's outstanding students was Rollins Adams Emerson, who received a B.Sc. degree in 1897. Emerson was born in upstate New York in 1873 and, from the age of seven, was raised on a farm in Kearney County, Nebraska, where his family settled when it moved west in search of better land.[8] His very first scientific paper, published the year he graduated, reported results on the internal temperature of tree trunks. As with Beadle decades later, Emerson had little interest in returning to the farm after his experiences at the university. With the likely help of Bessey's connections, he took a job as assistant editor for horticulture in the abstracting service of the Office of Experimental Stations, U.S. Department of Agriculture, in Washington, D.C. He didn't last long. Avoiding farming was one thing, but being stuck at a desk was not an attractive alternative. By 1899 he was back in Nebraska with two positions—assistant professor and chairman, in the department of horticulture, and horticulturalist at the associated Nebraska Agricultural Experiment Station. He immediately began research, conducting breeding experiments with the common kidney bean (*Phaseolus vulgaris*) as an experimental tool. Emerson's goal, like that of other American plant scientists and breeders, was to understand the rules by which desirable plant characteristics could be obtained by breeding.[9] It was an auspicious moment to begin such investigations.

The work of the Moravian monk, Gregor Mendel, published 35 years earlier, had just been remembered and resurrected. Although Mendel published his data and interpretations in a journal that was circulated across Europe, most scientists ignored it or didn't understand its significance.[10] Outside his local community of monks and others interested in natural science in Brno, he had no scientific standing. His Augustinian religious community prided itself on independence and even iconoclastic political and theological activities, and Mendel was not the only original thinker among the monks.[11] Even without modern communications, the import of Mendelian ideas spread rapidly from Europe to the United States after 1900. In the United States, botanists and plant breeders, motivated by their interest in artificial selection and hybridization, were especially quick to assimilate the news, and the Department of Agriculture encouraged the Mendelian approach through support at agricultural colleges and the agricultural experiment stations.[12]

Emerson had read Mendel's paper as early as 1896 while studying scientific German to pass a university language requirement. However, his German wasn't very good, and he, like virtually everyone else at the time, completely missed the significance of the work at this first reading.[13] That changed by 1902 when Emerson published a preliminary paper on the bean experiments incorporating Mendel's ideas and even reproducing, with beans, some of the monk's own experiments.[14] He was one of the earliest Americans to become a committed Mendelian, even though his mentor, Bessey, and others remained "doubters."[15]

Many students, even today, find genetics confusing and are relieved when the teacher turns to things other than the seemingly complex and abstract relationships between the color, texture, and size of pea plants in successive generations. In fact, Mendel's experiments were simple, and his conclusions have stood up after almost 150 years of experimentation. He himself extended his findings to include, among other plants, corn.[16] Over 12 years of work in the monastery gardens, he established a logical framework that is applicable to plants and animals alike and remarkably consistent with modern concepts of the chemical structure of genes and how gene structure determines the attributes of an organism.[17]

Mendel understood that to discover how plants transmit their traits from one generation to the next he would have to begin with individual plants that bred true for those traits he would study. The monk was a fine gardener and he started out by letting garden pea plants (*Pisum sativum*) self-fertilize, as they normally do, over several generations. This ensured that their different, distinct traits such as seed (pea) color or texture or size of plant indeed bred true. Then he cross-fertilized these inbred strains to obtain hybrid seeds and plants. For example, he crossed plants that produced yellow seeds with those that yielded green seeds, or smooth-seed producers with wrinkled-seed producers. To do this, he carefully removed the immature anthers, the source of pollen, from the tiny flowers of, for example, a yellow-seed producer to prevent self-pollination. Then he dusted the flower's female organ, the stigma, with mature pollen from a green-seed producer's flower. He also did the reverse. He took care that no pesky insects crawling or flying from plant to plant caused inadvertent fertilizations. Unlike plant breeders before him,

Mendel introduced a rigorous, scientific approach by carefully counting and recording the numbers of seeds of different colors and texture that were produced by the subsequent generations of plants.

From these and additional experiments, Mendel formulated several conclusions, one of which was that any particular characteristic of the pea plants could take alternate forms. He proposed that some unknown factors determined which of the alternate forms appeared; for example, yellow or green and wrinkled or smooth seeds, and white or purple flowers. Also, because self-fertilization of the plants grown from yellow hybrid seeds produced some green seeds, the factor that determined green seeds must have been present in some of the yellow seeds. Furthermore, the yellow color of the seeds harboring the green factor told Mendel that the yellow factor was dominant; similarly, smooth was dominant over wrinkled. The different factors (i.e., for yellow or green seeds) must be discrete because they did not give a blended seed color such as pale green or dark yellow. Mendel also recognized that each plant must carry two factors for each of the traits, the two factors specifying either the same or different versions of the trait. For seed color, for example, both versions could be yellow (yellow/yellow) or green (green/green), or there could be one of each (yellow/green). Furthermore, he surmised that progeny seeds and plants received only a single factor from each parent—the male pollen and the female egg.

Mendel carried out a variety of experiments to verify his ideas. Then he formulated his first major principle of inheritance, which states that the two factors for a particular trait separate (or segregate, as it is called in the literature on genetics) independently during the formation of eggs and pollen. Moreover, two different versions of a factor are not equivalent in effect, so that one usually dominates over the other when both reside in the same organism.

It was evident to Mendel that the looks of a plant or its seeds did not reveal which factors it carried; yellow plants could be yellow/yellow or yellow/green. This distinction was formalized in 1909 when the Danish botanist, W. Johanssen, proposed that the heritable factor needed to be distinguished from the trait it determined. William Bateson, an English zoologist who was an early enthusiastic and influential supporter of Mendel's propositions, had already called the new science of heredity "genetics" and so Johanssen introduced the word "gene" for what had been called factor. Johanssen also proposed the word "genotype" to denote the genetic constitution of the organism, that is, the versions of the factors (genes) (e.g., yellow/green or green/green). He called the observable traits or characteristics created by the genotype the "phenotype" (i.e., yellow or green seeds).[18] Bateson also introduced the word "allelomorph," generally shortened to "allele," to describe the variant versions of a gene such as green or yellow or smooth or wrinkled. This terminology is still universally used in genetics and will, from now on, be used in this book.

Mendel also wanted to know whether two different genes, each with its own pair of alleles, separated independently of one another during the formation of pollen and eggs. He chose to study the genes for seed color and seed texture. Again, he began by repeatedly self-pollinating plants until he could be sure that they bred true. This time, he carried out the preliminary breeding so that one of the two types of plants always yielded seeds that were yellow and round and the other seeds that were green and wrinkled. The genotypes of the first were yellow/yellow and round/round while those in the second were green/green and wrinkled/wrinkled. Using the same approach as before, he cross-pollinated the flowers from these two types of plants and recorded their phenotypes. He was not surprised when the hybrid seeds all had yellow and round seeds because he already knew that round dominated over wrinkled just as yellow dominated over green. But did seed color and texture go together in the next generation? To determine that, Mendel self-pollinated the plants that grew from the hybrid seeds. If there were no connection between the genes for seed color and the genes for seed texture it would be equally probable that a pollen grain or an egg made in the hybrid plants would contain either the yellow or green allele and either the round or wrinkled allele. When Mendel counted the seed phenotypes, he found just the ratio of yellow and round, yellow and wrinkled, green and round, and green and wrinkled seeds expected for random inheritance of the two pairs of alleles. These data led to Mendel's second general principle: Different genes are assorted independently of one another into eggs and pollen (i.e., germ cells), and thus into progeny.

Although Emerson's results with the beans were completely consistent with those of Mendel and others, he had not made enough individual crosses starting with purebred lines to report statistically significant results in his 1902 paper. Importantly, one aspect of his experiments was different from what Mendelian rules predicted. When he cross-fertilized purebred plants containing different alleles for bean color, the hybrid progeny occasionally had purple markings on a different color background rather than showing the pure dominant or recessive trait. When he made a more complete report in 1904, he described experiments with various traits including pod position, seed color, and pod stringiness. Again, he recorded occasional "intermediate" forms and used the term "blended" to describe some of these unexpected characteristics.[19]

By the time Emerson's first paper on beans was published, he was already working with corn, or maize, as it is usually called in the scientific literature and English-speaking countries outside the United States; the formal scientific name is *Zea mays*. Mendel himself had turned to corn after his pea experiments were finished. In an 1867 letter to Carl von Nägeli, Mendel reported that he had made crosses between a corn plant with dark red kernels and others with either white or yellow kernels. Three years later he wrote to Nägeli to say that the results were the same as those obtained with peas.[20]

Corn was attractive to Emerson, not least because of its growing importance in United States and Nebraska agriculture. Corn also had scientific advantages. Each of the hundreds of kernels on a corn cob is the result of a separate fertilization, and thus large numbers of offspring of as many independent matings can be counted, improving the statistical significance of the observations. In addition, the seeds (or kernels) stay put on the cob and have no tendency whatsoever to be dispersed. Another convenience derives from the separation of male flowers in the tassel at the top of the plant and female flowers well below on the stalk; the silks from the female organ capture the pollen. In a field, corn pollen blows with the wind and plants may fertilize themselves or be cross-pollinated by nearby plants. However, for experiments, it is a simple, if tedious, job to protect plants from random pollination by bagging the tassels and silks and then brushing the pollen from a selected "male" parent on the chosen silks.

There is another aspect of corn reproduction that made the plant convenient for genetic research. The properties (phenotype) of the bulky material that surrounds the embryo in the seed and provides nutrition to the maturing embryo have the genetic makeup (genotype) of the new seed, not that of the mature plant on which the corn cob appears. If the genes being studied affect the properties of the kernel, the results of genetic experiments can be seen in the kernels. It is not necessary to plant the kernel and let the plant grow before obtaining the experimental results, a great time-saver for the geneticist. The kernel can still be planted to continue to study the genes in question. Peas are similarly convenient for genetics, but for a different reason. The embryo inside peas is covered by swollen leaves. The color and texture of these leaves reflect the genetic makeup of what will be the new plant, not those of the plant on which the seeds appear.

When Emerson adopted corn as the organism for studying genetics, however, it was not because of these advantages. It was, instead, the puzzling and tantalizing results of an exercise he gave his Nebraska classroom students in the first years of the 20th century. The students received ears of corn from plants Emerson had grown for what was meant to be a demonstration of Mendelian principles. He had cross-fertilized two varieties of corn that bred true for traits called, respectively, starchy and sugary (or sweet). After planting the seeds from the hybrid offspring and letting the plants grow, he self-pollinated them and allowed them to produce mature ears. The students were to inspect the kernels and count how many were sugary or starchy. This is straightforward because the sugary kernels collapse and appear wrinkled while the starchy ones are plump and full. The students' numbers were not what Emerson expected. They were not the Mendelian ratios of starchy to sugary that should have occurred if the two phenotypes reflected two different alleles of a single gene, as Emerson thought they should. Emerson's first reaction was "Gad! They can't even count."[21] But when he counted himself, he saw that they were correct. It took him many years to discover the explanation for the discrepancy, but meanwhile, Emerson was sold on corn as a system for studying genetics because, for a great scientist like Emerson, puzzling results are always the most interesting kind.[22]

In the next dozen years, Emerson carried out many genetic experiments with corn and horticultural investigations with fruits, vegetables, trees, and ornamental plants. He collected interesting corn mutants at the "freak" displays common at Nebraska agricultural shows where his strong physical appearance—he was over six feet tall and sported a full beard—and firsthand knowledge of farming must have made him a notable figure. He also established himself as a leading and influential scientist and was appointed a full professor in 1905. In particular, he addressed head-on one of the phenomena that led some people to be skeptical of the Mendelian formulation: the inheritance of blended characteristics, such as the purple-spotted beans he saw in his early experiments. As he had already noted in 1902, not all phenotypes are simple, straightforward reflections of the genotype as is, for example, the seed color of peas (yellow or green). Emerson concluded that some phenotypes reflect the presence of two or more different genes (and their alleles) and vary quantitatively, depending on which alleles of genes other than the one being studied are present. Although genes are independent elements, they can influence one another to yield a phenotype that looks like a blend of several genes.

Although Emerson's data supported his staunch defense of the occurrence of blended or "quantitative" traits, he did not convince many others. Committed Mendelians thought he was throwing a shadow on established genetic concepts. Those still skeptical of Mendelism, such as Bessey, took his data to demonstrate the flaws in the new science of genetics. Thomas Hunt Morgan was another of the skeptics, although he would shortly change his mind and become one of the most important of 20th-century geneticists. Beadle loved to tell the story of a 1908 meeting of the American Breeders' Association at which Morgan derided the "jugglery" of Mendelism which "blinds us....to the common-place that the results are often so excellently 'explained' because the explanation was invented to explain them."[23] Morgan's example was that "If one factor [the term then still used, instead of "gene"] will not explain the facts, then two are invoked; if two prove insufficient, three will sometimes work out." Emerson followed Morgan on the program and, to the amusement of the audience, his talk illustrated Morgan's complaint. The experiments were about the mottled coat color of beans, a blended or quantitative trait, and Emerson interpreted his results in Mendelian terms by postulating the combined influence of two independent genes and their alternative alleles. As it turned out, Emerson was correct, although many at the meeting thought that Morgan had won the argument.[24]

E.M. East, a plant geneticist at Harvard, was not one of the skeptics. His own data and interpretations agreed with Emerson's proposal that the so-called "quantitative traits" in many plants likely resulted from the independent inheritance, by strictly Mendelian mechanisms, of several different genes and their alleles whose effects influence one another. Emerson spent 1910–1911 in East's lab at Harvard and obtained his doctoral degree (awarded in 1913) for work on quantitative traits done with East as his official advisor. In 1913, Emerson and East together published a huge compendium of data on a variety of quantitative traits in corn, a paper that became a classic and convinced many of the skeptics.[25]

The work on his Ph.D. convinced Emerson that he was much more interested in genetics as a science than in the very practical focus of the Nebraska College of Agriculture with its emphasis on improving farming in the state. In 1914, his growing reputation resulted in an invitation to join the faculty of Cornell University as head of the department of plant breeding in the college of agriculture. He welcomed the opportunity to move to surroundings more hospitable to fundamental research and accepted.[26] He, along with two of his graduate students, Ernest Gustav Anderson and Eugene W. Lindstrom, began what would become a Lincoln-to-Ithaca tradition that would eventually include Beadle. It was the start of an extraordinarily influential scientific family and a leading center for corn genetics. Although he was no longer on the campus, Emerson's insistence on and enhancement of Bessey's tradition of excellence and rigorous experimentation continued to shape the Nebraska College of Agriculture for decades.

This Bessey–Emerson tradition favored intelligent, hardworking young people such as Beadle, who could make the most of all opportunities. In his first semester, starting September 8, 1922, Beadle continued with chemistry, which he had begun in high school, and started studying botany, horticulture, and dairy and poultry husbandry. The next term was much the same with the addition of agricultural engineering and rural economics.[27] He performed brilliantly and won the Freshman Medal, given by the Agricultural Honorary Society, Alpha Zeta, to the freshman who received the best grade average. Campus recognition came with the medal and he soon joined Farm House, a social fraternity then restricted to agriculture students.[28] In group photographs, Beadle and the other Farm House members are dressed formally in jackets and ties that give no hint of the farms on which they had grown up. The picture for 1924–1925 includes G. Fred Sprague, who was then a graduate student and, like Beadle, destined to become part of the Lincoln-to-Ithaca tradition and a leading geneticist. As an undergraduate, Sprague had received many of the same honors that were later bestowed on Beadle. Life at the fraternity house was convenient, and Beadle promptly became active in Farm House affairs, including the track team that competed in interfraternity athletics. In his junior year, he was elected chapter historian for first semester (fall, 1924) and chapter secretary for the second (spring, 1925).

Although the occasional humor in the Farm House publications seems sophomoric, the articles mainly reflect a serious commitment to study and the university. Even so, life in the large fraternity house was not completely congenial for Beadle. The cost, about $45 a month for room, board, laundry, room cleaning, and fraternity dues, was not a problem for him although it was prohibitive for some.[29] Grandmother Albro's legacy and his father's help allowed him to have some financial independence. But he didn't like the noise and decided to move out. For a while, he lived at the home of his mentor, Professor Franklin D. Keim, whose family remembered Beadle as a likable if green country boy who rose early and, wearing an old stocking cap, left the house before anyone else was awake and returned after everyone had gone to bed.[30] He continued to enjoy Farm House's social life, athletic competitions, and participation in the large farmers' fairs and corn fairs that were major events in Nebraska. Campus activities, including some writing for the college's student publication, the *Cornhusker Countryman*, took up some of his energies. By the end of the 1924–1925 academic year, Beadle was not only elected to Alpha Zeta, but was the honorary group's chancellor, its top officer. In the next year, as a senior, he was elected to Sigma Xi, the national science honorary society, and to the National Honor Society of Agriculture, Gamma Sigma Delta. The honors were appropriate. For every semester as an undergraduate, his grades averaged 90 or above.[31] With all these undergraduate honors and activities, Beadle must have shed some of the shy country boy affect by the time he finished his studies in Lincoln.

Keim was part of Emerson's legacy at the university. He was 17 years older than Beadle and had a growing family, but there was not much difference between the educational status of professor and student. Born in 1886 and raised in Hardy, Nebraska on the Kansas border, he attended Peru State Teachers College and taught school before ever going to the University of Nebraska. He graduated in 1914, the year Emerson left, and promptly joined the university faculty as assistant agronomist. Four years later he was awarded a master's degree and promoted to professor of agronomy, a position he held until retirement in 1956.[32] The university honored his extraordinary contributions by naming a building for him. A straightlaced, conventional man who for years taught a celebrated Methodist Sunday school class, Keim was also a tireless and effective institution builder who concentrated on building a fine department of agronomy with the purpose of improving Nebraska agriculture. During his chairmanship of the agronomy department from 1932 to 1952, it became "the premier department in the University and one of the notable agronomy departments in the world."[33] One of his targets was the state legislature, where he lobbied for funds for crop and soil testing and for help to farmers for such things as the management of weeds. But his greatest accomplishments were as a teacher. Years later, George Beadle would remember him as an "inspired" teacher: "His complete candor and unpretentiousness was indelibly impressed on me by his habit of asking help of his students in solving textbook problems."[34] Beadle was never sure whether this was a deliberate teaching device or Keim's honest difficulty in problem solving. Either way, the student appreciated the feeling that he could do something that "has the Prof. stumped." Still, he realized that Keim was not a "very heavy scientist" nor "a brilliant teacher in the usual sense. He had an abundance of infectious enthusiasm and a fabulous understanding and judgment of students."[35]

During Beadle's undergraduate years, Keim was not only teaching but doing research for his Ph.D. thesis under Professor H.H. Love in Cornell's department of plant breeding. Although Keim did most of his thesis research in Nebraska, he and his family were in Ithaca during 1923 and their son, Wayne, was born there. Once back in Lincoln, Keim missed his contacts with Emerson and the lively and stimulating Cornell atmosphere, particularly the meetings of the geneticists' Synapsis Club. He also missed regular guidance because his research did not go smoothly and he had to juggle his time between his teaching and family responsibilities. Professor Love made Keim's life more difficult by ignoring requests for help. When Keim was trying hard to complete his degree by the end of the 1924–1925 academic year, he asked Love for the pedigree of a troublesome strain of wheat that was critical to his research.[36] Several months later, a desperate Keim wrote again requesting the pedigree.[37] The letter arrived in Ithaca two days after Love had departed on a prolonged trip to China without arranging for the handling and submission of Keim's thesis.[38] Nine months later, in early 1926, kindly Keim wrote to welcome Love back to Ithaca. The long effort finally succeeded when Keim was awarded his Cornell Ph.D. in 1927.

According to his own recollections, Beadle was inspired by the courses he took at the university and "found college a tremendously stimulating experience."[39] He recalled a phase during which he wanted to major in English, although he never spent much time on the humanities or classics.[40] Then, he "went all out for entomology...and was invited as a sophomore to do a checklist of all the ants in Nebraska...It was just too much for me." Later, after taking a course with an enthusiastic professor of plant ecology, he was still vacillating and "was pretty much persuaded that ecology was for me."[41] However, his introduction to genetics clearly made the deepest impression. The first hints about genetics came in the fall of his sophomore year (1923) in an agronomy course entitled Forage Crops given by Professor Homer T. Goodding. Goodding, like his other professors, must have recognized a promising student when he saw one and responded to Beadle's interest by giving him research papers to read. The young student was flattered by the attention, especially since the professor was a senior Farm House member.

When Keim returned from Cornell for the spring semester, he gave Beadle his first chance at genetic research by hiring him to collect data on the progeny of hybrids between spelt and club wheat. The work contributed to Keim's Ph.D. thesis and Keim acknowledged the assistance of Beadle as well as George Fred Sprague in data collection and calculations.[42] Beadle received 30 cents an hour for recording the kernel color, size, plant height, and other properties of the hybrids. To understand what he was doing, he "spent" his "spare time reading" about genetics.[43] Then, in the first semester of his junior year, he took a zoology course on the main "city" campus. The course was entitled Evolution and Genetics, and the professor, David Whitney, had studied with Thomas Hunt Morgan at Columbia University. Beadle found him an unforgettable teacher. Wanting more, he took Keim's genetics course in the agronomy department in the spring of 1925.[44] Keim was not as good a lecturer as Whitney, but he knew that doing research and paying attention to the latest literature were the ways to engage talented students. The following fall Beadle signed up for a weekly seminar course in which the students and Keim reviewed recently published research papers.[45]

Although Beadle was "hooked" by genetics, he would not realize it until after he finished at Nebraska and moved on. For reasons described in the next chapter, Keim assumed that his outstanding student was most interested in ecology, and Beadle, still an impressionable undergraduate, followed his mentor's lead.

CHAPTER 3

Genetics at the Quarter Century

When Beadle began studying genetics, the understanding of heredity had progressed well beyond Mendel's discoveries and his textbooks incorporated the most significant recent research. One of them was *Genetics and Eugenics* by W.E. Castle.[1] Beadle recalled that another text, *Genetics: An Introduction to the Study of Heredity* by H.E. Walter was "a simple book—just about right for me—and I was fascinated."[2,3] These books were first published only a little more than a decade after the rediscovery of Mendel and very shortly after the establishment of the chromosomal theory of inheritance. Besides studying Mendelian genetics from a textbook, Beadle also learned how to carry out experimental matings (genetic crosses) between plants with different characteristics. He learned how to predict the expected phenotype and genotype of the progeny and determine which alleles of the genes were dominant and which recessive.

Beadle also learned about the connection between genes and chromosomes. Microscopists working with many different plants and animals during the latter years of the 19th century had discovered that all cell nuclei contain small bodies that they named chromosomes because they were brightly colored by dyes. The number of chromosomes in each body cell is characteristic of the particular species. Corn, for example, has 20 chromosomes, and humans, as we learned only in the 1950s, have 46. However, the typical number of chromosomes is reduced by half during the formation of eggs and sperm (germ cells). Corn and human germ cells have 10 and 23 chromosomes, respectively. Then, when the egg and sperm meet during fertilization, the full number of chromosomes is restored.

In 1902, W.S. Sutton, then a graduate student at Columbia University with the great American biologist E.B. Wilson, advanced genetics by firmly establishing an idea that had been hinted at earlier by others and supported by Wilson.[4] Sutton (studying grasshoppers[5]) and others[6] learned that the typical set of chromosomes in each organism could be grouped into similar-looking pairs and that one member of each pair is derived from the mother and the other from the father. Sutton knew from earlier work that during the formation of germ cells, the chromosome number is reduced by half in a process called meiosis.

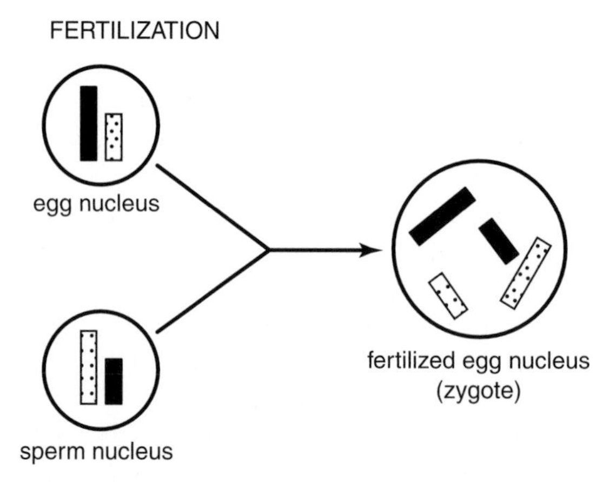

He recognized that at one stage of meiosis (synapsis), the pairs of similar chromosomes (called homologs) line up side by side, like two parallel lines. In the next step of meiosis, the two chromosomes in each pair separate and move away from one another within the cell. That is, for example, the ones on the right move to the right of the cell and the ones on the left move away to the left. The cell then divides into two down the midline between the two sets of chromosomes; each new germ cell has one of each of the original pairs and thus half the original number of chromosomes. Sutton hypothesized that if genes are carried on chromosomes, then homologous chromosomes might each carry one of the two alleles for the same gene.

About a year later Sutton developed his ideas more fully, stressing that the set of chromosomes lined up on each side could include a mix of chromo-

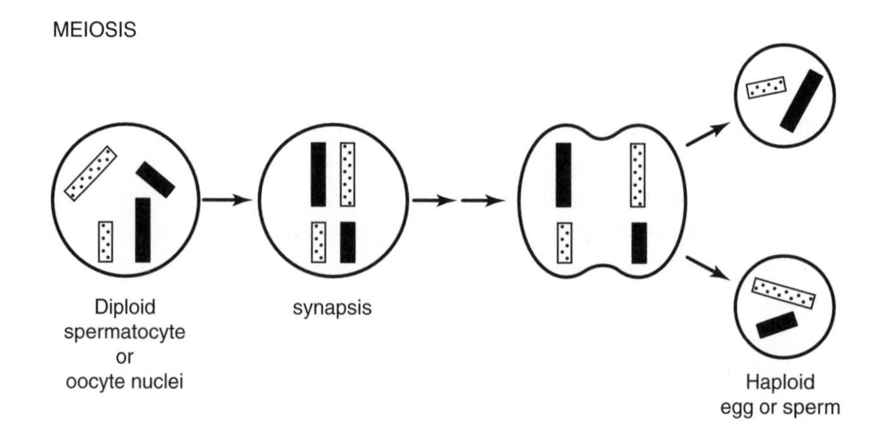

somes received from either mother or father. [7] Very importantly, he pointed out that the distribution of maternal and paternal chromosomes on the two sides was completely random. This meant that when the pairs separate at the division of the cell into two, each new cell receives one of each kind of chromosome, but whether each comes from the mother or father is a random selection. Each new germ cell will then contain a mixture of chromosomes derived from the parents of the individual making germ cells. Sutton's special insight was to recognize that Mendel's conclusions matched the behavior of chromosomes in meiosis if the chromosomes carried genes. Chromosome behavior during germ cell formation was the physical mechanism for separating the two alleles of a gene that were inherited originally from the mother or the father and sorting them randomly, and unchanged, into germ cells. Furthermore, if different genes were on distinctive (not homologous) chromosomes, then Mendel's observation that different characteristics, such as color or texture, were inherited independently would be expected. Ten years after Sutton's work, other scientists demonstrated conclusively that the maternal and paternal chromosomes do indeed sort randomly when the next generation germ cells are formed.

An astute student like Beadle would have recognized from his reading that 19th and early 20th century geneticists experimented with a great variety of organisms, including plants, invertebrates such as grasshoppers and sea urchins, and vertebrates such as guinea pigs and rabbits. Sutton and others seemed to assume, implicitly and correctly, that chromosomal behavior was similar regardless of the organism being observed. The details might differ somewhat, but the overall mechanisms were the same. Everything we now know, a century later, confirms this intuition. And Beadle would make extraordinary use of different organisms in his future genetic research.

Starting about the time of Sutton's work, geneticists began to obtain experimental results that were inconsistent with Mendel's second general principle. Some genes did not behave as though they were independently inherited; in fact, some alleles were always inherited together as if they were connected. Besides that, as more and more genes were discovered, it was plain that there were more genes in each organism than the number of chromosomes. How could these observations be explained if, as Sutton had proposed, the chromosomes carried the genes? As Beadle learned, the conundrum was resolved by Thomas Hunt Morgan and his research group at Columbia University in 1911 when they recognized that each chromosome carries many genes for many different characteristics. Only a few years earlier, Morgan had been a severe critic of Mendelian genetics (see Chapter 2), but by 1911 he was a confirmed believer because his own experiments, although designed for a different purpose, surprised him by confirming Mendel's observations. Genetics thereafter became Morgan's passion and the object of his research for years.

Morgan more than once quipped that he was "laid down" in 1866, the year Mendel published his seminal discoveries. His parents were from prosperous and venerable Southern families. Morgan's lineage in America started in 1636 and included the famous American banker, J.P. Morgan; his father, Charlton, fought with the Confederate army under the command of his brother John, leader of the infamous "Morgan's Raiders." When the war ended, Charlton married Ellen Key Howard, a granddaughter of Francis Scott Key, and the couple settled in Lexington, Kentucky, where their son Thomas was born and raised.[8]

Morgan's passion for nature emerged early, but it was not until he entered the State College of Kentucky that his interests in biology blossomed. To pursue that fascination, he enrolled as a graduate student in zoology at Johns Hopkins University, which had been founded only 10 years earlier but already achieved an international reputation. His Ph.D. thesis (1889) dealt largely with morphological aspects of the embryology and phylogeny of invertebrate sea spiders. In the fall of 1891, Morgan became an associate professor at Bryn Mawr College, replacing E.B. Wilson, who had just moved to Columbia University. Despite a heavy teaching load, Morgan continued his research, which, by then, had adopted experimental approaches to the embryology of marine organisms. He was particularly intrigued by the way that a single cell, the fertilized egg, differentiated into an array of cells with different shapes and properties during the development of an embryo into an adult form. One of Morgan's colleagues at Bryn Mawr, Lillian Sampson, became his wife in 1904, the year Wilson invited him to come to Columbia University as a professor of experimental zoology.

As an embryologist, Morgan was more concerned with how inherited information was transformed into traits than with how it was transmitted from one generation to the next. But he followed what geneticists were doing and he did not like it. His passionate opposition to Mendel's theory stemmed from his abhorrence of speculation. For him, only experiments and observations were significant. He regarded as sheer folly Mendel's and others' invention of "particles" to explain a process that was so poorly understood. He was equally skeptical of Charles Darwin's speculations on the origin of species, particularly the view that new species arose from even slightly variant organisms by a process of selection. He was more favorably disposed to Hugo De Vries' theory that new species were created by "mutations" that caused substantial enough changes in the organism's phenotype to prevent interbreeding with its progenitors. To collect evidence in support of De Vries' theory, Morgan decided to use *Drosophila* as his experimental subject.[9] Then, as now, the choice of experimental organism proved to be critical for discovering a new paradigm in biological research.

While still a young scientist at Harvard, W.E. Castle (the author of the textbook Beadle used) had learned about the rapidly breeding fly, *Drosophila*, from one of the graduate students, and then, from 1901 to 1905 carried out many brother–sister matings which established that inbreeding has very little effect on the fly's fecundity.[10] *Drosophila melanogaster*, now the most widely studied member of the Diptera order of insects, most probably originated in Southeast Asia, although many representatives of the species are now found in every part of the world. The flies are 2–3 mm long (about 0.1 inch) and easy to obtain, since they invariably congregate around decaying (fermenting) fruits and vegetables, where they feed on the yeast that abound on such plant surfaces; hence, their more popular but improper name, fruit flies. In the laboratory, *Drosophila* has a rapid reproductive cycle (about ten days) and produces abundant progeny (100–400 offspring per mating). A gauze-covered, narrow-necked bottle containing their food supply (usually yeast growing in a semisolid mixture containing corn meal and molasses) suffices as a laboratory habitat for scores of flies.

Male and female flies are readily distinguished by their size (the female is larger than the male), distinctive color markings, and the pattern of bristles on the thorax. In practice, flies are mated by introducing 5–10 appropriate virgin females, i.e., those that have not already been exposed to males, into a bottle containing 20–30 males of the desired genotype. After about two days, the females begin to lay eggs, and a day later the newly hatched larvae (about 0.5-mm long) crawl over the gelatinous surface consuming food. Then, the larvae grow substantially in size, necessitating several molts before they settle down and transform into what are called pupae, enclosed within a shell. By four days, most of the larval tissues are destroyed and replaced by new tissues and organs that are derived from buds of embryonic cells harbored in the larvae. These buds are called imaginal discs. Each bud contains the cells that can form a different and specific adult tissue, e.g., eye buds and wing buds. Four days later, the adult fly emerges from the shell and after another half-day, the newly hatched fly can mate. Progeny flies are usually examined with a binocular microscope after they have been anesthetized by exposure to carbon dioxide, from which they readily recover.

Drosophila's undemanding husbandry requirements appealed to Morgan. But he spent more than a year of frustrating and unsuccessful attempts to obtain with the fly the kinds of mutations that De Vries had proposed would, under appropriate selection conditions, result in a new species. Then in 1910, he had a bonanza, the result of which changed the course of genetics for all time; he spotted a fly with white eyes instead of the normal reddish-brown eyes. Repeated crosses of white-eyed flies with the wild-type parent yielded only flies with either reddish-brown or white eyes; none of the offspring had eyes of intermediate color.[11] Almost reluctantly, Morgan concluded that the white-eyed character was the result of a single mutation (*white, w*). It was clear that this mutation changed only a single trait and did not create a new species as De Vries had supposed.

Morgan also noted, however, that transmission of the white-eye character did not obey Mendel's law; indeed, only males inherited the mutant character. An explanation of the "sex-limited" inheritance of the white-eye trait was suggested by the earlier studies of his colleagues, Nettie Stevens at Bryn Mawr and Wilson at Columbia.[12] They had inferred from their examination of dipteran insect chromosomes in microscopes that females had a pair of chromosomes called X, whereas males possessed only one X chromosome. Morgan reasoned that if the *white* mutation occurred on the X chromosome, and a single mutant *white* allele was insufficient to alter normal eye color, then females carrying the *white* allele on one of their X chromosomes and the normal allele on the other would have normal-colored eyes. Only rare females that carried the defective *white* allele on both X chromosomes would display the white-eye character. In contrast, he reasoned, if the Y chromosome lacks the determinant for eye color, males would have white eyes if they inherited a mutated X chromosome from their mothers.[13] *white* was the first gene that could be assigned to a specific chromosome.

A few years later, Morgan discovered additional mutations affecting a wide variety of characteristics. Some were sex limited, such as *white.* Others affected both sexes equally frequently, and these were assumed to be associated with *Drosophila*'s other three pairs of chromosomes. Morgan then made another observation that was not predicted by Mendel's rules. Mutant alleles for several different sex-limited traits, all presumably associated with the X chromosome, were often inherited together while mutant alleles associated with different chromosomes (e.g., X and another of the chromosomes) separated from one another in the offspring. Traits that were most often inherited together were viewed as being "linked," while those that were inherited independently of each other were considered "unlinked." By that view, genes were organized into "linkage groups." Morgan made the intellectual leap that linkage groups and chromosomes were one and the same; that is, genes that are transmitted to offspring together are associated with the same chromosome and, conversely, those that pass to the offspring independently of each other are most likely associated with different chromosomes.[14] Among other questions, this explained why there could be more genes than chromosomes in a particular organism.

The physical similarity of homologous chromosomes reflects the underlying fact that they contain the same genes, though often different alleles, and in the same order. It's something like an encyclopedia. Each volume contains a particular set of articles, just as each chromosome contains a particular set of genes. All copies of, for example, volume 1 contain the same set of articles in the same order as long as it is the same edition. The same set of articles likely occurs in another edition. Some of the articles may be the same in both editions, like identical alleles. Others may have been altered by the editors, analogous to the different alleles of a particular gene. Mendel had been lucky. The genes for the seven traits he studied were, by chance, each on a different chromosome. If any two had been linked on a single chromosome, he would have had trouble figuring out what was going on.

As Beadle learned from his books and courses, it was not long after the discovery of linkage that Morgan and his research group realized that some of the *Drosophila* experiments produced results that were inconsistent with what linkage predicted. Occasionally, two alleles (of different genes) that were known to be linked on the same chromosome were not inherited together. The frequency at which they separated varied, depending on which pair of genes was being studied. Geneticists already thought of chromosomes as a line, with the different genes spaced along it, much like beads on a string. The two homologous chromosomes appeared to contain the same genes, in the same order, with occasional genes being represented by variant alleles. Morgan speculated, something he rarely did, that in the course of the meiotic process producing the egg there could be an exchange of alleles between homologous chromosomes. His idea was that alleles for two or more different genes that were originally associated with one chromosome became separated, some remaining with the original chromosome while others ended up with the other member of the pair.

Morgan imagined that when the pairs of homologous chromosomes lined up side by side during meiosis, each of the two chromosomes could break at some common point, say, between two genes, and then exchange pieces. The two new chromosomes would still be homologous, but they would contain one section with alleles from one of the original chromosomes and one section with alleles from the other. For example, if the alleles on one chromosome are designated ABCDEFG and the corresponding but different alleles on its homolog abcdefg, then breakage and exchange might form ABCdefg and abcDEFG.

But how could this breakage and exchange occur? Morgan was aware of the work of a Belgian cytologist, F.A. Jannsens, who had observed what appeared to be a physical intertwining between the paired, homologous chromosomes at the time of meiosis.[15] He surmised that in the course of or following the creation of Jannsens' intertwined chromosome pairs there was a physical exchange of segments between them. The process is now referred to as "crossing-over" or recombination. Morgan made still another inspired guess, namely, that the further apart two genes are from each other in the chromosome the more likely it is that they will be separated during the exchange process; conversely, closely spaced alleles are less likely to become separated during the crossover.[16] Many years later, Herman J. Muller acknowledged Morgan's genius when he wrote that the "evidence of crossing-over and Morgan's suggestion that genes further apart cross over more frequently was a thunderclap hardly second to the discovery of Mendelism, which ushered in that storm that has given nourishment to all modern genetics".[17]

Alfred Sturtevant, an undergraduate working in Morgan's lab, was intrigued by his professor's ideas about crossing-over, particularly the guess that it was more likely to occur between genes as the distance between them increased. One night, putting aside his regular homework, Sturtevant examined Morgan's data on the frequency with which linked alleles became separated in a new generation. Sturtevant realized that the variations in frequency offered the possibility of determining the order of genes along the linear chromosome. Assuming that the breaks and exchanges are equally likely to occur at any position along the chromosome, two genes that are far apart will be more likely to become unlinked than two that are close together. Thus, in our example, F and G are likely to stay together because there are five possible breaks that would leave them linked and only one that would separate them, whereas A and G are likely to exchange. By the next morning, he could show Morgan a linear "map" of the order and relative distances separating five sex-linked genes.[18] This was a remarkable and profound deduction, especially for an undergraduate.[19]

Years later in their 1939 textbook, Beadle and Sturtevant suggested a helpful analogy.[20] Knowing what time a train leaves New York and when it arrives in Philadelphia and Baltimore, it is a simple matter to deduce which city is closer to New York. Knowing in addition the length of time it takes a train to travel between all possible combinations of two cities, for example, New York and Baltimore, New York and Philadelphia, and Baltimore and Philadelphia, it is also possible to make a pretty good guess of the order and distance between the cities on a north–south map. The length of this whole map is the longest distance, that between New York and Baltimore. If the time it takes to go from Baltimore to Washington, D.C. is added to the data, the whole map would increase in size by the distance from Baltimore to Washington. Continuing with this, the ends of the map would become clear when no cities beyond Bangor, Maine, on the north, and Miami, Florida, on the south, were discoverable.

By the time Beadle learned all of this, the construction of genetic maps by the analysis of recombination frequencies was in the standard toolbox of animal and plant geneticists. Nevertheless, it would be some years before there was rigorous experimental proof for the association between genetic recombination and the physical exchange of material between chromosomes.

Besides describing what was known about heredity, the Walter textbook introduced Beadle to some of the unanswered questions and speculations that biologists discussed and which would inform their research agenda for the next 30 years. Walter had a modern view on the controversial issues of the time. He believed firmly in the centrality of the study of mutations in genetic research and clearly accepted Morgan's chromosomal theory of inheritance, while pointing out that not everyone was convinced. He affirmed that mutations and inborn factors are the cause of variation among organisms, not environmental factors, as claimed by the Lamarckian hypotheses about the inheritance of acquired characteristics. Walter identified a critical question that would remain unresolved until midcentury: "What are the determiners of hereditary qualities?" And he commented that "Whatever the answer...it may at least be affirmed that the determiner represents the adult structure without resembling it." In modern language, he was saying that the gene was not chemically similar to the chemical responsible for the observed phenotype. Moreover, he added, "It is not unlikely that the key to this whole problem will be furnished by the biochemists and that the final analysis of the matter of the heritage-carriers will be seen to be chemical rather than morphological in nature." In a section called "The Enzyme Theory of Heredity," Walter opined that "it is likely that heredity will be reduced to a series of chemical reactions dependent on the manner in which various enzymes initiate, retard, or accelerate successive chemical combinations occurring in the protoplasm." At the time, many biologists assumed an important, if obscure, relationship between enzymes and genes, and few of them believed that biochemistry would provide answers for biological questions. The chemical nature of enzymes was itself an unsettled question, and the relationship between enzymes and proteins was still a matter for speculation.[21] Nevertheless, the idea that enzymes themselves might be the key informational molecules—the genes—was a popular view then, and Beadle and others held to that belief for many years thereafter. The unsolved questions that Walter's text planted in Beadle's mind during his undergraduate years were the keys to understanding genetics and would become central to the student's own research when, 15 year later, he took up the question of the relationship between enzymes and genes.

Both the Walter and Castle textbooks discussed eugenics, which was then flourishing as a serious "scientific" endeavor among some geneticists in the United States and Europe and as an element of social policy. Early in his book Walter devotes substantial space to Francis Galton's "biometric" (statistical) method of analyzing breeding experiments and the phenotypic variations in traits observed among individuals of a species.[22] Working almost simultaneously with Mendel, and during the period when Mendel's findings were in eclipse, Galton, too, began with a study of peas. However, he soon concentrated on humans, especially such matters as the inheritance of criminal behavior and intelligence; he never achieved the insights that Mendel produced. The last chapter of Walter's textbook summarized the biometric and Mendelian approaches that were, by then, being used in what can now be called the pseudoscience of eugenics. Walter's conclusion about this debate is ambivalent: "It is possible that if some of the philanthropic endeavor now directed toward alleviating the condition of the unfit should be directed to enlarging the opportunity of the fit, greater good would result." Castle was less equivocal and criticized eugenic concepts in a reasoned manner.[23] These matters were not academic. By the time Beadle read these authors, Congress had passed and the president had signed the restrictive Immigration Law of 1924, which was based on eugenic principles and actively lobbied for by some geneticists.[24] Even as a young student, Beadle learned that genetics could raise social and political issues and that geneticists could be called upon to exercise their scientific judgment in the public interest. He would have many opportunities as a scientist to do just that.

All the while he was learning formal genetics, Beadle was also acquiring research skills and experience and even some income, thanks to opportunities provided by Professors Goodding and Keim. He taught the agronomy lab at the Agricultural High School one year and had a paying job organizing the collections of weed seeds and growing plants at the agronomy department.[25] Above all, the two professors made Beadle feel like a colleague.[26] Speculating years later on why small schools such as the University of Nebraska produced "an inordinate number of talented scholars," Beadle said "My theory is that they got good faculty members...[who] had to find their intellectual companionship with their students. To have that kind of a relationship with a professor is very stimulating for a student."[27] Keim knew just how to spot a talented young man and then immerse him in research. Professor John E. Weaver, who had taught ecology to Beadle, was also impressed by the student's talents and interest in that subject. But Keim and Weaver were rivals, and Keim wasn't about to let Weaver claim his star undergraduate student. Keim told Beadle that he could study ecology just as well in the department of agronomy as in Weaver's department on the central campus. He reinforced what he said by giving Beadle attention and financial support for summer work on ecological research.[28]

Beadle's work during the summers of 1925 and 1926 contributed to a formal, bound report on Nebraska hay.[29] Typically, the commercial value of hay was determined by federal hay standards that depended on the mixture of species found in hay lots. It was a matter of great consequence in Nebraska where, according to USDA and official Nebraska agricultural statistics, hay occupied the third largest cultivated acreage (it produced an average of 2.8 million acres in 1924–1935) and was the fifth largest crop in monetary value (it produced an average annual sum of $25.9 million from 1914 to 1925). Beadle had grown up with hay. It was one of the important crops on his father's farm and one that came in for particular praise in the 1908 USDA report.[30] Keim and Beadle noted that some farmers resisted the grading standards while many "more progressive farmers are enthusiastic advocates of the US grades." Their own data focused on the Elkhorn Valley region of Nebraska about 20 miles east of Wahoo near the Platte River, where hay is a main crop at least partly because the soil is not very good for much else. The native and cultivated species found in the Elkhorn Valley were catalogued, and the report concluded that as the region matures ecologically, true prairie species should be able to compete with the grasses that were then being grown. In the course of this work, Keim taught Beadle how to write a scientific paper, including a review of the background material and the careful reporting of their own data, enhanced by drawings and photographs.

Beadle accumulated sufficient credits to complete his bachelor of science degree in agronomy by the end of the first semester in the academic year 1925–1926; the degree was awarded on January 29, 1926.[31] Willa Cather could have used him to exemplify her concerns. He had applied for and been granted permission to skip taking required courses in sociology and history altogether, substituting instead courses in rural economics, entomology, agronomy, and horticulture. By graduation day, he was determined to go on with his studies and immediately matriculated as a graduate student pursuing a master's degree. In the 1926 spring semester he took several courses, including mathematics and a heavy load of German, his first study of a foreign language and at that time a necessity for a serious scientist. Research, particularly the continuation of the work he had begun with Keim the previous summer, was his primary activity. In the 1926 and 1927 summer sessions he registered as a research student, although by the fall of 1926 he was far away in Ithaca, New York, beginning his Ph.D. work. He finished writing the master's thesis while at Cornell and in it acknowledged Dr. H.P. Cooper, his first Cornell research professor, for comments on the first draft. The master's degree was awarded on July 15, 1927.

The subject of the master's thesis grew out of the report he wrote with Keim. The work in the Elkhorn Valley required identifying species of grasses for population counts. This was problematic because few hay grasses flower before they are harvested, and flowers are the easiest and most reliable way to identify species. A method for field identification based on the vegetative properties of green, growing plants was needed. At the time, the standard in use was a decade-old USDA publication. Beadle defined a set of vegetative characteristics such as hollow stem, hairy appendages, and pigmentation of several structures and described these for various species through drawings and photographs. Altogether, he examined about 25 species.[32] This was enough for the thesis, but it was a good many years before the work was really completed. Keim, along with another student, obtained additional data and finally, in 1932, long after Beadle had finished his doctoral degree at Cornell University, the three published two papers on this subject.[33]

Except for Farm House activities, an occasional tennis game and a girlfriend, Leona Davis, he took little time off.[34] He probably would never have entered the new football stadium if the students hadn't been pressured into buying tickets for the 1925 football game between Nebraska and Notre Dame. Having bought a ticket, he "had to use it"[35] and witnessed the decisive win by Nebraska that caused Knute Rockne to cancel any additional matches between the two universities. Occasionally Beadle worked in the campus blacksmith shop, practicing skills learned from his father. He couldn't avoid taking part in the huge annual agricultural fair on the Cow Campus. Later in life he remembered with pride doing his bit by painting big signs for the fair. But he had come to the university with the ingrained work habits of his father, and mainly he studied and worked. Leona and he were engaged to be married, and his life seemed to be on a promising track.[36]

During his last year in Lincoln, his sister Ruth arrived to start her own university studies. Like her brother before her, she was glad to leave Wahoo and the farm behind. He worried about her well-being and began his lifelong habit of looking after her and trying to help determine her directions. Ruth, however, was as independent as he was and, although she looked up to her big brother, she liked to make her own decisions. Beadle had arranged for Ruth to live with Leona Davis' family, but she didn't care for the situation and moved to another place where she supplemented her share of Grandmother Albro's legacy by working for her room and board.

Keim did more than inspire students through the pleasures of research; he also arranged their futures. He was, in Beadle's words, "a man who had an amazing facility to size up people; he knew what people should be professors, which should be farmers and which should be county agents."[37] Thanks to his sharp nose for talent, the most promising Nebraskans were sent on the Lincoln-to-Ithaca path for graduate work. He laid the groundwork for their acceptance at Cornell by keeping Emerson informed about the courses he was giving and his excellent students and by inviting Emerson to visit Lincoln to deliver talks. George F. Sprague received Keim's strongest endorsement as a brilliant student with exemplary work capacity, personality, and character as well as technical skills to rival those of "little Andy," referring to E.G. Anderson. He urged Emerson to take Sprague on as his own student when Sprague, like Beadle, went to Ithaca in the fall of 1926. In contrast, Beadle received just a passing reference in Keim's letters, indicating that he was a fine student, a fast thinker, and had an excellent capacity for work, but there is no request for Emerson to take Beadle as his own student when he, too, arrived in Ithaca that same fall.[38] Genetics was not Keim's plan for Beadle. Instead, impressed by Beadle's work on the grasses and believing, perhaps biased by his own interests, that Beadle was more enthusiastic about ecology than genetics, he sent him to Cornell to study ecology.

CHAPTER 4

The Corn Cooperation

When Beadle arrived in Ithaca in the fall of 1926, he registered as a graduate student in the agronomy department at the College of Agriculture in keeping with the arrangement made by Keim. The ecologist Herbert Press Cooper, assistant professor of agronomy, was to be his professor and provide essential financial support through an assistantship. Cooper's plan was for Beadle to concentrate on the ecology of New York State pasture grasses.[1] Genetics was to be his minor area of concentration. But Cooper gave Beadle little independence, and his new student balked. He had not come to graduate school expecting to carry out someone else's research ideas. As Beadle explained in an interview many years later, "It turned out that it [the New York State grass problem] was more his [Cooper's] problem than mine and I said, 'I'm going to change to genetics and cytology.' And so I did."[2] Years later he described the same event in more colorful language, recalling that he said to himself "To hell with it, I'm going to quit."[3] To Adrian Srb he recollected that the ecology department at Cornell was "dead."[4] In contrast, he sensed that the department of plant breeding where Emerson was breaking new ground in genetics was the right place for him.

When Emerson left Nebraska in 1914 to direct Cornell's department of plant breeding, he hoped to concentrate his efforts on fundamental research rather than applied agriculture. He was not disappointed. Between 1914 and 1926, his and his students' research in corn genetics made the department the leading laboratory worldwide in plant genetics. Cornell even exceeded Harvard as a center for maize genetics and magnet for fine students. In genetics in general, work on corn was second only to that on *Drosophila*, and some aspects of the science were best studied with plants. Most significantly, corn was a major factor in the productivity of U.S. agriculture, and corn breeding was well supported. Emerson's status grew within the scientific community where he, together with Morgan, became a leader in genetics. Within the university, too, he became an important figure and was appointed dean of the Cornell Graduate School in 1925, a notable achievement for someone associated with the School of Agriculture and a scientific discipline that was just beginning to be acceptable.[5]

The "Chief," as Emerson was dubbed by the students, had already trained several remarkable students, including the two who with him had started the Lincoln-to-Ithaca tradition, Ernest Gustav Anderson and Ernest W. Lindstrom. Lindstrom finished his Ph.D. at Cornell in 1917.[6] Anderson, a Nebraska undergraduate, completed his Ph.D. in 1920 and after spending some time in Morgan's lab and at Cold Spring Harbor, joined the faculty at the University of Michigan. Emerson expected that Anderson would become an outstanding scientist, and the two collaborated for years after "Andy" left Cornell. Many of his long letters to Emerson described his research results in detail, but others sought advice about how to handle the conflicts he consistently generated with collaborators and competitors. Anderson was a rigorous thinker and cautious interpreter of data, but not easy to get along with. In 1927, Emerson recommended him for a Guggenheim Fellowship, saying that of the four or five of his students whom he considered intellectually brilliant, Anderson ranked number one, or perhaps two, although "I have always noted some lack of easy adaptibility [*sic*]."[7] Soon after, when Morgan invited Anderson to join his new department at Cal Tech, Emerson seemed convinced that Anderson was in fact living up to the high expectations he had for this, "the keenest student who has ever worked with me. But," he goes on to say, "he was so temperamental that, if conditions were not to his liking, he accomplished little. I once told him that, unless he could get a grip on himself and quit worrying about things he couldn't help, I should wash my hands of him..."[8] Still, he continued to use Anderson as his gold standard when he had to recommend other students, including Beadle, for fellowships.

Emerson's evaluation of Anderson is particularly remarkable in view of the other young geneticists who were, by 1927, part of the Chief's scientific family and would be much more productive and influential than Anderson. The Yugoslav, Milislav Demerec, emigrated to the United States in 1919 and soon after became a student at Cornell. He completed his Ph.D. in genetics under Emerson in 1923 and joined the Carnegie Institution's department of genetics at Cold Spring Harbor, New York, where he remained, an increasingly important figure, until his retirement in 1960 after almost 20 years as director.[9] Barbara McClintock, who joined Emerson's group after completing her undergraduate and doctoral studies at Cornell, would also outshine Anderson.

From the time he arrived in Ithaca, the Chief consciously shaped his students and colleagues into a social and scientific community. For this, he revitalized the geneticists' Synapsis Club that had started as early as 1907 and held the club's dinner meetings at his home. When there were no special visitors, the members, both faculty and students, took turns reporting on their current research, recent scientific meetings, or the activities of the scientific community. Returned travelers talked about the countries and laboratories they had visited. New students were rapidly socialized as scientists and introduced in an informal setting to the frequent, well-known visitors such as George Shull, the first to demonstrate the importance of hybrid seed for corn productivity, and even Thomas Hunt Morgan. Beadle, who joined the Synapsis Club soon after arriving in Ithaca, remembered especially being introduced to the mold *Neurospora* by Bernard O. Dodge, a visitor from the New York Botanical Gardens on January 25, 1929.[10] "Dodge was at a loss for an explanation" for some of his novel genetic findings, but Beadle and others, who had been reviewing recent publications on crossing-over in *Drosophila*, could explain the data.[11] More than a decade later, Beadle would remember Dodge and *Neurospora* and realize that the mold was the right organism for understanding the function of genes.

The Synapsis Club was named for the side-by-side pairing of homologous chromosomes at the beginning of meiosis (see Chapter 3). Like all members, Beadle was dubbed a "chromosome" when he was inducted into the "cell" according to established ritual. By the March 28, 1927 meeting, he was on the "feed committee," which typically cooked the meat loaf, potatoes, salad, and pie in the Emersons' kitchen for the 25-odd participants. A special arm of the Synapsis Club, the Razzberry Club, met once a year and organized programs, outings, the printing of poems, spoofs of professors, and cartoons. The clubs encouraged romantic pairings as well as scientific friendships and Beadle, with his blue eyes, nice looks, and friendly manner, must have been attractive to the young women. He wasted little time making his own synapsis, although that brought on the teasing attention of the Razzberry Club. A cartoon headed "GENETICS 201, Special Problem #1000: Analyze the following genetic situation: (is it) a case of Synapsis or Non-disjunction?" shows the Strand Theatre's screen and the backs of a man and woman, with the man saying "Our Nebrass-ska scenery is prittier [*sic*] than that, Lilly!"[12] Lilly was Lillian Phelps, who also attended meetings of the Synapsis Club. Leona Davis back in Lincoln was apparently forgotten.

Beadle heard the geneticists' enthusiastic discussions at club meetings, and it was there that he began to realize that Emerson's group was the right place for him. The idea was reinforced by Professor A.C. Fraser, who gave the beginning genetics course Beadle took his first semester and "was a superb teacher, in the sense of organizing the facts of genetics known at the time." [13] In this, and all other courses, his grades were recorded as "satisfactory," the only designation used for those who passed. By June 1927, Emerson reported that he was making good progress on his thesis topic. This was likely *pro forma* because Beadle had not yet harvested his first corn crop or officially switched his major subject to genetics. The official change was made by the start of the spring semester in 1928. His minor subjects became cytology with Professor Lester W. Sharp in the department of botany, and plant physiology with Professor Lewis Knudson.[14]

When Beadle switched departments, he lost the financial support of his assistantship, and "prepared for an even more Spartan life."[15,16] Emerson came to the rescue with a new, part-time assistantship that Beadle supplemented with income from his Grandmother Albro's legacy. Eventually he saved enough money to acquire "Ophelia Bumps," a Model A roadster that replaced the Model T car he had left behind in Nebraska.[17] He was happy with the new arrangements, and Emerson's approach to students suited him well. Although a professor and dean of the graduate school, Emerson was readily available to students. Despite all his responsibilities in the corn genetics community and the university, he carried out his own research. Research was also his primary approach to teaching his graduate students; he rarely gave formal lectures. In contrast to the outstanding lecturer, Professor Fraser, who "was little concerned with what remained to be discovered," Emerson "was little interested in what was already known but fascinated with what remained to be discovered."[18] He set a high standard because "His persistence and objectivity were great and he never published until he had extracted the truth from his experimental material and verified it not once but many times in many ways."[19]

Generations of Emerson's students recollected him with awe. None even hint at any dissatisfaction or perceived flaw. He is always portrayed as a deeply sympathetic person who was dedicated to science and his students. Geneticists worldwide admired him and trusted his leadership. Yet, he was not an easy mentor. According to him, it was not the professor's job to pamper students either intellectually or emotionally. He treated them as mature people, encouraged their independence, and expected them to take the initiative. The approach succeeded admirably, producing scientists who went on to have major impacts on genetics, and it became Beadle's model when he had students of his own.[20]

A great deal of the group's scientific fervor was fed by the exciting and novel marriage of maize cytology and genetics that was occurring at Cornell literally before their eyes. Cytology employs microscopes as a way to study the insides of cells. The marriage produced what came to be called cytogenetics. The senior cytologist in Ithaca was Lowell F. Randolph. But it was Barbara McClintock who made the brilliant technical innovations and highly original insights that established corn cytogenetics on firm ground. By studying both inheritance patterns and the unusual properties of chromosomes associated with the patterns, experiments could be interpreted with more depth of understanding.

McClintock was a year older than Beadle.[21] A descendant, through her mother, of Mayflower voyagers, she had been a maverick since childhood. Her parents supported this self-styled "tomboy" in a range of interests unusual for a young girl in those days. But her mother balked at the idea of her going off to Cornell to college when she finished at Brooklyn's Erasmus Hall High School in 1918. It was only her father's intervention after his return from World War I service in France that made it possible for her to go to Ithaca in the fall of 1919. After McClintock finished her undergraduate studies at the College of Agriculture in 1923, she decided to remain and study for a doctoral degree in the college's botany department. She might have preferred to work under Emerson, but the plant breeding department did not accept female students.[22] Lester W. Sharp, who had already taken her under his wing, became her thesis director. Sharp knew that she did her best when left on her own and gave her the freedom to pursue her own research ideas.

Needing, like most graduate students, to support herself, McClintock took a position as research associate to Randolph. He was on the staff of the USDA's Agricultural Experiment Station at Cornell, where he worked to understand corn chromosomes so as to maximize the utility of genetics to plant breeders.[23] With McClintock, Randolph described the first example of a maize plant with three rather than the normal two sets of chromosomes.[24] Little else came of their collaboration, which ended shortly after it began because they couldn't get along. Randolph was a careful, methodical scientist. McClintock too was careful, even methodical, but she also had brilliant, lightning-quick insights into the significance of complex cytological observations and was adventurous in adopting new techniques. As fine as that was for advancing science, it robbed Randolph of the intellectual pleasure of interpreting his own results, not to mention the admiration of his colleagues. For her part, McClintock chafed at any perceived threat to her independence. Randolph was only the first to find her a problematic colleague, although others would temper their annoyance by recognizing that McClintock was indeed a person of rare talents whom many would eventually call a genius.

One important question about corn chromosomes eluded Randolph's skills and dedication. He wanted to establish a way to tell each of the corn chromosomes apart and count them. No one had been able to do this, although the four *Drosophila* chromosomes, while small, had long since been distinguished from one another, and each of the four linkage groups of genes was associated with a particular chromosome. Similarly, the corn linkage groups needed to be assigned to specific chromosomes, but to do this, the chromosomes needed unique identities. Randolph tried, but he could not succeed. Worse yet, McClintock did.

McClintock realized that developing germ cells showed the individual chromosomes more clearly than the cells from plant roots that Randolph used. She also improved a recently described technique for making microscope slides and staining the chromosomes. With these methods, she associated each of the corn chromosomes with a unique set of physical properties, such as length and the relative size of the two chromosome arms extending from the constriction known as the centromere, and determined that there were ten chromosomes in germ cells (and thus ten pairs, or twenty, in all normal cells). She numbered the chromosomes one through ten from the longest to the shortest. Randolph only learned of her success in 1929 when her paper diagramming the ten corn chromosomes was published.[25] The full extent of the work became the basis of her Ph.D. thesis.[26] Almost immediately afterward, she began to use offspring of the "triple dose" plants to assign the known linkage groups to particular chromosomes. This was possible because some of the offspring had nine normal double sets of chromosomes and one triple set. The phenotypes of such plants gave clues about which linkage group was associated with the extra chromosome. By 1931, she and others had assigned the ten known groups of linked genes to particular chromosomes. Altogether, these were accomplishments that established McClintock as a leader in the corn genetics community and the preeminent corn cytogeneticist. Randolph, however, "was embittered, since she, unknown to him, had reached his long-sought goal."[27]

By the time Beadle and Sprague went to Cornell in the fall of 1926, the College of Agriculture was world-famous. It was a center for the burgeoning fields of genetics and cytogenetics and had an excellent faculty, good facilities, and a community of outstanding, hardworking students. These students had been well trained primarily in public colleges, and many had the practical biology of farming in their hands and a yen for research in their heads. The young scientists gathered around Emerson were as talented a group as could be found anywhere. Friends as well as colleagues, they shared the excitement of their individual projects, their knowledge, and ideas.

Although the departments of botany and plant breeding were in separate buildings, the spirited group of students and research assistants had their work tables in the graduate student common room in the attic of Stone Hall, the home of botany. Cubicles fashioned from bookcases marked off the individual spaces on the wooden floor. When Harriet Creighton arrived in 1929 to begin her graduate work under McClintock, she was assigned to share a cubicle with Beadle and his dog.[28] It was so crowded that anyone sitting in the cubicle had to get up from his or her chair to let another person in or out. Under such conditions, anybody's business was everybody's business. Creighton thought the crowding advantageous; it made it easier to learn from one another. Excitement, talk, and argument were incessant. Anything interesting that turned up in one person's microscope was quickly viewed by the others. For all its advantages, the intense scientific intimacy was also the cause of some dissension among these smart, ambitious, and independent people. McClintock, for example, could be irksome when she peered into other people's microscopes and quickly interpreted their results, usually correctly, before they had a chance to think themselves.

Camaraderie continued at the several-acre experimental field where corn was planted. Called the "Hole," its surrounding banks provided (and still do) some protection from late spring or early autumn frosts. Although each person had an assigned piece of the plot, they all helped one another with the hard work of planting, cultivating, and watching over the plants. It was not unusual for the students and the Chief to have several thousand plants growing at one time. Spring and summer were hectic because of the short window of time in which the work had to be done. Emerging tassels had to be wrapped or removed so that pollen did not fly in the wind and make unwanted fertilizations, ruining everyone's carefully planned experimental matings. Similarly, the stigmas had to be protected so that they did not receive the wrong pollen. The right pollen had to be applied to the right plants and the stigmas and tassels covered again. Later, when the corn matured, the ears had to be collected and the kernels carefully labeled and stored. Meticulous records had to be kept.

During corn pollination seasons, they "all worked seven days a week from dawn to dark, with Emerson setting the pace and presiding over lunch and rest period bull sessions."[29] Hard as the fieldwork was, it was also a time when the whole group was together sharing experiences and when Emerson did most of his teaching. The shed in the middle of the Hole has been preserved and carries a sign reminding visitors that two Nobel Prize winners worked there: Beadle and McClintock. Years later, Beadle looked back fondly on the intense spirit that pervaded his graduate student days. Charles Burnham, who was working at the Bussey Institute at Harvard, the other center of corn genetics in the United States, spent the summers of 1929 and 1930 at Cornell learning cytogenetics from McClintock. He recalled that Emerson engendered a "sense of teamwork" there just as he fostered a "remarkable esprit de corps among maize workers" worldwide.[30] Among other things, he made his stocks of maize mutant seeds available to all, thus establishing a uniquely cooperative community.

Beadle and another of Emerson's Ph.D. students, Marcus M. Rhoades, developed strong ties with McClintock. Rhoades, a Kansan, arrived in 1928 from the University of Michigan with a fine recommendation from E.G. Anderson.[31] The two students realized that McClintock's unique findings and methods would be helpful for their own research. Rhoades, in particular, appreciated early her very special talents. He knew that Randolph had complained to Emerson about McClintock's independent ways and, hoping to counter any disapproval, made sure that Emerson knew about her experiments.[32] Rhoades never seems to have clashed with McClintock's competitive spirit, perhaps because he came to terms early with her superior gifts. Cytogenetics became Rhoades' primary scientific interest, and he and McClintock remained close colleagues and friends until they both died close to their 90th birthdays. She and Beadle developed a similar relationship, but their road was rockier and the ties less compelling. Beadle initiated correspondence with her only occasionally in the coming decades, usually when he needed her help or advice. Perhaps he always remembered her as they were as students, serious competitors in the lab and on the tennis court.

Early in his stay at Cornell, Beadle established two independent lines of research. One concerned the relationship between corn and a wild, Central American plant, teosinte. Because the teosinte work was performed as a paid research assistant to Emerson, it could not be used toward the requirements for his degree. The other, which became the core of his graduate thesis, focused on mutations that were associated with sterility of corn plants. Such mutants had been collected by earlier workers, including Emerson, who noticed that occasional plants lacked well-developed anthers on their tassels and produced little or no pollen. The implication according to Mendelian principles was that the normal parents of the mutant plants carried one functional and one nonfunctional allele of a gene essential for the production of viable pollen. Those offspring that had the bad luck to inherit two nonfunctional alleles produced infertile pollen (and often infertile eggs as well). Beadle planned not only to investigate the formal inheritance patterns of sterility, but also to use cytogenetics to study the mechanism behind the inability of the mutant plants to manufacture functional eggs and sperm. Although he had done well in Professor Sharp's cytology course in the summer of 1927, he needed McClintock's help. She was willing to teach her techniques to anyone who wanted to learn, provided she didn't think the person a fool. Beadle passed the test.

Beadle self-pollinated plants known to harbor one such mutant allele and planted the kernels produced from that mating. He and McClintock examined the pollen-producing cells on the new plants in the microscope. In some of the plants, something was seriously awry in the normal development of the germ cells; the chromosomes were not lining up in pairs as they should in the early stages of meiosis. This meant that the process for making functional pollen aborted early, which explained the plants' sterility. The gene was named *asynaptic* because the mutant allele interfered with normal meiotic chromosome pairing; that is, with synapsis. Later, Beadle would report that the maturation of eggs was also aberrant in these plants. His first Cornell paper was a one-page report on *asynaptic*, coauthored with McClintock.[33,34]

asynaptic was a significant discovery because it was the first identification of a mutation that affected chromosome behavior. Beadle would "never forget the incomparable thrill of discovering the asynaptic character. My enthusiasm was shared—so much so in the case of Barbara McClintock that it was difficult to dissuade her from interpreting all my cytological preparations. Of course, she could do this much more effectively than I."[35] McClintock, however, recalled the situation differently: "He was very good, very fast. So then he was on his own."[36] Beadle felt strongly enough that he complained to Emerson and then did go ahead on his own. Although he had learned McClintock's "smear technique" for preparing microscope slides of chromosomes, he, like the others, remained dependent on her stock of "good" acetocarmine dye. Without the right dye preparation, the whole cell could take up the color, not just the chromosomes.[37] When it came time to write his Ph.D. thesis, Beadle made a point "especially to express appreciation of the help so freely given by Dr. McClintock both in matters of technic [*sic*] and of interpretation of material."[38]

Beadle did all the genetic experiments himself, including carrying out the fertilizations, collecting the seeds, and planting them to grow the next generation. This work went on largely during the summers of 1927 and 1928, but also in a greenhouse during the 1927–1928 and 1928–1929 winters. He must have found this "farming" a familiar comfort because no matter what he did scientifically later in life, he always found a reason to grow corn. Two years after the paper with McClintock, he published on his own the full story of the genetic and cytological experiments.[39] He had studied the offspring of the self-pollination of plants harboring the mutant *asynaptic* allele. Out of 144 kernels planted, 109 produced normal plants and 35 produced sterile plants. This was comfortably close to the expected Mendelian outcome (a ratio of three to one) if the parent plants each had one normal allele and one defective allele and the defective allele was recessive. Beadle confirmed his straightforward interpretation of the results by application of Mendel's methods. He self-pollinated fertile plants that grew from his first breeding experiment, planted the seeds the plants produced, and analyzed the seeds produced by the new generation of plants. He could not easily breed the infertile ones, which must have had two copies of the sterility-causing allele. Four of the eleven plants that he self-pollinated bred true; that is, all their progeny were fertile as expected if the four parent plants had only normal alleles. Seven of the eleven produced seeds that resulted in some plants that gave no pollen; this is what is expected if the parents contained at least one copy of the mutant (infertile) allele. Beadle's data were pretty close to perfect, which was especially surprising because he had only eleven plants to begin with.

Beside the central discovery that a gene could control chromosome behavior, Beadle posed an insightful and original question in the major paper: "How a single gene in one pair of chromosomes can so alter physiological conditions that there results a failure of synapsis of all the chromosomes, is a question on which one can only speculate." Eleven years later he, with Edward Tatum, would discover the answer. They learned that what a gene does is to provide a cell with a particular protein. The protein corresponding to the asynaptic gene is presumably needed by all chromosomes for correct synapsis. Today, we know that many genes and their corresponding proteins are required for this complex but critical pairing. Beadle was not just a good farmer and experimentalist, but he also knew how to think about the significance of his data.

Besides his own research, Beadle was busy with his courses and Emerson's teosinte project. During his first year, he took a course given by the chemist James Sumner. It was a memorable experience even though Sumner "wasn't such a hot lecturer."[40] The chemist had that year astonished the world of chemistry and biology by announcing that he had obtained crystals of the enzyme urease. Up until then, there were great disputes as to whether preparations of enzymes, that is, proteins, could possibly be pure collections of identical molecules, and crystallization depends on just such purity. Many people were skeptical even to the extent of questioning whether urease was indeed a protein. It wasn't until the 1930s, when John Northrup crystallized the enzyme pepsin, that people generally accepted the fact that enzyme proteins were real molecules, with precise structures. And, although it would be another decade before the actual connection could be established, Sumner's accomplishment provided an essential fact for thinking about the relationship between genes and proteins.

The Synapsis Club was an opportunity for scientific and social interactions; Beadle rarely missed a meeting. In the fall of 1927, the beginning of his second year in Ithaca, the club's records show a new member, Marion Hill. She was a native Californian who was 2 years younger than Beadle and had received her bachelor's degree at Pomona College. She registered at Cornell as a master's student in botany, the subject of her undergraduate major. Besides plants, she loved birds and was interested in ecology.[41] Soon she would replace Lillian Phelps as Beadle's girlfriend. Meanwhile, Ruth Beadle had also arrived in Ithaca. Her brother had convinced her that Cornell was a better college than Nebraska and that she should come there to study nutrition so that she could be a dietician, one of the few professions open to women. Ruth, who was already interested in painting, wasn't terribly excited by nutrition, but her world centered on her brother and she took his advice. She moved into George's apartment and did the housekeeping, but, as in Lincoln, she had her own ideas about where she wanted to live and the arrangement didn't last long. Ruth wanted to be in the dormitory, and, in any case, it was obvious to her that George and Marion were in love and would soon be married.[42] The wedding took place in the presence of Cornell friends on August 22, 1928, at one of the beautiful waterfalls feeding the Finger Lakes in the green, hilly surroundings of Watkins Glen, New York, near Ithaca.[43] Ruth was there to represent the family. Beadle must have been busy beginning to harvest his corn at that time, but Marion and he appeared to friends as the most romantic couple in Ithaca. In October of 1928, Marion Hill began to sign the Synapsis Club roster as Marion Beadle.

Like many women in that era, Marion Beadle put aside her studies after her marriage. She obtained a position as a research assistant in pomology (fruit cultivation) and found time to help Beadle in the lab. Their son David Beadle recalls her as worldly, and others thought her outspoken and not very sociable.[44] She shocked Ruth Beadle when she lit up a cigarette at their first meeting. McClintock too must have been a startling surprise to Ruth; a cigarette dangles from her hand and a grin decorates her face in a photograph taken while she was probably still an undergraduate.[45] Marion was broadly interested in music and art. In contrast, Beadle seemed always to be the unworldly, outgoing Nebraska farm boy who was interested in little besides the lab. His favorite author was said to be H.H. Munro (Saki), although there seems little in those mannered tales of early 20th-century upper-class British society that could appeal to him.[46] Perhaps it was a case of opposites attracting. Marion probably entered the relationship on what used to be called a "rebound." She had been disappointed when a California college boyfriend ended their relationship. Mementos from this young man, whom she called "the cowboy," remained with her for the rest of her life: a pair of spurs, a Colt 38 revolver and holster, and a ring fashioned from a silver dollar. As for Beadle, perhaps he saw in Marion a good-looking and sophisticated woman, a shadow of his memory of Bess MacDonald. In any case, the marriage did not slow the pace of his research.

Besides *asynaptic*, he was studying a different sterile, recessive mutation of corn, one that he eventually named *polymitotic* because of the many abnormal cell divisions late in meiosis. Again, he first published a brief preliminary report followed by a full description a year later.[47] He self-pollinated hybrid plants containing one normal allele and one *polymitotic* mutant allele; there were three normals to one with the *polymitotic* character in the offspring, the result typical for a simple recessive character. As with *asynaptic*, Beadle analyzed the developing pollen grains in the microscope to determine the actual defect. This time, he did the cytology on his own without McClintock's collaboration. He discovered that the *polymitotic* mutation interrupted normal meiosis at a step after synapsis.

The cell division that occurs after synapsis (called the first meiotic division) produces two cells, each with what appears to be a single set of the ten distinct corn chromosomes. As seen in a microscope, however, each of the ten is actually a doublet consisting of two, identical sister chromosomes. This is because prior to synapsis, each chromosome is duplicated and the duplicates remain closely associated. In the second meiotic cell division, the sisters separate into separate cells, so that four cells, each with ten single chromosomes, are produced. Thereafter, these cells, regardless of whether they will develop into eggs or pollen, undergo one or more additional cell divisions and then stop dividing when mature germ cells are produced. Prior to each of these divisions, the chromosomes are again duplicated and the two duplicates separate when the cell divides into two daughter cells. Each daughter cell, and thus eggs and pollen, receives one copy of each of the ten chromosomes. This type of cell division, in which each daughter cell has the same number of chromosomes as the parent cell, is called mitotic.

In corn plants with two mutant *polymitotic* alleles, pollen maturation started off normally. The chromosomes synapsed properly, and the cells divided once and then again in a normal second meiotic division. Then the trouble began. Instead of ceasing to divide after a few additional mitotic cell divisions as happens normally at this stage, the cells continued to divide again and again. Moreover, the mechanism that precisely sorts the full complement of ten distinct chromosomes into each daughter cell failed. The ten chromosomes typical of normal corn pollen and eggs at this stage were divided randomly so that many of the daughter cells wound up with fewer than the ten chromosomes needed for effective fertilization. Beadle noted that, unlike many recessive alleles in which something appears to be missing (e.g., synapsis in *asynaptic*), here something was added; namely, supernumerary cell divisions. He speculated that the mutation involved the absence of some factor whose normal function was to regulate the nature and timing of cell division. Here, as with the *asynaptic* paper, Beadle had thought deeply about his data and proposed a novel mechanism. We know now that there are many genes that regulate biological processes rather than impart some obvious characteristic. The last sentence in the 1930 paper is prescient: "The fact that the *polymitotic* gene seems to be of the nature of a factor regulating the time and character of cell division may be of interest in connection with the hypothesis that certain tumorous growths are due to mutations in such genes as have to do with the regulation of cell division and growth." He credited the hypothesis to a 1929 paper by Leonell C. Strong.[48] Modern research has indeed identified genes, called tumor suppressor genes, whose normal function is to regulate cell division and when missing or present in mutant forms, they lead to uncontrolled cell division and human tumors.

Beadle also carried out experiments that allowed him to associate *polymitotic* with a linkage group, that is, a particular chromosome. He found that *polymitotic* was inherited independently of many genes, but was linked to two that were themselves known to be linked. One was *P*, whose mutant allele gives purple-colored plants (not kernels), and the other *Y*, whose mutant form gives yellow instead of the normally white endosperm (the bulky material filling the kernel). One hybrid plant that he self-pollinated had obtained the normal alleles for *polymitotic* and *Y* from one parent and the mutant forms from another. Of almost 1200 offspring Beadle scored, about 88% still showed either the two normal or the two mutant characteristics, indicating that the pairs of alleles that were together in the original parents were still linked two generations later. He obtained similar results with plants carrying *P* and *polymitotic*. Thus, all three genes were linked on the chromosome numbered three by McClintock.

Most of this work was finished late in 1929 and was sufficient to meet the requirements for a Ph.D. In October, Beadle presented a major, summarizing seminar entitled *Genes Affecting Meiosis in Maize* to the plant breeding department.[49] His 28-page thesis described the work on asynaptic and discussed related recent work by others on mutations that affect meiotic chromosome behavior in other organisms, including *Drosophila*.[50] His Ph.D. was awarded on February 5, 1930. Even many years later, Rhoades affirmed that Beadle's demonstration of the genetic control of chromosome behavior was among the major accomplishments of the explosive period that followed the development of corn cytogenetics.[51]

The Beadles remained in Ithaca for almost another year. During this time, his official title was experimentalist in the plant breeding department and his annual salary was $1380. There was a substantial advantage to remaining in Emerson's circle and having access to the mutant seeds that had been collected, catalogued, and stored over many years along with detailed information about the plants themselves. This material provided many opportunities for original work and Beadle made good use of it. He finished some projects and began others. One mutant he studied was called *yellow stripe* because otherwise normal plants develop yellow streaks on their leaves. Emerson had detected it in 1926, and it seemed to be a recessive mutation. Beadle concluded, from examining leaves in the microscope, that the yellow color was caused by unusual chloroplasts that were yellow, not green.[52] The streaks typically occurred at the midpoint between the major vascular bundles on the leaves. He speculated correctly that the problem was inadequate transport of some important factor from the main plant circulation into the body of the leaf. The factor is now known to be iron; the *yellow stripe* gene encodes a protein that facilitates the transport of iron into cells.[53]

As happy as he was to continue working in Ithaca, Beadle was also ambitious and realized that he had to think about his future. The comfort provided by his legacy disappeared when the banks failed in the 1929 crash, and even Ruth was forced to take a job if she was to continue her studies.[54] There was one way to postpone having to look for a real job while advancing his career and remaining under Emerson's wing within the community of colleagues he valued. That was to apply to the National Research Council in Washington, D.C. for a National Research Fellowship in the Biological Sciences.[55] Both Emerson and Beadle hoped that Beadle would continue to work at Cornell, the first choice indicated on the fellowship application. The California Institute of Technology, Cal Tech, was the second choice. Regardless of where he worked, Beadle proposed to work on two problems in corn genetics: genetical and cytological aspects of Mendelian functional sterility and of hybrids between corn and teosinte, the wild Mexican grass closely related to *Z. mays* he had investigated as Emerson's research assistant.

The plan was for him to continue to work under Emerson's guidance if he stayed at Cornell or with E.G. Anderson if he went to Cal Tech. Emerson preferred that Beadle remain in Ithaca to complete the work on sterility mutants because he assumed that he needed continuing help with the cytology from Randolph and McClintock.[56] Evidently, Emerson didn't realize how competent Beadle himself had become with McClintock's techniques; perhaps he simply did not pay enough attention. The Chief suggested Anderson as an alternative because of his high regard for his former student and expectation that he would provide a supportive environment for corn genetics. Also, he must have thought that Anderson was more likely to welcome collaboration with Ithaca than would one of the Cal Tech fly geneticists. Much as he wanted Beadle to remain at Cornell for the fellowship period, Emerson also recognized that a period at Cal Tech might be advantageous. He was hoping to recruit Beadle for the permanent staff at Cornell, and if that was possible it would be "a compelling argument for sending him to Pasadena" because the proposed appointment would be more favorably received by the university if the candidate was not at Cornell.[57]

Early in January of 1930, Frank R. Lillie, chairman of the NRC board, requested evaluations from Beadle's three major professors. Lillie wanted their judgments on his attainments and "promise in research as indicated by his ideals, ability, originality, judgement, enthusiasm, industry and personality." He also wanted them to rank Beadle among the other graduate students they knew.[58] Sharp, Knudson, and Emerson wrote back promptly.[59,60] Their letters were unanimous in their excellent appraisals of Beadle. They all agreed that he was industrious, enthusiastic, and good at formulating research plans. Knudson and Sharp ranked him at least equal to the best students that they had seen. Emerson's letter recounts Beadle's switch of departments by way of explaining why it had taken him so long to finish his Ph.D. work.

"In the fifteen years since I came to Cornell I have had a good many graduate students working with me. Only a few of them, however, gave great promise of future achievement, and an equally small number have ever done noticeable work since leaving here [Emerson then lists those outstanding people]. It is perhaps too early as yet to say just where in the above group Mr. Beadle belongs, but unless I have wholly lost my ability to judge of future promise by one's graduate work, he distinctly belongs somewhere in that group. Mr. Beadle is energetic and a persistently hard worker, and he directs his energy effectively. He has shown marked originality and initiative in his work. In fact, the surest way to prevent his accomplishing anything worthwhile is to attempt to direct his efforts in minute detail. This was tried by a Professor with whom he was formerly associated and it just didn't work. This experience, I feel sure, was in part responsible for his changing his major to me. I put him at once on his own responsibility, as I do all my students in order to let them show what they have in them or what they have not, and he has come through without over much floundering. I have felt closer to Mr. Beadle than to many of my students because he has come to me for discussion more than is ordinarily true of students. I have no notion that he has done this because he sensed a need of my advise, but rather because he is naturally so enthusiastic about his work that he must talk over any new developments. This may seem to you an unduly enthusiastic estimate of Mr. Beadle, but I am willing to let it stand. If he does not accomplish really important things in the next few years, I shall be not only surprised, but disappointed."

Emerson had also had an inquiry from the fellowship board, requesting Beadle's course grades. He wrote back, "It is quite impossible for me to give you this information. The Faculty of the Graduate School at Cornell is strongly opposed to the recording of grades made in courses by graduate students or the evaluating of graduate work at all in terms of grades or of hours taken in courses."[61]

The Board of Fellowships reported favorably on Beadle's award a few days after its meeting in early February.[62] However, it had a problem with the proposal because they favored having a fellow obtain additional training in a new environment. Beadle recalled that his "fondness for Cornell and Professor Emerson was so great" that he had wanted to stay on after his degree.[63] Emerson wrote back to the board, explaining why he wanted Beadle to remain at Cornell.[64] He acknowledged that Beadle was very desirous of going to Pasadena and that perhaps it was he himself who was trying too hard to have him stay at Cornell. Emerson pointed out that Beadle was an assistant in the department of plant breeding, with an appointment until July 1. Marcus Rhoades would return from his year at Cal Tech thereafter and assume the assistantship, relieving Beadle. However, if Beadle stayed until then, it would be too late to plant corn on the west coast where March 1 is the common planting date. Finally, Emerson again argued, as he had to Anderson, that Beadle needed the advice of the Cornell cytologists for his proposed work. He recommended a solution proposed by Beadle: begin the fellowship at Cornell on July 1, 1930, stay until September or, at the latest, January 1931, and then complete the fellowship year at Pasadena. September 1930 was the earliest departure date because the corn kernels had to be dried, shelled, and wrapped in packets for travel to avoid spreading the corn borer infestations present in the Ithaca region. This plan was approved by Allen, and the first NRC check was sent out on June 5, 1930.[65]

Not only was Beadle's immediate future in research assured, but his annual stipend would increase to $2300, a relatively generous stipend for the time.[66] According to the fellowship application, the only other income Beadle had was the $400 annual interest on real estate mortgages. Presumably this was part of his Grandmother Albro's legacy that was not lost in the 1929 crash. As the year went on, he did consider an offer from the botany department at the University of Washington in Seattle because of the attractions of a permanent professorial position. Finally, his concern about whether corn would grow in the Seattle weather decided the issue and he declined the offer.[67] By mid-October 1930, the Beadles were ready to leave Ithaca. They drove west in their Model A Ford, stopping in Ames, Iowa, and visiting with Beadle's father in Wahoo and Marion's family in Claremont, California, before arriving in Pasadena about November 1.

Beadle left Cornell feeling satisfied with what he had accomplished. He had proven to himself and others that he had the hands and the head to do original genetic research, notwithstanding Emerson's puzzling doubts about his ability to do cytology on his own. He must have recognized that his self-confident decision to change to genetics from Keim's plan that he study ecology was fully vindicated and in the future he could trust his own, independent judgments.

During the four years that Beadle was in Ithaca, the lead in cytogenetics had passed from *Drosophila* to corn because of McClintock's techniques and the continuing difficulty in visualizing the fly's small chromosomes. Soon after Beadle went to Cal Tech, the advantage would return to *Drosophila*, and Beadle would be there to participate. Years later, Beadle acknowledged the wisdom of Allen's insistence that he change institutions and he urged his own and other students to do the same after receiving their Ph.D. degrees.[68]

CHAPTER 5

The Fly Group

When Beadle arrived at Cal Tech, the institution was less than 15 years old, although it had its roots in an earlier small provincial technical school. The astrophysicist George Ellery Hale was the force behind the conversion, and he had a grand vision: Cal Tech would become the MIT of the west.[1] Hale never had small thoughts. He had come to Pasadena from Chicago in 1904 to establish on the nearby Mount Wilson an astronomical observatory for the Carnegie Institution of Washington and, by 1918, he had completed construction of the two largest telescopes in the world. By dint of his own personality, he had recruited Robert Millikan, a Nobel Prize-winning physicist from the University of Chicago, and A.A. Noyes, a distinguished MIT chemist, to help him establish Cal Tech. Research and teaching in physics, chemistry, and engineering dominated the institute's early days, with Millikan serving as chairman of the executive council (the closest thing Cal Tech had to a president) and Noyes chairing the chemistry division. Hale's dream required that a fine city surround the great university, and he saw to the civic development of Pasadena with the same zeal he applied to Mount Wilson and Cal Tech. All of Hale's plans required a lot of money, but he was as astute, tireless, and successful a fund-raiser as he was an astronomer, so he pulled it all off. Within a decade, Cal Tech was producing first-rank research, and the graduates of its small undergraduate school and graduate programs were in demand by academic institutions and the growing southern California industrial sector.

Although Cal Tech was new by midwestern and eastern university standards, its biology division was younger still. Noyes had realized that a program in biochemistry was important for his own emerging research interests and the burgeoning California agricultural industry. There were also inquiries from the Rockefeller Foundation, then one of the very few sources of financial support for research, about Cal Tech's interest in starting a medical school. Fearing the financial burden of a medical school, Hale favored a program in biology with an emphasis on biochemistry. Millikan agreed that a program in biological research would enhance Cal Tech's role. To lead such a program, however, they needed to recruit a person who matched their own scientific statures and could share their vision for the future of the institute. Thomas H. Morgan was the obvious candidate. Millikan was assigned the challenging task of persuading him to leave Columbia and come to Pasadena.

As Beadle had learned as an undergraduate, Morgan was a legendary figure, the preeminent geneticist in the world. Over the years, his books and lectures had emphasized a need for a coalescence of biology with physics and chemistry that fit perfectly with the Cal Tech vision. In the introduction to his book *Experimental Embryology,* Morgan wrote "The working hypothesis is...an attempt to find an answer to some feature of a complex situation in terms of accepted chemical or physical principles. This is conceded in physics and chemistry and enough has been done even in biology to warrant their employment."[2] The trio, "the thinker, the stinker, and the tinker," as Hale, Noyes, and Millikan, respectively, were dubbed locally, knew Morgan well through their common memberships in scientific organizations, most notably the National Academy of Sciences. The Columbia professor had been elected president of the Academy in 1927. At first, Morgan was cool to the idea of coming to Pasadena. He was only a few years away from the statutory retirement age of 65 and was wary of undertaking major administrative responsibilities because he preferred to do research. Still, the opportunity to implement his dream of creating a new breed of biologists grounded in physical, chemical, and mathematical principles was a strong lure. Overcoming his early reservations, he accepted the Cal Tech offer in July of 1927. He would officially become chairman of what he named the biology division in the summer of 1928. He eschewed the title director in favor of chairman to emphasize that he would not be directing the faculty's investigations. Rather, he hoped to recruit independent scientists who would create a highly interactive environment for teaching and research.

In the year following his acceptance, Morgan made plans for recruitments and facilities. He traveled widely in Europe, renewing acquaintances with European biologists and looking out for possible appointments to the Cal Tech faculty. His extensive, often transatlantic, negotiations with Millikan about laboratory space were simplified when two southern California businessmen and the Rockefeller Foundation pledged the resources for construction of a home for the biology division, the William G. Kerkhoff Laboratories. Finding the scientists who could assure success for the new venture was more complex. Morgan persuaded his two former students, Alfred H. Sturtevant and Calvin B. Bridges, now senior colleagues at Columbia, to move with him to Cal Tech. These two appointments would assure *Drosophila* genetics a major place in the new division, especially since three of the Columbia group's graduate students, Albert Tyler, C.C. Lindegren, and Jack Schultz, also came along. Morgan wanted the new division to include studies on evolution and recruited another Columbia associate, Theodosius Dobzhansky, for this area.

R.A. Emerson's students were an important additional source of talent for genetics. Morgan recruited from the University of Michigan Sterling H. Emerson, R.A. Emerson's son, who had just received his Ph.D. in genetics, as well as E.G. Anderson. He hoped that Anderson would develop a program in plant physiology and biochemistry, although he must have known from Emerson that Anderson could be difficult.[3,4] Emerson was as pleased about Anderson's appointment as he was about that of his son, Sterling.[5]

As important as the people Morgan brought to Cal Tech was the interactive and collaborative style of research that had evolved under his guidance at Columbia. Although it resembled, somewhat, the close-knit relationships Beadle was accustomed to at Cornell, the fly group's sociology was unique. Its style had developed in the small and crowded Columbia laboratory that was widely and affectionately known as the "Fly Room." With Morgan's encouragement and occasional chiding, his young acolytes set a pace that few other labs could match. From 1910 to 1920, they elevated *Drosophila* to new prominence as the preferred model organism for genetic research. From 1915 onward, the Columbia fly group was the mecca to which budding geneticists flocked.

The fly group's success owed much to Morgan's early recognition of *Drosophila*'s unique qualities for genetic analysis. When the fly made its way into the geneticist's repertoire, no one knew that crossing-over during meiosis occurs only in female flies, a property that greatly simplifies the interpretation of mating results. Nor was it suspected that the fly's entire genetic complement is contained in only four chromosomes, which considerably eases the task of mapping genes to their chromosomal locations. Corn, the second most important organism used for genetic studies, and the one that Beadle was most familiar with, has ten chromosomes, and crossing-over occurs in both male and female meiosis. Corn also produces abundant progeny (an ear of corn contains hundreds of kernels), but successive generations are rarely obtained more frequently than once a year. Additionally, maintaining *Drosophila* is simpler and cheaper than growing corn or maintaining colonies of animals for genetic studies.

Although Morgan often bragged that he carried out all experiments with his own hands, he realized early that he needed help with the burgeoning fly work. As important as *Drosophila* to the very rapid progress the group achieved was his astute choice of Sturtevant, Bridges, and H.J. Muller to assist him. Each of these three extraordinary students had heard a challenging vision of embryology and the future of genetics in E.B. Wilson's undergraduate course in experimental zoology. Each had also attended Morgan's lectures, and, although they recalled that his presentation was not especially stimulating and lacked any mention of his recent discoveries, they were attracted by the opportunity to work in an emerging field and by his unpretentiousness.

Sturtevant joined as a lab assistant about four months after the discovery of the white-eyed fly in 1910. The youngest of six children, he was born in 1891 in Illinois and grew up on a farm in Alabama. After he graduated from high school, his older brother, Edgar, who was teaching Latin and Greek at the university's Barnard College, urged him to enroll at Columbia. Sturtevant excelled in the entrance exam and was admitted to study zoology in 1908. While growing up, he had taken a keen interest in the pedigrees of his family's horses, and at Edgar's urging he began to read what was then known about heredity. When he heard Morgan's lectures, he saw in Mendel's hypothesis a possible explanation of the pedigrees he had constructed. Edgar encouraged his brother to seek Morgan's views on a manuscript describing his deductions. At the time, Morgan was still skeptical of Mendel's propositions, but he encouraged Sturtevant to publish his findings and invited him to be his assistant. Even as an undergraduate, Sturtevant revealed precocious intuition and insight into *Drosophila*'s secrets as exemplified by how he constructed the first chromosome map (see Chapter 3). After completing his undergraduate degree (1912), Sturtevant remained in Morgan's laboratory as a graduate student, receiving his Ph.D. in 1914. During the ensuing years, he became the mainstay of Morgan's team and was the first to be invited to help found the new enterprise at Cal Tech.[6]

Over decades of research, Sturtevant initiated extensive investigations on some of the most fundamental aspects of chromosomes. One of his most far-reaching efforts concerned the *Bar* mutation, which results in flies with small eyes. The mutation is not very stable and can revert to produce normal-size eyes or mutate further to produce even smaller eyes than *Bar* itself. Sturtevant imagined that the reversion reflected the loss of the mutation, and the enhancement a duplication of the mutant gene, with both phenomena arising from unusual recombination events he called "unequal crossing-over." Later, when it became possible to see the details of *Drosophila* chromosomes, Muller and Bridges confirmed Sturtevant's hunch. Now, it is clear that unequal crossing-over occurs in all organisms with profound consequences for the building and evolution of chromosomes. Another universal genetic property that Sturtevant discovered is the influence on gene activity of the position of the gene on the chromosome. Although laboratory genetics occupied much of his time, Sturtevant always reserved time for field biology and natural history, both of which kept him constantly concerned with evolutionary processes and the role of genes in evolution.[7] His interest in the taxonomy of various *Drosophila* species, for example, helped him to establish the occurrence of inverted segments of chromosomes.

Bridges, the second of the undergraduate threesome, was born in 1889 in upstate New York and was orphaned early in life. Because he insisted on supporting himself throughout, he did not complete high school until he was 20, but his excellent academic record led to a scholarship at Columbia, where he began a year after Sturtevant. Coincidentally, Bridges and Sturtevant were enrolled in the same zoology course, and Bridges, too, was turned on by Morgan's personality and by what he heard about the promise of genetics. He petitioned Morgan for a place to work in the lab and was given a job as a bottle washer. When Bridges spotted a fly with an abnormal eye color through a thick glass bottle, a character that most others would have had difficulty detecting even with a microscope, he was promptly promoted to lab assistant.[8] Over and over, Bridges displayed an uncanny talent for detecting mutant flies amid overwhelming numbers of their wild-type brethren. Of the 365 *Drosophila* mutants listed in the Morgan collection in 1925, Bridges was responsible for the discovery of 240![9]

After completing his undergraduate work (1912), Bridges initiated a series of cytological studies of *Drosophila* chromosomes that provided detailed and convincing proof of the chromosome theory of heredity. In the course of this work, he identified the Y chromosome as being morphologically distinct from the considerably larger X chromosome, thereby overturning the widely held belief that the X and Y chromosomes were indistinguishable. He also explained an earlier anomalous observation by Morgan. As expected, crosses between females with white eyes and males with wild-type eye color gave rise predominately to males with white eyes and females with normal-color eyes; the males inherited one mutant X chromosome and a normal Y chromosome (which lacks the relevant gene for eye color) and the females one normal X from the father as well as a mutant X from the mother. The surprise was occasional sons with normal eye color and daughters with white eyes. Bridges explained the surprise by suggesting that during meiosis leading to eggs, the two X chromosomes occasionally failed to separate so that some eggs retained two X chromosomes and an equal number received none.[10] Bridges surmised that fertilization of eggs with two mutant X chromosomes by sperm containing a normal X chromosome would yield females with two mutant X chromosomes and one normal X chromosome and normal eye color. However, when eggs with two mutant X chromosomes were fertilized with sperm carrying the Y chromosome, females with two mutant X, one Y, and white eyes would result. Fertilization of eggs lacking any

X chromosome by sperm carrying the father's normal X chromosome would give rise to the males with normal eye color because they were what Bridges termed an XO male. No viable offspring were expected if an egg lacking an X chromosome was fertilized by a sperm carrying a Y chromosome because the X chromosome carries many essential genes.

Bridges then confirmed experimentally that meiosis does indeed occasionally go awry in just that way. He referred to the normal separation of two paired chromosomes during meiosis as disjunction. The term "nondisjunction" was adopted for unequal chromosome segregation, the rare occasions when both homologs of a chromosome pair end up in only one of the two germ cells.[11] Bridges' astute cytological observation not only explained Morgan's anomalous findings, but also provided the most definitive evidence at the time that X chromosomes bear the genes for the sex-limited characters. Furthermore, it put to rest any remaining doubts about the chromosomal basis of Mendelian inheritance.

Bridges is remembered to this day because of his iconic series of drawings indicating the positions of the various bands (maps) in the four chromosomes, drawings that are still the reference standards. These maps provided a direct way to localize mutations resulting from a variety of chromosomal rearrangements to specific bands of the affected chromosome. They were made possible by the recognition that chromosomes in the fly's salivary gland cells (as well as a few other tissues) are unusually large and easily seen in resting cells at the usual magnification.[12] Such "polytene" chromosomes, as they were named, result from repeated rounds of chromosome duplication without the usual intervening cell divisions. As a consequence, more and more copies of each chromosome accumulate in a giant bundle and have a characteristic and reproducible banded appearance when viewed with even a low-powered microscope. Both Bridges and T.S. Painter developed methods for staining the bands in polytene chromosomes, which made it possible to couple cytological observation with genetic information. At the time and continuing to the present, such maps provided an indispensable foundation for the genetic analysis of *Drosophila*. For example, Morgan, Sturtevant, and Muller had earlier inferred the existence of chromosomal anomalies such as deletions and inversions as well as unequal crossing-over. Using Bridges' banding patterns, the genetic data could now be correlated with visually demonstrable structural changes in the respective polytene chromosomes. There was no longer any doubt that chromosomes were the carriers of the genes.

Aside from his own investigations, Bridges contributed to the lab's efforts by his remarkable skill at constructing flies with unusual genotypes, many of which made a variety of novel experimental studies feasible. Bridges' remarkable work formed the basis of his Ph.D. thesis submitted in 1916. Some years later, Sturtevant noted that Bridges' achievements required great patience, accurate observation, technical skill, ingenuity, and an understanding of what was important.[13] Bridges' interests in biology were more limited than those of Morgan and Sturtevant. He was first and foremost an experimentalist. His renowned observational gifts enabled him to focus attention on his subjects and carry out manipulations that frustrated others. The fly group's research at Columbia and Cal Tech could not have succeeded so well without Bridges' devoted and meticulous stewardship of the lab's mutant fly collection. Dobzhansky described Bridges with an expression which exists both in Russian and German but not in English: "Gotterfunken, a person who has in him a spark of God."[14] He recalled that "Once in a while, he [Bridges] would have this inspiration flying from Heaven to him, and for two months, he would do the work of genius. Then he would relapse back into his routine and gadgeteering. Otherwise he was a man of the utmost naivete." Many years later, Max Delbrück commented that he had never met such an easygoing person in Europe and he was especially fascinated by Bridges' unorthodox and utterly unpretentious way of life.[15]

Bridges' almost scandalous lifestyle fascinated his colleagues and even outsiders. He was a staunch supporter of the political left and defended its actions against all comers, especially since that stance was unpopular. He was also an outspoken advocate of free love and practiced what he preached, to the consternation of Morgan and Sturtevant. Being married and the father of three children did not stop his escapades as women flocked to his side. He made advances to every woman he ever met, anywhere, even to Dobzhansky's wife, but he readily accepted "no" for an answer when rebuffed. On a planned visit to Cold Spring Harbor, he was warned by Demerec, the lab's director, to stay away from the young women there. But his unaffected manners, outgoing personality, and stunning good looks went a long way to forgiving his chosen lifestyle.

Bridges enjoyed living to the fullest, and was relatively unconcerned that he was not invited to be a Cal Tech faculty member; he continued as a research associate at the Carnegie Institution of Washington until he suffered a heart attack in December 1938 while attending a meeting at Cold Spring Harbor. Although he recovered sufficiently to make it back to Pasadena, he died from a massive heart attack three weeks later. Many years later, Beadle admitted to being tortured by the possibility that he was partly responsible for Bridges' death. He recalled "how on a cold evening in Palo Alto I let him have my Model A roadster with the top down to take Kitty out in the Stanford hills. He came back pretty chilled and it wasn't too much later that he died."[16]

Morgan particularly, but all who knew him (including Beadle), realized that science had lost an incredible scientist, an invaluable colleague, and a free spirit.[17] After Bridges' death, Morgan found a catalog of his "women," together with dates, and had it destroyed promptly.

Herman J. Muller was arguably the most brilliant of Morgan's early students, but he was also the most contentious and mercurial. He was born in New York City in 1890 and, in keeping with his lifelong leftist political views, he often called attention to the fact that he and Stalin shared the same birthday. Like many New York children, his interest in science was sparked by frequent visits to the city's Museum of Natural History. Muller's intellectual brilliance was already evident in high school, and he won a scholarship to enter Columbia College in 1906. Like Sturtevant and Bridges, who arrived 2 years later, Muller was "turned on" by Wilson's masterful one-semester course on chromosomes and heredity. Reading R.H. Lock's book on *Variation, Heredity and Evolution*[18] convinced Muller of the generality of Mendelism in explaining heredity. He was particularly persuaded that genes were true physical entities and that they were associated in some way with chromosomes. Convinced that he would be a biologist, Muller founded a biology club for Columbia undergraduates where he "lectured" frequently on what in retrospect were exceptionally mature views on genes and heredity. It was at the biology club meetings that Muller learned about the exciting developments in Morgan's laboratory from Sturtevant and Bridges. Muller sought to join Morgan's group but was rebuffed, ostensibly because of a shortage of laboratory space. Nevertheless, Sturtevant and Bridges kept him abreast of the fly group's work.

Upon graduation in 1910, Muller accepted a fellowship in physiology at Cornell Medical College, not because the subject interested him, but because he needed an income. Within 2 years, however, he tired of neurophysiology research and reapplied to join Morgan's laboratory. This time, Morgan accepted him as a Ph.D. student and gave him space at the table at which everyone in the budding fly group worked side by side. Even at that stage, Muller harbored more advanced ideas about the nature of genes and their action than did the others, particularly Morgan. Muller chose not to work on the problems that Sturtevant and Bridges were exploring, in large part not to get caught up in their "wake." Nevertheless, he kept abreast of what they were doing, and he was prone to making suggestions when he thought he could help.

For his own project, Muller sought to understand more fully how crossing-over between homologous chromosomes caused linked alleles of different genes to become unlinked during formation of the germ cells. He constructed flies whose chromosomes were marked by many mutations along their length and determined the frequency of crossing-over in different chromosomal regions. Surprisingly, he found that there were striking differences in the frequency of crossing-over between certain sets of alleles and between the different chromosomes. He recognized that the sum of the distances between closely spaced alleles did not add up to the distance between the alleles that were at the ends of the cluster. Muller surmised that if, occasionally, there were closely spaced double crossovers instead of a single crossover, it would seem as if no crossover had occurred. In that event, one would conclude that the distance between alleles is less than it really is. Furthermore, he reasoned that anomalous structural features, e.g., inversions, insertions, or deletions in one of the homologous chromosomes, would suppress crossing-over because continuous pairing of two homologous chromosomes is required before exchanges can occur. His studies showed clearly that Sturtevant's mapping strategy, which was based on the frequency of crossing-over, was too simpleminded.

Unlike Sturtevant's and Bridges' relative lightheartedness, Muller's manner was intense, and he took himself and all he did very seriously. His relations with Morgan were somewhat strained, and he decided that it was best for him not to remain in Morgan's shadow after he received his Ph.D. in 1916 but to accept a position in the new biology department at Rice University in Houston, Texas. Although he rejoined the fly group at Columbia in 1918, his stay was short-lived because he was denied promotion. Muller attributed this action incorrectly to a lack of support from Morgan, but it was Wilson, the head of the department, who made the decision.[19] Once again he moved, this time to a post as professor of genetics at the University of Texas in Austin. It was there, in 1927, that he made the dramatic discovery that exposure to X rays was a potent means for inducing mutations in flies.[20] This was convincing evidence that genes were physical entities and that their structure could be altered by radiation. At about the same time, L.J. Stadler at the University of Missouri independently discovered that X rays were highly mutagenic in corn and barley plants.[21] Later, he discovered that ultraviolet (UV) radiation was also mutagenic. These artificial means for producing mutations greatly accelerated the study of genetics in a variety of organisms in which spontaneous mutations were difficult to find. For this breakthrough alone, Muller received the Nobel Prize in physiology and medicine in 1946. Both X-ray- and UV-induced mutagenesis turned out later to be critical in Beadle's revolutionary studies on gene function.

In the early 1930s, Muller became involved with leftist movements and radical student groups at the university. In time, his political activity on campus became more and more problematic, and he decided to join N.I. Vavilov, the leading geneticist in the USSR. By 1937, however, Muller's frustration with Trofim Lysenko's heavy-handed tactics and destructive influence pushed him to leave the Soviet Union, and eventually he accepted a position at Amherst College in Massachusetts where he stayed until 1945, when he became a professor of biology at Indiana.

The fly group's effort from 1910 to 1915 transformed the field of genetics. Their identification of chromosomes as the physical locus of genetic information and their explication of rules for the way in which that information is transmitted from one generation to the next were electrifying discoveries. C.H. Waddington wrote that "Morgan's theory of the chromosome represents a great leap of the imagination comparable to Galileo and Newton," and C.D. Darlington added that "the chromosome theory begins to appear as one of the great miracles in the history of human achievement."[22] But Morgan understood, perhaps better than most, the extent to which their success was attributable to Sturtevant's, Bridges', and Muller's unique personalities and distinctive way of approaching problems. The dynamics of their interactions owed much to the backgrounds, intellectual gifts, dissimilar values, and lifestyles that each of the members brought to the enterprise. Morgan, of course, was the guiding force in setting the tone. He was imposing without being formal. His age and station never interfered with his scientific dialogue with his junior colleagues. He was anything but the "Herr Geheimreit Professor" who demanded homage and blind obedience. When his colleagues referred to him as the "Boss," it was more out of respect and affection than recognition of his lofty status.

The rapid and remarkable achievements of the fly group intimidated other *Drosophila* geneticists. It seemed that the most important problems had been sequestered to the Fly Room. Although the Columbia group's work overshadowed that of other labs, they were generous in making their mutant strains and information freely available to others through the *Drosophila* exchange network. Of course, the Columbia fly group benefited handsomely from these exchanges since they could keep abreast of the findings made by the recipients of their "generosity." Nevertheless, such token "collaborations" did not alleviate the resentment toward the Columbia group's monopoly and its influence in monitoring competitors' publications.[23]

The intellectual energy supplied by Morgan and his students overcame the shortcomings of their physical surroundings. Located on the sixth floor of Schermerhorn Hall on the Columbia University campus, their work space was a single room about 16 x 23 feet crammed with eight closely packed desks. Columbia was not yet enveloped by large residential apartment houses, and the view from the main lab included goats grazing in the nearby pasture. Visitors were struck immediately by the room's messiness and makeshift appearance. Gauze-topped milk bottles with flies flitting within vied for space on the tables and shelves with papers and debris from completed experiments. Part of the fly group's mystique derived from the suspicion that the milk bottles used to house the flies were harvested from the front steps of nearby houses. While the flies were confined to their assigned milk bottles, cockroaches resided in every nook and cranny, roaming freely to feast on fly food and other edible leavings. Mice, too, had a field day finding sustenance in the leftovers accumulated in the room's filth. The room reeked of yeast and the fruity odor of overripe bananas. Occasionally, friends and colleagues in the building came to sample the stalks of bananas that adorned the walls.

Morgan himself set the tone with his unpleasant laboratory decorum. His unread mail accumulated on his desktop, and when the pile threatened to inundate his current work, it was shuffled onto a nearby desk, only to be returned or thrown out when the occupant of that desk returned. Whereas others disposed of their flies in a jar of oil, Morgan squashed his on the porcelain observing plate that was already layered with the leavings from earlier experiments.[24] He was also reputed to appear unkempt, occasionally even being mistaken for a janitor. Once, just before a lecture he was to give, someone noticed that his collar had been ripped away from his shirt, but Morgan brushed all concerns aside by saying "never mind, Bridges will fix it with adhesive tape," which he did.[25] However, in all the pictures extant of Morgan in the lab, he wears a white shirt, tie, and vest, a style that would be considered anything but informal today.

Cross talk among the Fly Room's occupants created a nearly continuous din to which Morgan contributed as he talked constantly while counting and examining flies. The conversations were lively, occasionally boisterous, and peppered with jokes and bursts of laughter. One foreign visitor was bewildered by it all, having "expected something wonderful in God's country." But she noted "how different everything was from the European universities" where "the professor was generally aloof, at a distance, who made a round in the lab now and then to guide and answer questions." She likened the setting to a gathering of students having a good time with the professor and thought "Well this must be the way in the U.S." Only later did she realize "that the atmosphere in the Morgan group was unique" and resembled a family where Morgan knew about everybody's life, family, money, and even love affairs.[26] But the Fly Room was more than just a place for fun and games. Morgan encouraged and treasured this almost carefree, exciting style as the way he thought science should be done.

Sturtevant attributed the atmosphere to Morgan's attitude, strong critical sense, generosity, open-mindedness, and remarkable sense of humor. He also acknowledged the "unfailing support and appreciation of the work" by Wilson, then chairman of the zoology department.[27] The Fly Room was a place where "everyone did his own experiments with little or no supervision but each new result was freely discussed by the group...we discussed, planned and argued all day, every day. I've sometimes wondered how any work got done, with the amount of talk that went on."[28] Like a father, Morgan supported his young charges and even fed their growing cockiness. Morgan, Sturtevant, and Bridges forged a special bond based on the synergy of their scientific efforts, their shared commitment to science, and their mutual admiration and respect.[29]

A unique feature of the fly group's sociology was the practice of depersonalizing the source of ideas or suggestions—both were presumed to be group property. Moreover, group members were not expected to stake out problems for their exclusive study, an understanding to which members did not always adhere. Sturtevant recalled this "give-and-take atmosphere" as one in which as "each new result and each new idea came along, it was discussed freely by the group. The published accounts do not always indicate the source of the ideas. It was often not only impossible to say, but was felt to be unimportant, who first had an idea...I think we came out somewhere near even in this give-and-take, and it certainly accelerated the work."[30] Authorship was usually credited to the ones who carried out the experiments. Muller was the only one who bristled at this practice.

From the time he joined the fly group, Muller's penetrating mind was quick to provide theoretical interpretations for the frequent unexpected findings, often before the group had digested the results. Not surprisingly, his ready explanations and suggestions for experiments to confirm his hypotheses tested people's patience, as did his boasting about having provided the solutions to problems that others had defined. After leaving Columbia, he was outspokenly critical of Morgan, always believing that the Boss and Sturtevant had taken credit for achievements that were rightfully his own. Morgan, however, acknowledged the importance of Muller's intellectual contributions to the *Drosophila* group's discoveries by including him as a coauthor in the monumental publication that summarized the chromosomal basis of inheritance.[31]

One of Morgan's less admirable qualities was that he was stingy to a fault, a quirk resented by the members of the fly group. Much of their equipment was makeshift, and Bridges was frequently called on to improvise and design items that were readily available commercially. Microscopes were rare, and hand lenses were deemed sufficient for examining flies; Morgan himself preferred a jeweler's loupe. Part of this frugality stemmed from the early days of the *Drosophila* project, when only scant funds were available from the department budget. Subsequently, as the work blossomed, Morgan sought outside funding and received grants from the Carnegie Institution of Washington from 1914 until he retired. Still, he was proud when he could return unspent grant funds at year's end, even when it meant denying valid expenditures.

On occasion, the students vied to concoct ways for convincing the Boss to approve the purchase of needed items. Soon after Beadle came to Cal Tech, he needed a rather inexpensive balance but was assured that Morgan would not approve the $10 expenditure. He persuaded James Bonner, a plant physiologist who could justify such equipment, to request it and then give it to him. Sometimes, Morgan's response to requests for equipment depended on his mood. Speculating that Morgan was most likely to be agreeable on Sunday mornings, Beadle succeeded, perhaps during one of their occasional tennis matches, in getting his approval for the purchase of a new microscope oil immersion objective. Most considered that a triumph.[32]

Shortly before the move to Pasadena, Jack Schultz joined the group as a Ph.D. student. Born in New York City in 1904, Schultz's intellectual interests and brilliance surfaced early, and he made his way through high school with such distinction that, like Bridges, he received a scholarship to attend Columbia. He loved the life of learning but also needed to earn some money and answered an advertisement to wash bottles and make fly food for Morgan's laboratory. Driven by his innate curiosity, he quickly absorbed the basics of the genetic experiments. Sturtevant and Bridges recognized his potential, and within short order he was making contributions to the discussions in the Fly Room. Later, Morgan accepted him as a graduate student. His Ph.D. thesis, which was completed in 1929 after the move to Cal Tech, dealt with a genetic curiosity: A large class of mutations dubbed "*minute*" produced nearly the same phenotype even though they mapped to many different chromosomal locations.[33] That posed a serious conceptual paradox: How could a variety of mutations exert so similar a phenotypic effect during embryonic development? These experiments provided the foundation for Schultz's lifelong interest in understanding the molecular basis of gene action during fly development and for his close association with Beadle at Cal Tech.

At Cal Tech, Schultz initiated a spectrographic comparison of the eye-color pigments in wild-type flies and those in the many eye-color mutants that had been accumulated over the years. These studies were fundamental for Beadle's later work on the development of the eye pigments (see Chapter 7). Beadle was also influenced by Schultz's perceptive review article that laid out the principal conundrums concerning gene function.[34] In time, Schultz became more and more concerned with learning the chemical constitution of the genetic material and the way in which it functioned to produce an organism's phenotype rather than probing further into the mechanisms governing the transmission of genetic information. To pursue this interest, Schultz spent two years in T. Caspersson's laboratory in Sweden applying new analytical techniques in microspectrophotometry to chromosomal nucleic acids. His return to Pasadena when World War II started was disappointing because Morgan failed to provide funding for the equipment he needed to continue those studies. Instead, he insisted that Schultz take on the care and feeding of the *Drosophila* stocks, as Bridges had done before his premature death. Schultz's relations with Sturtevant and Morgan became more and more troubled, and he left to start a genetics unit at The Lankenau Research Institute in Philadelphia. Although Schultz's publications were relatively sparse, they were provocative and visionary. Even among the most ardent conversationalists in the fly group, Schultz held his own, offering interesting ideas to all who would listen.

At Columbia, Morgan attracted several additional gifted postdoctoral fellows and graduate students who went on to have outstanding independent research careers. Especially notable was Curt Stern, a visitor from Germany, and Theodosius Dobzhansky, who had come from Leningrad on a Rockefeller Foundation fellowship. Dobzhansky was born in 1900 in a small town near Kiev where he lived through the turbulent years of the Bolshevik revolution.[35] Early on, he developed a love for biology through collecting and characterizing butterflies and ladybird beetles. After studying biology and graduating from Kiev University, he visited Kozlov's institute in Moscow where he began to experiment with the collection of *Drosophila* mutants brought there years earlier by Muller; he published the first Russian papers on *Drosophila* genetics. This attracted the attention of Y. Filipchenko, the head of the new department of genetics at the University of Leningrad, who invited Dobzhansky to become his assistant and encouraged him to continue working with *Drosophila* and subsequently to go to Morgan's laboratory. Dobzhansky imagined Morgan as "next to God," the fly group as "semi-divine beings," and the lab as "next to heaven." This idyllic view was shattered when he arrived and found that "Morgan's laboratory was very small, poorly equipped and positively filthy." He had expected that "any laboratory in the United States must be luxurious and, of course, Morgan's laboratory should be the top of luxury."[36] It was, however, less commodious and certainly far dirtier than the laboratories at the University of Leningrad.

Sturtevant and Bridges helped Dobzhansky acquire the vocabulary and skills to initiate his own projects. Within a few years, he had made significant contributions to understanding the mechanism of crossing-over and clarified certain anomalous features of the genetic maps disclosed by unexpected frequencies of crossing-over. Using mutant strains containing one or another anomalous chromosome that resulted from rare crossovers between nonhomologous chromosomes,[37] Dobzhansky proved that the linear arrangement of genes based on linkage relationships was, as Sturtevant had proposed, colinear with the genes' physical order in the chromosome. Dobzhansky also extended Muller's findings on the anomalous estimates of crossing-over frequency in different parts of the chromosome, particularly near the spindle attachment site.[38] This phenomenon of reduced crossover frequency in certain chromosomal regions later served as Beadle's introduction to *Drosophila* genetics.

Dobzhansky worked incessantly at the bench and was obsessed by the need to publish frequently and widely. James Bonner, Dobzhansky's colleague at Cal Tech, is reported to have overheard him say "a month gone by without a paper sent to press was a wasted month."[39] But Dobzhansky's principal claim to fame was his role in formulating the modern synthesis of evolutionary theory by expanding the potential of *Drosophila* for experimental studies of evolution. Initially, he collaborated with Sturtevant, who was also interested in the genetic basis of speciation. Through their common interest in evolution, their relationship became less that of student and mentor and more that of colleagues. They relished their joint forays into the field to collect flies. To distinguish the different races of flies, they studied the characteristic features of polytene chromosomes as markers and combined genetic, cytological, and field investigations to explore issues of hybrid sterility and the evolutionary history of regional populations. Using different patterns of chromosomal inversions among several *Drosophila* species in their natural habitats, they explored the natural distribution of related races and attempted with some success to infer the sequence in which the inversions arose and even to construct phylogenies based on these features. Sturtevant took the lead in developing an extensive plan of work for a large-scale study of the genetics of natural populations but, after 1936 when the two close colleagues had a falling-out, it was left to Dobzhansky to implement.[40] There are many different recollections about the cause of the rift between Sturtevant and Dobzhansky, but the most likely explanation appears to be that Sturtevant cooled when he suspected that Dobzhansky tended to overinterpret his findings.[41]

Dobzhansky's book, *Genetics and the Origin of Species*, published while he was still at Cal Tech, is considered one of the most important and influential 20th-century theoretical works concerning evolution.[42] It provided a conceptual framework that stimulated experimental work for many years. Dobzhansky remained in California until 1940, when he returned to

Columbia to become professor of zoology. There he turned his interest to the evolutionary history of humans. He became a champion of human individuality, arguing that it was the pervasiveness of genetic variation that provides the biological foundation of individuality. Since there is more genetic variation within any human genetic race than there are genetic differences between races, Dobzhansky argued that individuals should be evaluated by what they are, not by the genetic race to which they belong. Now, with confirmation of Dobzhansky's view by the sequence of human genomes and enlightened social attitudes about genetic groups, this seems obvious, but at the time it was novel, even revolutionary.

It's not surprising that during the years in which the Fly Room was in operation its occupants were the subject of untold numbers of anecdotes and reminiscences by visitors and those who experienced its ambiance. Taken together, they tell of a very special group of individuals operating at the very edge of existing knowledge, all the while adjusting to and often competing against each other. Serious, lofty science, yet a joyous adventure! The fly group's social structure, novel for the time, was a precursor of a new kind of group interaction that has since been adopted in many areas of biology and indeed throughout science.

Morgan's fly boys, as the members of the *Drosophila* group were often called, were undeniably brilliant, but more importantly, the talents they brought to the lab were complementary. Although precise and creative in his thinking, Morgan was uncomfortable dealing with mathematical and quantitative aspects of the work. But Sturtevant, Bridges, and especially Muller were at home with these matters, and they frequently had to guide Morgan over such difficulties. Although Morgan had always disdained speculation that was not strongly backed by experiments, his young associates relished the challenge of theorizing on the implications of even preliminary results. As the evidence for the gene's physical reality became more widely appreciated, Morgan had difficulty accepting the molecular or even physiological explanations of gene action that were being advanced by his young colleagues.

Morgan believed in a simple, uncomplicated, unpretentious way of life in science, a belief that was fully shared and practiced by Sturtevant. The Boss was most comfortable with close friends and colleagues and was always ready for serious discussion of either scientific or personal matters.[43] Being closest to Morgan in background, scientific interests, and philosophy, Sturtevant was privy to Morgan's thinking. Neither Morgan nor Sturtevant had any patience with "established dogma" or pretentiousness. The Boss was expert at banter and liked to tease people, especially those who he believed couldn't take it. Like Morgan, Sturtevant was easygoing, self-possessed, and witty when he chose to be, but his outwardly gentle manner also hid a sharp-edged sense of humor that was provoked particularly by those of whom he disapproved. His occasionally blunt criticism of a scientific work or presentation offended some colleagues, but most often his perceptive insights were welcomed. Dobzhansky characterized Sturtevant "as a very steady worker, no flashes, no sparks, just good steady work every day including Christmas day, New Year's day, always in the laboratory counting flies or reading the literature."[44]

Such was the culture Beadle entered when he arrived at the Cal Tech biology division. They welcomed him as they had done each new recruit before. And like those who preceded him, he was expected to participate fully in the group's activities and to formulate his own research program. As usual, he was confident that he was intellectually and temperamentally ready. Having planned to continue his work on the sterile mutants of corn meant that most of his time would be spent at the Cal Tech experimental gardens known as the "Farm," a 20-acre parcel at Arcadia, about a half-hour's bicycle ride from the Cal Tech campus. The mentor named in his fellowship application, E.G. Anderson, lived and had his laboratory and experimental fields on the Farm. As a bachelor, a "loner," and the only maize geneticist in the division, this suited him well.[45] He was also a proficient *Drosophila* geneticist, and Muller credited Anderson with the insight to suggest, as early as 1920, that the occasional duplication of small segments of chromosomes was an important evolutionary process, an idea that contemporary biologists are inclined to think has a more recent provenance.[46] Among the buildings at the Farm, Anderson occupied the larger house rent-free. In return, he was expected to provide a helper who would cook and keep house for him and for other Cal Tech people who stayed over when they came to work in the fields.[47] Cal Tech students went out to the Farm for his genetics course that met in the loft of the barn, but "Andy was hopeless when it came to teaching and lecturing."[48]

Although Anderson kept to himself and rarely participated with the others on the main campus, Beadle was strongly attracted to the fly group and quickly became part of its scientific culture. Initially, corn research was his main occupation, but before too long the excitement of *Drosophila* genetics would prove irresistible. As at Cornell, where he changed from ecology to genetics, he would not hesitate to change his scientific direction and place himself where the science was most compelling.

CHAPTER 6

From Corn To Flies

Southern California was Marion's home, but it was to all new to Beadle. The climate and surroundings were unlike anything he had known. Although Pasadena itself was flat like the countryside around Lincoln and Wahoo, a glance upward revealed majestic mountains nearby to the north. Unlike the green hills surrounding Ithaca, these mountains were brown and only occasionally snow-capped in winter. The palm trees and other semitropical plants were new to him as well, and the prospect of gardening year round was a welcome change.

Beadle moved his family into a small house on the Farm. He and Anderson had a lot in common, including childhoods on eastern Nebraska farms, undergraduate years in Lincoln, and graduate work with Emerson at Cornell. However, although they seemed to get along, they never became close friends and never published any joint papers.[1] In contrast, Beadle easily established friendship with Sterling Emerson. Charles Burnham, another midwesterner and occasional Cornell colleague, was at the Farm working with Anderson when he arrived. Burnham had spent two summers in Ithaca (and the intervening winter at Harvard) tending his corn in the Hole and learning cytogenetic techniques from McClintock.[2]

A year after the Columbia group moved to Pasadena, Morgan had added to his all-American genetics faculty two English-born scientists, Henry Borsook, a physician and biochemist, and Kenneth V. Thimann, a plant biologist. Borsook's scientific interests covered a wide range of biochemical questions, including vitamins, nutrition, and the energetics of enzymatic reactions. Thimann was already at Cal Tech as a postdoctoral fellow working on plant growth hormones when he was appointed an assistant professor. Morgan tried but failed to entice Milislav Demerec, another former Emerson student, away from Cold Spring Harbor or Curt Stern, who had returned to Germany after his earlier postdoctoral training period at Columbia.[3]

The Morgan who came to Cal Tech was not the geneticist of the *Drosophila* group's heyday, although he was still affectionately and respectfully called the "Boss" by his colleagues. Some people thought that fly genetics had, by the late 1920s, become too quantitative, even too complex for

Morgan's intuitive scientific style.[4] He no longer kept up with the experimental advances, and leadership of the fly group's work fell increasingly to Sturtevant and Bridges, who were also deputized to look after the students. However, Morgan still participated actively in national and international scientific affairs and continued to produce scientifically influential talks and publications. A large part of his time was taken up with recruiting faculty for the new biology division and raising the funds needed for completion of the Kerckhoff Laboratories. But even these burdens and the passing years did not quench Morgan's devotion to scientific work and his belief in the primacy of experimental results over theorizing and speculation.[5] One of his favorite expressions was "let's try it," where trying meant doing an experiment.[6]

After coming to Cal Tech, Morgan picked up his old interest in embryology. He spent weekends at the Institute's Kerckhoff Marine Laboratory at nearby Corona del Mar and summers at the Marine Biology Laboratory at Woods Hole on Cape Cod. Albert Tyler, who was appointed a faculty member after finishing his Ph.D., worked with him, and the others, too, were welcome to spend weekends at the seashore with the Boss. But his withdrawal from much of the day-to-day activity in the division meant that his interactions with students and postdoctoral fellows such as Beadle were only occasional. Students were considered a group responsibility, and official assignments of students to individual faculty members were largely ignored except for administrative purposes. To Sturtevant and Bridges, students were interesting colleagues, not commodities who would carry out the faculty's research. It followed that the students were responsible for their own research ideas and findings, and their own destinies.[7] The spirit and style of the Fly Room had been transported to California, including an inviolable open-door custom to encourage and facilitate exchanges, interactions, and collaborations.

Schultz and Dobzhansky, particularly, helped newcomers such as Beadle adapt to the unusual laboratory culture. Decades later, Beadle wrote to Schultz, "As a National Research Council Fellow, I suppose I may have designated Andy (Anderson) as my sponsor. But almost immediately I felt I was a member of the group and in fact I worked with Dobzhansky, Sterling (Emerson), Sturtevant and others with no formalities in the various shifts. Through Morgan, Sturtevant, Dobzhansky, Albert (Tyler), Bridges, you and others, the Columbia fly Lab spirit was, it seems to me, directly transferred to Cal Tech Biology and although it became diluted in other areas...it persisted in the genetics group throughout Morgan's time and beyond."[8] Schultz always felt that Morgan set the spirit in the lab by insisting on the paramount importance of getting on with the work; cooperation was always taken for granted and taught by example. The fly group's lab and lifestyle were comforting for

Beadle since they closely resembled the way Emerson's group at Cornell functioned. Many years later, he acknowledged how this free and easy, largely noncompetitive, style influenced the kind of setting and spirit he tried to create in his own research group.[9]

Soon after arriving in Pasadena, Beadle presented a seminar, as was expected of a new fellow. To everyone's surprise, Morgan, who usually fell asleep at seminars, stayed awake and even took notes. People thought that Morgan found something especially interesting in Beadle's presentation. When Beadle was called into Morgan's office, the Boss said that indeed the seminar was good and then handed him his notes; they contained a list of words that Beadle had mispronounced. Perhaps all the farmers in Ithaca never noticed his Nebraska accent, but Morgan had.

Letter and paper writing took up at least some of Beadle's time in the early months in Pasadena. He even worked with Keim on the description of work he had done as an undergraduate at Nebraska.[10] He and Emerson were still collaborating on the teosinte work and they corresponded frequently, exchanging information and seeds for various projects. As Emerson's research assistant, he had been responsible for the teosinte project, and now the work had to be prepared for publication. Emerson wrote, telling him about related experiments being done by Randolph and by McClintock and her graduate student, Harriet Creighton, hinting that he worried about a three-way competition among them all.[11] Two papers by Emerson and Beadle that were published in 1932 are the first chapters in the long tale of Beadle's work on the origin of corn (see Chapter 18). Beadle was especially interested in experiments concerning crossing-over between corn and teosinte chromosomes in hybrids between the two plants. These experiments were pertinent to general questions about the mechanism of crossing-over, a topic of great interest to the *Drosophila* group at Cal Tech and one that Beadle investigated in a collaboration with Sterling Emerson—his first foray into *Drosophila* genetics.

Besides the writing, Beadle got right to work completing experiments he had begun at Cornell. In March, four months after he arrived in Pasadena, he was planting seeds for the final phase of his investigation of a third mutation that caused sterility in corn, one he named *variable sterile*. By the time he wrote the work up in November of 1931, he had interacted enough with the *Drosophila* group to realize the many common questions inherent in studying the genetics of the two organisms. The paper on *variable sterile* starts out with a brief review of genes that affect chromosome behavior in several plants and animals, including the fly.[12] Anderson may have been content to stick to the Farm, but Beadle preferred the stimulating scientific conversations with Morgan and the group on the main campus. And he found the Boss a challenging competitor on the tennis court despite the 36-year age difference. Morgan continued to lecture in the general biology course for Cal Tech students, and Beadle assisted in the laboratory. Years later he remembered how hard Morgan labored over his lectures. He also wondered whether Morgan's occasional errors were a deliberate device for keeping the students interested, but concluded that Morgan probably just made mistakes.[13] Other exciting things went on at the main campus. Einstein visited during the 1931–1932 academic year, and Linus Pauling, already famous and a professor of chemistry at age 30, was beginning to be interested in the relationship between chemistry and biology and sometimes attended the weekly general biology seminar.[14]

The seminar was held after dinner. Morgan, who presided, and his wife, Lillian, had only to cross the street between their home and the lab to attend. She, too, was a geneticist and had been a student at Bryn Mawr when Morgan taught there. After raising the Morgan children, she returned to an active role in the lab. One of her important contributions was the discovery of a strain of flies in which the two X chromosomes were attached to each other.[15] This *attached*-X mutation was a convenient experimental tool and was used by Beadle and Sterling Emerson in their study of crossing-over.[16] Often, after getting the seminar program under way, Morgan would doze off for a while. Sometimes he awoke again spontaneously and sometimes only after Mrs. Morgan prodded him, but he promptly rejoined the discussion with a penetrating question or witty comment. Like the Synapsis Club meetings that Beadle knew well, the programs ranged from the latest results in the lab to interesting items that Morgan found in letters received or the *New York Times* that arrived a week late in Pasadena.[17]

Variable sterile was another of the corn mutants that Emerson's sharp eye had identified and saved because it produced very few anthers and had a low pollen yield. The "*variable*" in its name refers to the fact that both the number of anthers and the pollen yield differed from plant to plant. This made it a headache to score plants as either mutant or not. Observations had to be made repeatedly over several days to minimize misclassification. With Marion giving him some technical help, he carried out cytological analyses on the pollen-forming cells. The *variable sterile* chromosomes did everything right until it came time to divide into two cells, each with ten chromosomes, that is, one of each pair. However, at that point there was no sign of the plate that normally forms between the two halves of the cell, and the cells did not divide. In the absence of cell division, the cells collected too many chromosomes and, "sensing" that something was very wrong, died. A few developing pollen cells managed to escape the tragedy and developed normally. Thus, *variable sterile* is an example of a mutation whose effect is only partial. Using the few viable seeds, Beadle demonstrated that a single recessive gene was involved; he also determined that *variable sterile* was linked to genes already known to be on corn chromosome 7. The paper reporting these results, like all of Beadle's corn papers, has both camera lucida drawings and photomicrographs illustrating the behavior of the chromosomes.[18]

In the investigation of *variable sterile*, Beadle went one step further in the genetic analysis than he had with *polymitotic* and determined the relative position on chromosome 7 of the genes in the linkage group. That is, he constructed a map based on recombination frequencies according to the principles he had learned as a student in Lincoln. Using *variable sterile* and two other mutant genes on chromosome 7, he measured the deviations from the results expected for complete linkage (called recombination frequencies). He started with plants that contained the three mutations on one chromosome of the homologous pair and normal alleles on the other, and he self-pollinated the plants. In the majority of the offspring, the parental distribution of genes was retained. The next most abundant forms reflected crossing-over. In one, the *variable sterile* mutant allele had changed places with its normal allele, but the other two mutant alleles remained linked. This suggested that the other two were closer together than either was to *variable sterile*. In the third most abundant form, another of the recessive mutations, *yellow seedlings*, exchanged places with its normal allele, leaving the other two mutant and normal alleles in place. These results suggested that the order was *variable sterile–glossy seedlings* (the third mutant allele)–*yellow seedlings*. These sorts of exchanges should also yield the reciprocal products normal–*glossy seedlings–yellow seedlings* and *variable sterile–glossy seedlings*–normal, and these combinations were also found in the chromosomes of some progeny. From his data on the relative frequency of the several different exchanges, Beadle concluded that the distance on the chromosome between *variable sterile* and *glossy seedlings* was about the same as that between *glossy seedlings* and *yellow seedlings*.

By the time Beadle published this work, there was finally rigorous evidence for the 20-year-old assumption that genetic recombination reflects the physical exchange of segments by homologous chromosomes. McClintock and Creighton had proved that deviations from fixed linkage are indeed associated with the exchange of chromosome parts—the "crossing-over" that occurs when the paired homologous chromosomes twist around one another during synapsis.[19] Almost simultaneously, Curt Stern demonstrated the same association of genetic and cytological crossovers in *Drosophila*.[20]

Beadle was on the lookout for other mutations that would contribute to understanding chromosome function. In a paper published in 1932, he described no fewer than 18 different sterile strains collected from many sources.[21] He examined the pollen-forming cells of many of these strains microscopically but could not pinpoint the defects. Many seemed normal until they suddenly disintegrated before forming fertile pollen. This must have been a disappointment, because his previous work had revealed a way to use genetics to study the complex process of chromosome function in meiosis. If a series of mutants, each defective in successive steps of the process, could be assembled, then a great deal about this critical process could be learned. This approach, using a series of mutants each causing a different defect in the same multistep process to define a biological mechanism, foreshadows the pathbreaking work with *Drosophila* eye color and *Neurospora* metabolism that would be Beadle's great contributions within the next decade. Was he aware of the significance of this approach? Apparently not in 1932, because neither his published papers nor personal correspondence mentions it.

When Beadle arrived in Pasadena near the end of 1930, it was already time to apply for a second year of the National Research Council fellowship. His application for an extension proposed continuing the search for new mutations affecting meiosis, further work on *variable sterile*, on the unexpected finding that crossing-over appeared to be normal in *asynaptic* plants, and on a recently examined mutant with "viscous" chromosomes (what would eventually be called *sticky chromosome*). The proposal also emphasized the extension of his work on the corn-teosinte hybrids. Anderson[22] and Emerson[23] both wrote of their enthusiastic support for a second year, emphasizing that Beadle had a lot of ongoing projects to finish as well as many additional ideas. Emerson was prescient when he compared Beadle to Anderson. "I am not certain that he [Beadle] has the brilliancy of Dr. E.G. Anderson, tho [*sic*] he lacks little in that respect, but I predict that in a period of ten years he will accomplish more than Anderson has in an equal period." Besides the letters from his mentors, the board on fellowships received an enthusiastic report about Beadle and his situation from one of its members, E.J. Kraus, who had visited Beadle and Anderson.[24] All went well, and the fellowship was extended for another year.

The *sticky chromosome* mutation, like *asynaptic, polymitotic*, and *variable sterile*, had a unique effect on meiosis.[25] In cells carrying two mutant alleles, homologous chromosome pairs synapsed properly, but then, at the point when normal paired chromosomes began to move to opposite sides of the cell, the sticky chromosomes instead stuck together, resisting segregation into separate cells during the first meiotic division. Also, the chromosomes tended to break, as if they were resisting the forces acting to separate them. The mechanism of cell division was disturbed in all the plant's tissues, not just in the precursors to the germ cells, so that the effect was on both mitosis and meiosis. Even more unexpected, all the chromosomes were sticky and could break, not just chromosome 4 on which Beadle had mapped the *sticky chromosome* gene. Using genetic experiments, he confirmed that the mutation affected genes on at least three chromosomes besides chromosome 4. Apparently, the *sticky chromosome* mutation was a general troublemaker for corn.

The discovery of *sticky chromosome* provided Beadle with an "occasion on which my morale was given a substantial boost."[26] It was the first known of a whole group of genes—which are now called mutator genes—that occur in many different organisms. Mutant alleles of these genes cause a general increase in the rate of mutation. To Beadle's delight, Sturtevant was impressed enough to ask "May I show this to the boss?"[27] And the observations were particularly pertinent to McClintock's later interest in spontaneous chromosomal aberrations in corn. She was struck by the fact that the rate of such aberrations and of spontaneous mutation could be controlled by genotype, that the chromosome sticking and subsequent rupture occur not only in meiosis but in mitosis, and that unusual phenotypes were observed in all parts of the plant.[28]

McClintock had been awarded a National Research Council fellowship starting in the summer of 1931, and, while her headquarters remained at Cornell, she traveled extensively and avoided the frigid Ithaca winter by visiting Cal Tech from November through the following March. Unlike the situation at Cornell, a woman was an oddity at Cal Tech, whose faculty and students were all male. Even an invitation from Sterling Emerson to lunch at the Athenaeum, the Cal Tech faculty club, provoked some turmoil; Morgan had to argue with the Cal Tech trustees themselves before Emerson was given permission to take her for a meal. Years later, McClintock recalled that everyone stared as they came into the dining room.[29] No wonder that she reserved her comings and goings to the biology laboratories and the Farm, where she visited with Burnham and arranged for Anderson to make plantings for her that spring.[30] She also talked with Linus Pauling, whom she thought of as belonging socially to the genetics group.[31] At that time McClintock was doing experiments leading up to another major discovery, the identification of certain chromosomal regions as nucleolar organizing regions or NOR, which are required for cells to form the prominent nuclear structures called nucleoli. Besides working in the lab, McClintock did have time for old friends. In her car one day, with Marion Beadle driving and Barbara in the passenger seat, the only place for Beadle was the running board, which was illegal in

Pasadena. When a cop stopped them to make out a ticket, Barbara spoke up saying, "Don't worry, that's her husband, she can see right through him." And he let them go.[32]

Marion and George Beadle's son David was born in December of 1931. This event made it more necessary than ever for Beadle to worry about what kind of job he might have when his National Research Council fellowship ran out the following June. Although these fellowships were rarely extended much beyond 2 years, Anderson thought that Beadle might obtain a third year, and an application was made. Meanwhile, Emerson, who assumed that his student would continue with corn genetics, wrote to say that Demerec was interested in offering Beadle a job at Cold Spring Harbor and that he himself would like to have Beadle return to Cornell. He worried that Demerec could offer a better salary, $1800 compared to $1500 a year.[33] Beadle declined Demerec's offer, indicating that he was eager to continue his corn work at Cal Tech.[34] Demerec persisted, actually offering a smaller stipend than Emerson had assumed and promising freedom to work independently as well as providing him with field and greenhouse space for the corn.[35] But Beadle's real hope was to return to Cornell where he would find a large group of corn colleagues, and so he again turned down the offer with profuse thanks and apologies, as well as news of David's birth.[36] Shortly before Beadle wrote this to Demerec, Emerson had told him that an assistantship was available on September 1.[37] His idea was that Beadle would come for the August International Genetics Conference in Ithaca as planned, and then stay on.[38] Demerec did not give up easily. In mid-January of 1932 he wrote asking Beadle to reconsider.[39]

By early February, Beadle was informed that his fellowship would be extended by three additional months, through September 1932.[40] He decided to take his chances and stay at Cal Tech, especially since the job possibility at Cornell did not offer much more security, being only an assistantship, not a faculty appointment. By spring, Beadle seems to have been assured of support at Cal Tech for at least another year, although the source of his support is unknown. Disappointed as he was that Beadle would not return to Cornell, Emerson advised him to remain in California.[41] Sterling Emerson told his father that there might even be a permanent job for Beadle at Cal Tech, but the Great Depression put a damper on that and other plans.[42] Morgan was not allowed to make any new appointments and had to limit technical assistants to half-time work because of the severe financial restrictions. Knowing that the Rockefeller Foundation's help was needed if a broad approach to biology was to be developed, he discussed the emergency with Weaver several times in 1933.[43] In the fall, he sent Weaver a formal proposal.[44] It emphasized the connection between genetics and physiology and the importance of tracing the chemical changes associated with a genetic trait to the gene on which the trait depends. This was exactly what Beadle would attempt with Ephrussi a few years later. Although the Rockefeller Foundation responded with a grant of $50,000 for two years,[45] Beadle was not offered a permanent appointment; indeed, although Weaver talked to several of the young biologists when he visited Cal Tech in the spring of 1934, surprisingly he made no record of even talking with Beadle. He found the group largely "disappointing" and singled out McClintock, who was visiting, as "quite first class."[46]

When Morgan went to collect his Nobel Prize in 1934, he searched in Europe for potential Cal Tech faculty who were in fields other than genetics. He had little luck finding "the kind we are looking for"[47] and appeared "desperate" to Harry M. Miller of the Rockefeller Foundation staff, who was traveling in Europe at the time.[48] The Rockefeller resident staff member in the Foundation's Paris office noted that Morgan was "combing England and Scandinavia for a physiologist who is not Jewish."[49] Was it Morgan or Cal Tech that had a bias against the many fine Jewish scientists seeking to escape the Nazis by emigrating to the United States? Finally, Cornelis Wiersma and his colleague Anthonie von Harriveld in Utrecht, specialists in the electrophysiology of nerve cells, were given positions. They broadened the scope of the Division's activities, but neither one would add much luster to Cal Tech. Miller proved correct when he wondered "whether Morgan might not have secured equally good or even superior young Americans, if he had made half the effort here that he did in Europe."[50]

Beadle's agenda was full. Besides his growing attention to *Drosophila* and the corn genetics projects, he was becoming interested in the evolutionary relationship between corn and teosinte, its wild, Central American relative. He also had a central role in a huge collaborative project with Emerson and another Cornell geneticist, R.A. Fraser. Soon after his arrival in Ithaca, Emerson had appointed Beadle "secretary" for the lab's corn chromosome collection, and he was still acting in that capacity for the expanded effort when he went to Cal Tech. Besides collecting the data, seeds with interesting chromosomes and mutations were stored in small bags until someone wanted to explore their genetics. Collecting the seeds, keeping careful descriptive records, and responding to requests from corn geneticists for these precious materials must have kept him busy. Emerson believed in freely sharing both materials and information with the whole community, and there were constant demands on Beadle. Typical were Demerec's many requests made at the last minute before planting time and with urgent pleas for a prompt response.[51]

To keep the collection useful, the catalog required constant updating and had to be made available to all interested workers. This was done informally for some years. In a letter apparently sent to all concerned in late 1929, Beadle referred to a letter that Emerson had mailed out the previous winter announcing ambitious future plans and enclosing a mimeographed summary of gene linkage.[52] The idea was to collect all data, published and unpublished, on corn genes and genetic maps and to collate and publish them in a compendium. It would include all the known "genetic factors" in corn, whether they were dominant or recessive, a brief description of the characteristic changes in the plants, and, to the extent known, the linkage relations. Emerson (and Beadle in quoting Emerson's letter) promised to give proper credit to all. He also said that while this mammoth undertaking was going on, revised mimeographed information would be distributed. In fact, Emerson had been "threatening" to bring the compendium out for some years. A "cornfab" had been held to lay the plans in Emerson's hotel room during the 1928 meeting of the American Association for the Advancement of Science. In the late fall of 1930, the work was still in progress, with the hope that it would shortly be ready for publication.[53] In the intervening years, Emerson and Beadle collected information and sent repeated apologies for dragging their feet.[54] Some of the letters suggest feisty reactions to would-be competitors. Emerson asked at one point for a sketch of the chromosome maps Beadle had been constructing from the collected data, saying he wanted to "impress the crowd" at a talk he was to give in East Lansing.[55] Meanwhile, they initiated a mechanism for disseminating the news on a regular basis.

On October 5, 1932, the first, mimeographed Maize Genetics Cooperation was issued by Marcus Rhoades, who was the secretary.[56] It reported on a meeting called by Emerson on August 6, 1932, during the Sixth International Genetics Congress in Ithaca, to establish a clearinghouse at Cornell for data, stocks, and the standardization of gene names. About 45 people attended. The group assigned each chromosome to one or more investigators, with the idea that they would be responsible for receiving relevant data from all and collating and sending the information out through the Cooperation. Emerson was assigned chromosome 1; Beadle and Rhoades, chromosome 2; Paul Mangelsdorf at the Agricultural and Mechanical College of Texas, among others, chromosome 4; and so on. This was the first organized effort by geneticists to manage the large amounts of data required to understand a whole genome. Its utility was quickly recognized, and in 1934 the Cooperation received a 5-year grant from the Rockefeller Foundation in support of its efforts. Issue 3 (November 13, 1933) of the Cooperation announced that the drosophilists were going to do the same thing. Rhoades expressed "hope that the *Drosophila* community will find the same generous spirit of cooperation which you maize workers have shown."

The Corn Cooperation still functions and stands as a model for all genome studies.[57] In more recent times, human geneticists followed suit, with large committees responsible for each of the 24 different human chromosomes. The latest examples of such cooperation among geneticists are the international, organized efforts that yielded the sequences of the entire genomes of yeast, *Drosophila*, the mouse, and humans, and the publicly available databases for the complete DNA sequence of these organisms. Also, just as Beadle and Rhoades made mutant corn seeds available, centralized repositories of mutants for important model genetic organisms are now common. Another important community action was proposed in issue 5 of the Cooperation: the adoption of a system for naming genes, mutations, and rearrangements of chromosomes that had been devised at a Boston meeting (January 25, 1934). This may seem a trivial matter, and it is certainly not an exciting one. But then, and even now, it is not uncommon for the same gene to be discovered and named in different laboratories; the different names and often distinctive phenotypes of the observed mutants are at best confusing and at worst downright misleading. The system that was proposed was, by mutual agreement, consistent with (although not identical to) the one adopted by the drosophilists. Altogether, the Corn Cooperation established exemplary scientific practices which have advanced knowledge through the sharing of fundamental information.

The full compendium included more than 300 genes when it was finally published in 1935.[58] Besides a brief description of many corn genes, it cataloged all the available data on linkage and the frequency of crossing-over for the different genes. The earliest reference in the compendium is to the work of C. Correns in 1901. Correns was one of the original rediscoverers of Mendel's work and recognized the particular features of corn reproduction that made it so convenient for genetics: the manner of germ cell development, the double fertilization process, and the fact that a kernel's endosperm has the genetic makeup of the new generation, not the parent plant (the phenomenon called xenia). Long tables summarized and synthesized the cytogenetics and the formal genetic analyses. The tables were used to construct simple cartoons of each of the ten corn chromosomes. Their relative lengths were determined from the sum of all known intervals of crossing-over from one end of the chromosome to another (e.g., 128 for chromosome 1 and 32 for chromosome 10), and the order of the known genes on each chromosome was shown. The crossing-over intervals confirmed the relative physical lengths defined by McClintock 5 years earlier. Although the publication came 15 years after Bridges and Morgan had made similar maps for *Drosophila* chromosomes, it was still a triumph that corn, with its complex genetics, ten chromosomes, and long generation time, yielded up the second actual chromosome map ever known.

The authors of the compendium acknowledged the accomplishment as "an almost unique example of unselfish cooperation" by the whole corn genetics community, and at the time it surely was. The enormous task was in the self-interest of the three authors because it gave them prompt access to everyone else's findings, but it also advanced the interests of all corn geneticists and breeders. However, compared to the commercialization of genes today, the community did not seem concerned about the economic potential of the corn mutations although corn had already emerged as a major U.S. crop; in 1935, for example, 99,974 acres of corn were planted with a total production of about 2.3 million bushels.[59] The growing productivity of U.S. corn farmers resulted from the introduction of hybrid corn seed, an innovation arising from early 20th-century research. By the late 1920s, four companies that grew and sold hybrid corn seed had been formed. Henry A. Wallace, then a farm-magazine editor who had proselytized the importance of hybrid corn seed, founded the most important of these in 1926—Pioneer Hi-Bred. Wallace was later Secretary of Agriculture and Vice President of the U.S. during the third Franklin D. Roosevelt administration. He lost a bid for the presidency in 1948, when Harry Truman was elected.

Beadle came into his own early during his years at Cal Tech. His connection with Emerson facilitated his growing role. Rhoades believed that the linkage maps were finished because of Beadle's enthusiasm and effectiveness.[60] He even took on a distinguished older scientist. The British geneticist, C.D. Darlington, had, in 1932, reinterpreted Beadle's work on *polymitotic* to suggest that the supernumerary divisions after completion of meiosis were in fact like additional meioses and involved not only synapsis but even homologous chromosome exchanges. Beadle struck back, marshaling new and older data, and arguing that Darlington's view was untenable and unsupported by experimental evidence.[61] By this time, Darlington had visited Cal Tech, and the bloom was off his distinguished reputation, at least for some in Pasadena. Anderson, blunt as usual, is reported to have said after hosting Darlington at the Farm that he would "never believe another thing that man says."[62] Most important, however, was the fact that Beadle was becoming increasingly excited about working with the fly group and they were impressed with him.

Beadle was familiar with the formal and theoretical aspects of *Drosophila* genetics from his undergraduate studies and the many discussions during his years at Cornell. But *Drosophila* genetics, with its own idiosyncratic lexicon, curious strain and mutant nomenclature, and experimental methodologies, was foreign to a corn geneticist. Most newcomers turned to Bridges for an introduction, but with the help of his Cornell mentor's son, Sterling Emerson, Beadle quickly became comfortable with the lore. Then, although still uncertain whether he could keep up his end of the conversation with such a "distinguished and brilliant geneticist," he sought Bridges' views on a theory of genetic crossing-over that he and a colleague, H.S. Perry, had developed at Cornell. Much to his relief, Bridges turned out to be very friendly, quite patient, and methodical during their discussion. "He insisted on going over each point carefully and fully before going on to the next one."[63]

Knowing about Beadle's interest in the mechanics of crossing-over in corn, Sturtevant urged him to examine those events in *Drosophila*. The issue at the time was whether the probability of crossing-over between homologous chromosomes was equal over all regions of the chromosomes; that is, does crossing-over occur preferentially in some regions of the chromosomes, and is it inhibited in others? This question was fundamental if the frequency of crossing-over during meiosis was to be used for mapping the relative position of alleles in chromosomes. Indeed, in his 1913 landmark experiment, Sturtevant had assumed but not proven that the likelihood of a crossover event was the same over the entire length of the chromosome.[64] Over the years that this method of estimating the arrangement and spacing of alleles was applied to chromosomes other than the X and to related fly strains, there were more and more indications that the principal assumption underlying the mapping procedure was suspect. Muller's supposition of double crossovers explained why the distances between distant alleles were invariably underestimated. He also found that structural aberrations in one member of a homologous chromosome pair gave rise to unexpected frequencies of crossovers in the region of the anomalous structures. Dobzhansky had extended this work and learned that the crossover frequencies in the region containing chromosome translocations were markedly reduced.[65] He also noted that alleles near the spindle-fiber attachment site (the centromere) appeared to be much closer together, based on the lower frequency of crossing-over between them, than the actual physical distance between them measured cytologically.

With Dobzhansky's encouragement, Beadle set out to compare the likelihood of crossing-over between two homologous chromosomes in the vicinity of the centromere with that at locations distant from that site. His experiments were unequivocal; the frequency of crossovers was considerably lower near the centromere.[66] Lillian Morgan's *Drosophila* strain, whose *attached*-X chromosomes stay together during meiosis, provided a clever approach for further exploration of the phenomenon.[67] Using female flies that contained three X chromosomes, two attached to one another and the third free, Beadle found that crossing-over was more frequent between the two attached X chromosomes than between the free X chromosome and either of the two attached chromosomes. This indicated that proximity to one another was an important parameter influencing the frequency of crossing-over.[68] Later, Beadle teamed up with Sturtevant to measure the frequency of crossovers in a region in which relatively small chromosomal segments had become inverted. Such anomalous structures could be easily identified in polytene chromosomes because the bands in the region of the inversion were in the opposite orientation compared to the normal. They found, as Muller had suggested earlier, that crossovers were relatively rare within and near small inversions but they increased in frequency as the length of the inversions increased.[69] These and other experiments led Sturtevant and Beadle to the novel hypothesis that the frequency and nature of crossovers, which depend on the extent of the homology between the chromosome pairs, greatly influence the separation of genes during meiosis.[70]

In August of 1932, when Beadle traveled back to Ithaca for the Sixth International Genetics Congress, he was an accomplished, independent scientist. Apparently he realized that on a farm and in the laboratory alike, problems have solutions that can be found through thinking and hard work. In Ithaca, besides catching up with the work of American and foreign scientists, he, Emerson, and Fraser took advantage of being together to straighten out some matters concerning the compendium of corn mutants. Morgan was the president of the Congress and Emerson the chairman of the local organizing group. Beadle, more than most young geneticists, was importantly connected to both these leaders, and both were looking out for his welfare. After hearing Beadle's talk about the crossing-over experiments in *Drosophila*, Morgan once again found fault and told Emerson "That was fine work, but you ought to tell Beadle to improve his presentation."[71] Back in California after the Congress, Beadle, eager to master everything, arranged to practice his lecture style in a public series at the Los Angeles Public Library. He speculated that, although the audience asked intelligent, even penetrating questions, they were less interested in enlightenment than in keeping warm.[72]

The Congress was also an opportunity for Beadle to meet geneticists from abroad and learn about the international scene. While genetics was flourishing in the United States in the late 1920s, and plant genetics was making a significant contribution to U.S. agricultural productivity, the situation in the USSR was a negative mirror image. The failure of the Soviet winter wheat crops in 1927–1928 and 1928–1929 had made Stalin desperate for agricultural improvements. He turned to Trofim Denisovich Lysenko, who, in 1929, became the most powerful influence in Soviet agriculture and remained so until 1964. Lysenko promised spectacular results, and his ill-informed, unscientific approach appealed to Stalin, who distrusted the professional scientists, although there were then outstanding plant geneticists in the U.S.S.R. Chief among these was Academician Nikolai I. Vavilov, president of the Lenin All-Union Academy of Agricultural Sciences and a person of international renown. Vavilov attended the congress in Ithaca. At that time, he was still defending Lysenko's ideas and he presented them to the assembled international community. As reported in the second issue of the Cooperation (January 23, 1933), Vavilov described Lysenko's 1929 "discovery" that the growing period of plants can be significantly shortened by soaking and chilling the seeds in the dark, a process called jarovization (or vernalization). In fact, the adoption of this technique led Soviet agriculture into desperate situations for decades. Also, because of Lysenko's political influence, modern genetic research and its teachings were suppressed for more than a generation while genetics advanced enormously in Europe and the United States. The Cooperation says, in a comment presumably by Rhoades, "If the claims of the workers investigating this problem are justified, this discovery is of great importance to plant geneticists and to plant breeders." It's impossible to know with certainty what Rhoades actually thought, but it seems likely that his comment, for the record, was motivated by politeness and respect as well as a sense of the tragic implications that honest criticism might have for Vavilov. After years of trying to accommodate Stalin and Lysenko, Vavilov in 1937 incurred Stalin's wrath by his devastating scientific critiques of Lysenko's corrupt ideas. By 1940, Vavilov had lost his institute and, in 1943, his life.[73]

On the return trip from Ithaca to Pasadena, Beadle stopped to visit Herman Muller's fly group at the University of Texas at Austin. He left them with the impression that he was going to turn full attention to *Drosophila*, which offered major advantages over corn for genetics; namely, a generation time of ten days.[74] Back in Pasadena, he completed several major reports on his corn work, including the genetic and cytogenetic studies on corn–teosinte hybrids. Altogether, he published nine papers in 1932, including six on corn and his first paper on *Drosophila*. After that, except for the 1935 corn compendium and a brief report on teosinte in 1939 (see Chapter 18), he published no more corn papers for almost 40 years. Beadle had turned to flies.

CHAPTER 7

Three-Eyed Flies

Beadle's decision to shift his focus from corn to *Drosophila* was not so much a bold move as an opportunistic one. The Morgan fly group, albeit beyond its heyday, was still one of the world's leaders in genetic research with *Drosophila*. Moreover, much of the intellectual energy in the biology division was in fly work, and the people who flocked there came for that purpose. There was still much about the action of genes to be learned, and doing so with the fly in the midst of those who had founded the field was an opportunity not to be missed. Also, the rapid experimental results with *Drosophila*, as compared with corn, were compatible with the intensity and impatient style Beadle brought to research. Although tending to shyness in social settings, he was anything but retiring with those who shared his passion for research. In Sturtevant, Bridges, Dobzhansky, Emerson, and Schultz he found not only the intellectual energy that could sustain him, but also company who shared that passion.

Beadle's close contact with Sturtevant during their collaboration in the crossing-over study created a lasting and influential impression. He was particularly taken by Sturtevant's prodigious knowledge of biology and genetics and by his soft-mannered yet insightful commentaries on research presentations and publications. Sturtevant had a distinctive way of working, spending mornings doing his experimental crosses and counting their offspring. Afternoons were invariably spent reading almost every newly received journal in the library or updating his plant and animal collections by forays in the field. Although he taught undergraduate and advanced genetics courses, his favorite teaching format was afternoon teas, where he would bring up a paper that had recently captured his attention.[1] Visitors who were likely to be awed in Sturtevant's presence were quickly put at ease by his broad smile accentuated by the rarely missing curved pipe between his lips. Jack Schultz recalled an impression that most people had when they entered Sturtevant's laboratory: "...he would look up from his microscope or clipboard and begin (before you had opened your mouth) to tell you what he had just found out. 'Sturt' was most interested in the excitement of discovery, and all other things came afterwards no matter how important they might be considered."[2] To those who knew him,

Sturtevant seemed to possess mythical qualities. E.B. Lewis, a student and later a colleague, commented on his remarkable intellect and encyclopedic knowledge of public events and history, all of which were invaluable for his passion for doing challenging crossword puzzles. Lewis noted that "It was as if his memory were composed of a plethora of matrices waiting to be filled with any data that lent themselves to classification into discrete categories. Whatever form the data took, the observations fell into the appropriate matrix in his memory from which they were retrievable to a degree that was truly phenomenal." Sturtevant liked to refer to this as the "blockhead approach."[3]

Many years later, Beadle reminisced about the satisfaction he shared with Sturtevant in "interpreting a seemingly complex situation in a relatively simple way" and "how this experience gave me some small measure of appreciation of the excitement of the earlier *Drosophila* days at Columbia University."[4] Beadle never acquired Sturtevant's passion for literature and the arts, but they shared another passion—science. In an interview for Ann Roe's study on the making of a scientist, Sturtevant recalled an aspect of his collaboration with Beadle during 1935–1936. "Yes, we just said to ourselves, well, there's a lot of material here and it isn't understood and what we've got to do is collect a lot of cases and put them together in this diagram and see what comes out of it...and then it came out, largely I think because of Beadle's imagination and ability to think in three dimensions, which is what you have to do." When asked if he could think in three dimensions, Sturtevant answered, "Yes but not as well as Beadle."[5] When told that his ability to think in three dimensions was well above the mean of others that she had tested, he replied "but I still think that Beadle was much better and it had to be for us to get this one worked out." The mutual admiration created in these years would ripen over time.

All the while that Beadle was engaged by his experiments on crossing-over, there were ongoing discussions within the fly group about the role of genes in embryonic development. One of the most active centers working in that area, a group headed by Charles M. Child at the University of Chicago, questioned whether genes had anything to do with development. Child believed that development was driven by morphogens emanating from specific sites inducing variable types of differentiation in a concentration-dependent manner in the cytoplasm; he dismissed a role for genes in the nucleus.[6] In reviewing that work, Sewall Wright, who was one of the creators and leaders of physiological genetics, was quick to point out that "...it is to be regretted that the author has made no attempt to bring the facts of genetics into relation with his theory." Indeed, Wright argued that genetics needed to be included in any comprehensive treatment of development.[7] At the time, however, there was no experimental approach to explore that connection.

By the end of the 1920s, Morgan, too, had recognized that embryological development and evolution would eventually be explained in terms of genetics. Accordingly, he alerted his research sponsors to the likelihood that major discoveries in development would soon be forthcoming.[8] In a lecture entitled "Genetics and the Physiology of Development," Morgan foresaw the need to link development to the action of genes, but that prescient view was never translated into experimental efforts,[9] very likely because he preferred to focus the *Drosophila* research on the transmission rather than the expression of genes and was reluctant to speculate beyond existing empirical evidence.[10] By 1932, however, in his presidential address to the Sixth International Congress of Genetics at Cornell University, Morgan emphasized how genetics had become integrated with cytology, physiology, and embryology and that among the most important problems he saw facing genetics in the immediate future was "the relations of genes to characters: the way the information of genes is translated physiologically into adult characters." Solving that problem, Morgan mused that it would take "industry, trusting to luck for new openings, by the intelligent use of working hypotheses" and "by a search for favorable material, which is often more important than plodding along the well-trodden path, hoping that something a little different may be found."[11] Little did he suspect at the time that the breakthrough for solving the physiological basis of gene action would have its origins with George Beadle at Cal Tech.

Morgan's extraordinary contributions in biology were recognized the following year (1933) by the Nobel Prize in Physiology or Medicine for discoveries concerning the hereditary function of the chromosomes. He was in fact the first nonphysician to receive the prize in that category. Morgan was characteristically calm, even reserved about the news, although he was pleased that the Nobel committee had acknowledged the importance of genetics to physiology and medicine, a recognition they had ignored when he was nominated earlier.[12] As is often the case on such occasions, there was considerable excitement and very likely celebrations at the Institute and in the fly group. Reluctantly, he submitted to press interviews and even relented to a reporter's request for a photograph, but only after it included the children in the vicinity.[13] Morgan did not attend that year's ceremony in Stockholm, pleading that he was in the midst of important organizational business in Pasadena. He went the following summer to collect the award, the monetary portion of which he shared with Sturtevant and Bridges as an acknowledgment of the magnitude of their contributions to his achievements. Morgan provided $4700 for each of his four children, and comparable amounts for Sturtevant's[14] and Bridges' children.[15]

In his Nobel Prize paper, Morgan reflected on his ambivalence about the nature and function of genes. He raised a hotly debated issue: "What are genes? Now that we can locate them in the chromosomes are we justified in regarding them as material units; as chemical bodies of a higher order than molecules?" He noted further that "there is no consensus of opinion among geneticists as to what the genes are—whether they are real or purely fictitious—because at the level that genetic experiments lie, it does not make the slightest difference whether the gene is a material particle. In either case, the unit is associated with a specific chromosome; if it is a fictitious unit, it must be referred to a definite location in the chromosome, the same place as in the other hypothesis. Therefore, it makes no difference in the actual work in genetics which point of view is taken."[16] It was another decade before Oswald Avery and his colleagues provided evidence that the gene was a physical reality, in fact, a component of the chromosome, deoxyribonucleic acid (DNA).[17]

Morgan also used this occasion to reflect on his earlier thoughts on the genetic control of embryonic development. He confessed to the sad state of affairs when he acknowledged that "The biochemical and cellular processes that link gene action with embryonic development remain unexplored, a wide gulf separating the two fields."[18] But his book, *Embryology and Genetics*, published about the same time, failed to bridge the two subjects as its title suggested. He was content to discuss the two processes without relating one to the other.[19]

Sewall Wright, however, was not content to let embryology and genetics remain unconnected. At the 1934 Cold Spring Harbor Symposium, Wright speculated that "...the specificity in gene action is always a chemical specificity, probably the production of enzymes which guide metabolic processes along particular channels." He went on to suggest that a given array of genes determines the production of a particular kind of protoplasm with particular properties and, depending on the particular kind of chemical environments, the protoplasmic patterns become more complex and differentiated. This, he believed, brought additional genes into play that could lead to any degree of complex pattern in the organism.[20] Though prescient in outline, Wright provided neither ideas on how this control might be achieved nor guidance on how to go about exploring those processes.

By the mid-1930s, most biologists conceded that genes are the determinants of developmental processes, the principal evidence being the effect of mutations on the end results or on various stages of embryological development. Richard Goldschmidt cataloged many of these mutational effects, including abnormalities affecting morphology, physiology, and metabolism, and concluded that "the mutant gene must control or produce a deviation in the series of developmental processes leading to the visible character." He foresaw development as a highly precise series of steps where "each step is dependent upon the normal appearance of the preceding one" and that each one could be blocked or altered by mutation. The question remaining, however, was, How do genes influence or determine developmental processes?[21] Jack Schultz saw the relation of genes to characters as including all the processes of development, from the gene in the chromosome to the end result. But he believed that "If we knew enough about development, and about cellular physiology in general, it would be possible to use the effects of genes on development to find out something about the nature of the gene. Contrariwise, were we familiar with the properties of the gene from independent evidence, we might proceed to unravel the peculiarities of the developmental system. But our problem is knowing neither, to solve both at once."[22]

Sturtevant, as early as 1932, was also thinking about genes and development when he wrote "one of the central problems of biology is that of differentiation—how does an egg develop into a complex many-celled organism?" This traditional problem of embryology, he said, "also appears in genetics in the form of the question, how do genes produce their effects?"[23] Beadle, too, recognized that "the two great bodies of biological knowledge, genetics and embryology, which were obviously intimately interrelated in development, had never been brought together in any revealing way." Always perceptive about experimental organisms, he thought that the obvious difficulty "was that the most favorable organisms for genetics, *Drosophila*...were not well suited for embryological study, and the classical objects of embryological study, sea urchins and frogs as examples, were not easily investigated genetically."[24]

In the midst of all this speculation and uncertainty, experimental efforts were few. Sturtevant sought to study the genetic control of development by exploiting a discovery that Morgan and Bridges had made years before.[25] Bridges' uncanny ability to spot flies with structural irregularities was key. He was observing heterozygous female flies with distinguishable mutations on their X chromosomes. Unexpectedly, certain tissues contained patches of cells that were genetically different from the cells that surrounded them. Closer examination revealed that the cells in the patches had morphological features that could be explained if the unusual cells contained only one of the two X chromosomes. Bristles in the patches appeared to be genetically male, while most of the cells were, as expected, genetically female. Such mosaic individuals were called gynandromorphs.[26] Because the cells in the patches were genetically male, Bridges surmised that they arose during development from cells that had lost one of the two X chromosomes, an occurrence now known to be common. Such cells are genotypically XO and behave as if they were male cells. Bridges had shown earlier that flies whose genotype is XO are viable males. He inferred that in the gynandromorphs the size and location of the mosaic patches depend on when during development the X chromosome is lost.

One particularly interesting question was whether the phenotype of a patch of XO cells that revealed the mutation on the retained X chromosome was influenced by the surrounding cells or tissues that contained both X chromosomes and were female in their appearance. The experiments showed that in most instances the neighboring XX cells did not influence the mutant character of the genetically male (XO) cells in the patches. Such mutations were referred to as "autonomous" because the phenotypes are determined by the cells' genetic constitution and not by influences external to the affected cells.

Some years later, Sturtevant noted a different outcome while studying an X-linked eye-color mutation named *vermilion* (*v*). That mutation causes the eyes of affected males to appear bright reddish-orange instead of the dull reddish-brown characteristic of wild-type flies or heterozygous female flies. He was following the development of female embryos that carried two recessive mutations on the same X chromosome, one *vermilion* and the other a mutation whose phenotypic effect in the eye is also readily discernable by visual inspection. As expected, most of the cells in the eye were wild type in color and normal with respect to the other mutation; that result was consistent with the eye cells having retained the two X chromosomes. There were, however, occasional offspring that had patches with the mutant trait characteristic of the second X-linked mutation but wild type in color.[27] Sturtevant explained this initially puzzling finding by assuming that the cells in the patch arose from a cell that had lost the X chromosome containing the normal alleles of both the marker and *vermilion* genes sometime during development. Why then wasn't the eye color in the patch *vermilion* instead of wild type in color? Sturtevant surmised that the *vermilion* mutation was "nonautonomous," that is, the surrounding "female" cells enabled the mutant cells to manufacture the normal eye pigment, possibly by secreting a substance that bypassed the block caused by the *vermilion* mutation.

Sturtevant believed that such "nonautonomous mutations" might provide a route to analyzing the genetic control of selected developmental processes. However, he and generations of students failed in their attempts to use mosaic flies to determine when and where in the embryo such nonautonomous genes acted. Finally, he abandoned the approach.[28] A few years later, Beadle and Boris Ephrussi picked up on the nonautonomous *vermilion* mutation and developed a novel approach that succeeded in advancing the genetic analysis of development.

Cal Tech had many visitors who spent variable periods of time in the biology division. One of these, the geneticist and biometrist, J.B.S. Haldane, was aware of studies in England with mutants that appeared to be blocked in different chemical transformations responsible for the formation of characteristic flower colors. He deemed it most likely that the varied colors resulted from genetically determined changes in pigment structure. Arguing further that the chemical alterations responsible for creating the various pigments were almost certainly catalyzed by protein enzymes, he surmised that genes were enzymes or were in some way responsible for their formation or action.[29] As protein enzymes were more and more linked with the catalysis of biochemical reactions, it became fashionable to consider that genes were either enzymes themselves or involved in the production of enzymes. Morgan was skeptical of that explanation: "Enzymes might, for all we know to the contrary, be many stages removed, in a chemical sense, from the genes that initiate the chain of reactions that come to an end in the final reaction."[30] Notwithstanding Morgan's views, Beadle may well have stored the notion of a gene–enzyme relationship in his mind as he engaged in discussions about how genes could influence or direct development. He may also have recalled this idea from his undergraduate work at Nebraska (see Chapter 3). Then when Boris Ephrussi, a well-trained French embryologist, arrived at Cal Tech to learn genetics, the speculations were transformed into experimental plans.

Boris Ephrussi emerged from a very different background, culture, and scientific tradition than Beadle. He was born on May 9, 1901, in Moscow, Russia, to a prosperous Jewish family that enveloped him in a liberal intellectual tradition and environment. At one of Moscow's finest schools, he indulged his passion for classical Russian literature and the arts. Following the Russian revolution and about a year after the end of World War I, Ephrussi spent a year at Moscow University, but soon left to take charge of the family's property in Bessarabia (formerly Roumania). He rapidly became disenchanted with the bucolic life and returned to Moscow. As a youngster, Ephrussi had displayed a talent for art and, in 1920, with a Roumanian passport and fellowship he departed for Paris to pursue his artistic ambitions. On his arrival, he enrolled in the Sorbonne where he met Andre Lwoff, also of Russian heritage, and the two became lifelong friends. Although there is no evidence that Ephrussi had previously shown more than a passing interest in science, Lwoff persuaded him to study zoology. The two spent many summers at Roscoff, a marine biological laboratory in Brittany that provided a summer meeting ground for

France's leading biologists, much the same way the Woods Hole Marine Biological Laboratories on Cape Cod, Massachusetts, served American biologists. While there, he met many of the leading French biologists of the day, including Louis Rapkine, a biochemist of some renown from the Pasteur Institute, and Jacques Monod, a promising protégé of Andre Lwoff.

While at the Sorbonne, a mutual friend introduced Ephrussi to a young refugee who had fled Russia at the time of the revolution to continue her medical studies in Paris. Boris and Raya Ryss were married in 1923 and the next year they had a daughter, Irène (now Mme. Irène Barluet). Because it was difficult for a woman refugee to pursue a career in medicine, it wasn't until 1935 that Raya completed her training and was able to practice her chosen career. Irène recalled that her mother "was a very charming woman but not made for daily life; she never mastered French and spoke with a distinct Russian accent," something that troubled Ephrussi, because he took pride in his use of his adopted language. "My father was very meticulous...imaginative...living for the future and taken up with his work, while my mother was a dreamer and a very poor housekeeper, which my father resented very much."[31]

Within a year of receiving his certificate in zoology (1922), Ephrussi entered the Laboratory of Comparative Embryology in the College de France to pursue graduate research on the early embryogenesis of sea urchins. Working under the guidance of Emmanuel Faure-Fremiet, one of France's leading experimental embryologists, Ephrussi explored various aspects of sea urchin development. Rapkine encouraged Ephrussi to consider the chemical basis of differentiation in his research in embryology, advice he followed during much of his scientific career. Ephrussi's restless but adventurous scientific spirit drew him to adopt Alexis Carrel's new techniques of tissue culture to examine the patterns of differentiation of embryonic tissues in culture. The work in these areas resulted in two theses for which he received a doctorate in science from the Sorbonne in 1932.[32]

Ephrussi became widely known and respected in France as the only one familiar enough with tissue culture techniques to deal with fundamental biological issues. His growing visibility led to his appointment as assistant to Faure-Fremiet in the experimental cytology laboratory at the Rothschild Foundation's Institut de Biologie Physico-Chemique on rue Pierre Curie. Despite the then-existing antipathy, even hostility, to genetics in France,[33] Ephrussi became more and more persuaded that genetics was the key to understanding embryological development. He examined three mutations in mice, each one known to cause death of the embryo early in development. His goal was to determine whether these genes specified a "death program" that could be manifested in embryonic tissues propagated in culture. To his surprise, the tissues from the mutant mice survived and differentiated normally, leaving the question of the mechanism of the mutation-induced lethality a mystery.[34] These findings made Ephrussi even more determined to learn genetics because the expression of genes clearly held the answer to understanding development.

With the assistance of a Rockefeller Foundation fellowship, Ephrussi chose Morgan's laboratory at Cal Tech as the place to start. On his arrival in Pasadena in the autumn of 1934, he was drawn to Sturtevant largely because they shared a common goal of linking gene action to embryology. As a start, Sturtevant encouraged the young Frenchman to examine two *Drosophila* mutants that failed to survive embryonic development and to determine when and how death occurred. He hoped these mutations might be responsible for identifiable steps in the fly's embryonic development. Ephrussi decided to use Sturtevant's trick of constructing mosaic embryos for investigating this question. Each of the *Drosophila* mutations Sturtevant had assigned to him was in a gene known to be on the X chromosome. Males carrying either mutation in their single X chromosome, and females carrying either mutation in both their X chromosomes, died early in embryonic life; female flies with either of the mutant alleles on only one of their X chromosomes developed normally. Ephrussi generated embryos carrying the lethal genes in one of the two X chromosomes and examined offspring that were mosaic for a cluster of genes located on the same X chromosome. Among the female offspring, Ephrussi found frequent patches of cells on the fly's abdominal segments that were genetically "male," that is, they had only one X chromosome, the one bearing the lethal mutation. Ephrussi reasoned that if cells carrying the lethal mutations died during development they would have been absent from the mosaic patches; such mutations would have been autonomous. But these cells were viable and had differentiated normally to produce the appropriate cuticle structures! Apparently, the lethal effect of these mutations was either not manifested in the fly tissues he examined or they were overcome by the action of the adjacent normal cells; that is, these lethal mutations might be nonautonomous. Possibly, something being produced by the surrounding normal cells was preventing the mutant male cells from dying.[35]

Although working on quite different problems, Ephrussi and Beadle were drawn together and spent a great deal of time during the late evening hours speculating on why it was that geneticists and embryologists had not yet succeeded in discovering the way genes influence embryological development.[36] Beadle was well aware of the amusing encounter Ephrussi had with Morgan at the Marine Biological Laboratories at Woods Hole soon after he arrived in the United States in the summer of 1934. As Ephrussi related it, "Morgan told me that his book *Embryology and Genetics*[37] had just come off the press and he offered to give me the copy he had on his desk, provided that I promised to read it and give him my frank opinion about it. After a day or so, I said I found the book very interesting but that the title was misleading because the book did not try to bridge the gap between embryology and genetics as he had promised in the title. Morgan looked at me with a smile and said 'you think the title is misleading! What is the title?' 'Embryology and Genetics,' I replied. 'Well', Morgan replied, 'is there not some embryology and some genetics?'"[38]

As the end of Ephrussi's stay at Cal Tech drew near, he and Beadle became increasingly frustrated by the lack of experimental opportunities for identifying genes controlling embryonic development. No suitable experimental system was available for that purpose. *Drosophila*, the "reigning queen" of genetics, seemed very poor for studies of development, largely because of the complex transformation of its larva into an entirely different-looking adult fly. Because Beadle believed that the gene–development conundrum was more important than the crossing-over studies in which he was engaged, he and Ephrussi agreed to do something about it; but what would it be? They agreed to gamble a year of their lives to develop a system with which they could explore the functional role of genes in development and decided to exploit the rich and varied collection of *Drosophila* mutants for that purpose. But that meant that they would have to learn a good deal more about the embryology of the fly, and Beadle agreed that Ephrussi's tissue culture approach was the way to begin. Ephrussi persuaded Beadle that his laboratory at the Institut de Biologie Physico-Chemique in Paris would be an ideal place for that purpose.

To Beadle's delight, Morgan endorsed his plan and approached the Rockefeller Foundation requesting that they provide Beadle with a special fellowship for that purpose. The request was turned down, but Morgan assured Beadle of support by appointing him a Cal Tech research associate. Being in the midst of the depression, the annual salary was only $1500, down from the $2700 he had been receiving, but he was not dissuaded. Many years later, Beadle speculated that the money was actually provided by Morgan's personal funds.[39] Given Morgan's generosity when it came to personal matters, this was very likely true.[40]

Beadle set off for Paris in May of 1935 leaving Marion and his 4-year-old son David behind in a house at the Farm that Cal Tech had provided rent-free and that he had renovated. His welcome by Ephrussi's family was warm and friendly, and Ephrussi's daughter recalled that "My father was very happy when Beadle arrived...they had very good relations and lots of fun together."[41]

The Ephrussi family's friends welcomed Beadle, making him comfortable on his first visit to Paris. After a brief stay at Ephrussi's home, Beadle obtained room and board near the Institut in the vicinity of the Sorbonne, for about $1.25 a day. The Institut de Biologie Physico-Chemique, where Ephrussi and Beadle were to work, had been founded only 5 years earlier by Baron Edmond de Rothschild. At the time, it was arguably one of the best-equipped biological research laboratories in France. We can only imagine Beadle's thoughts as he walked along the Boulevard St. Michel and the streets leading to rue Pierre Curie (now rue Pierre et Marie Curie) where the gates of the Institut were located. The ambience and history of the area were certainly a far cry from Wahoo, Ithaca, and Pasadena!

Although Ephrussi loved music and art as well as the vibrant Paris night life, he does not seem to have enticed Beadle into sharing those aspects of Paris's charms, nor is there any indication that Beadle tried to learn to speak French. With the exception of the trip they made to the Marine Biological Station at Roscoff on the Brittany coast to meet and talk with some of the key figures of French biology, we know only that the two spent most of their waking hours working at the Institut. Ephrussi was already a well-established personage in French science, and he considered Beadle as a scientific equal rather than as his student.[42] Their doggedness, patience, and determination paid off, and within a few months of Beadle's arrival they achieved a breakthrough that produced major repercussions in genetics.

Beadle's farm-bred work ethic did not allow dalliance when there was something to be done, so he and Ephrussi started their experiments without much delay. They were well aware of the existence of "imaginal discs," the groups of embryonic cells ("buds") in the larvae that are transformed during development into the tissues of the newly hatched fly. Remarkably, the embryonic buds can be transferred from one larva into the abdominal cavity of another, and their developmental potential is conserved. Beadle and Ephrussi set out to determine whether such embryonic buds could develop into the corresponding fly structures outside the animal's body, that is, in tissue culture. Lacking experience in distinguishing or dissecting the miniscule embryonic buds, they sought the advice of Professor Charles Perez of the Sorbonne, who was the world's greatest authority on the development of the blowfly. Perez warned them that even with the blowfly, which is 10–20 times larger than *Drosophila*, their approach was unlikely to succeed, so how could they expect to succeed with the tiny fruit fly! Perez's advice notwithstanding, they went ahead and managed to master the problems about which he had warned them. Nevertheless, the tissue culture approach failed; none of the various kinds of embryonic buds developed outside of the animal into anything resembling the adult organs. But they were neither stymied nor discouraged since they had another idea. Because they were intent on examining the development of *Drosophila*'s eye pigments, their new plan was to determine whether the embryonic eye buds would develop into adult eyes if they were transplanted from one larva into another. Kohler cites an exchange of correspondence between Curt Stern and Sturtevant indicating that such work was in progress elsewhere, but there is no evidence that either Ephrussi or Beadle was aware of such efforts.[43]

The *Drosophila* eye consists of about 700 structurally identical facets (ommatidia), each containing 14 cells. Each group of cells, which constitute the fly's optic apparatus, is surrounded by two sets of pigment cells, one on top of the other. In normal eyes, the pigment cells contain purplish-red and ocher-yellow granules. The reddish-orange appearance of eyes is the result of a mixture of these two differently colored granules.[44] Over the years, Morgan's fly group had collected mutants whose eye color varied from the norm. The various mutations, each of which was genetically distinct and mapped to specific chromosomal locations, were named for the color that the eyes most closely resembled, e.g., *vermilion, cinnabar, claret, apricot, brown, white*, etc. The *vermilion* and *cinnabar* mutants that were to play a central role in Beadle's and Ephrussi's breakthrough experiments have bright red eyes and produce primarily the red granules. A mutant called brown, whose eye color appears brown, as the name indicates, produces only the yellow-ocher granules. The *white* mutant, which got Morgan into *Drosophila* genetics, fails to produce either of the pigmented granules. The two pigments have different physical properties, the red being soluble in water and easily extracted from the granules, whereas the yellow ocher is insoluble in water and cannot be removed from the granules except with alcohol or other organic solvents. Thus, it is easy to determine which of the pigments fails to be made by any of the mutants affecting eye color.

Embryonic eye buds acquire the two colored pigments during the pupal phase of development. Ephrussi and Beadle set out to determine whether embryonic eye buds transplanted from one larva into the abdomen of another larva or pupa would develop into a third eye in the newly hatched fly's abdominal cavity. They were emboldened to try that approach because a German group had succeeded in transplanting a variety of different embryonic buds from the larva of a mealmoth, *Ephestia*, into another moth larva's body cavity with the result that each was transformed into the corresponding adult structure.[45] The German workers also discovered that when embryonic testes buds from a mutant moth whose adult testes were pale instead of darkly pigmented were transplanted into normal larvae, the transplanted testes developed dark pigmentation. In essence, the mutation causing pale testes was nonautonomous because the mutant characteristic was overcome by something from the normal environment into which they had been transplanted. Furthermore, when normal testes buds were transplanted into mutant larvae whose own testes were pale, the newly hatched moth's own testes were normally pigmented. In this case, the German workers surmised that the transplanted testes provided what was missing in the mutant. These experiments demonstrated the feasibility of the transplantation approach and indicated that the genetic deficiency affecting pigmentation of the testes could be overcome by implantation in a normal tissue background. But *Ephestia* is "gigantic" compared to *Drosophila*, and Beadle and Ephrussi realized that transplanting such tiny bits of tissue without killing the transplanted larvae would be far more difficult. At Roscoff, they obtained the advice of an embryologist who had some experience with such procedures, and when they returned from the coast they sought to master the procedure.

Beadle and Ephrussi soon learned that what was relatively easily managed with the naked eye for the larger *Ephestia* larvae proved to require a microscope and demanding eye–hand dexterity with *Drosophila*. Using two binocular microscopes arranged so that they could both peer at a common object, one of them dissected out the embryonic bud from a donor larva and the other injected the bit of tissue into the abdominal cavity of another larva or early pupa. Predictably, removing the miniscule bud intact was very difficult, and their initial attempts ended in failure. Beadle complained in a letter to Demerec about "the small size of the beasts."[46] But Beadle's handworking skills, honed by the necessities of making do on the farm, saved the day. He devised a special micropipette joined to a microinjection apparatus that enabled them to remove an embryonic bud whole and deliver it intact into the recipient larva. More often than not, however, the injected larvae died. The lack of progress was frustrating, and only after many days of practice did they succeed in transplanting a variety of embryonic buds, each of which developed into the corresponding recognizable and functional tissue.[47] Remarkably, transplanted ovary buds produced adult ovaries that were fully functional in that their eggs could be fertilized and yield viable offspring. In time, as many as 200 embryonic buds for ovaries, testes, antennae, wings, and legs could be transplanted into larvae in a day, each yielding the relevant supernumerary tissues in the abdomens of the newly hatched flies.

Fully confident of their technique, Beadle and Ephrussi transplanted eye buds, with the result that the mature fly contained a third eye visible through the abdominal wall. As Beadle recalled it, "we came to the lab one morning to find an adult fly with three eyes, two at their normal position in the head and one in the abdominal cavity. We were tremendously excited and encouraged... and spent most of the day at a nearby sidewalk café celebrating and making plans for future moves."[48] Although the implanted embryonic eye buds developed into normal-appearing eyes, they were unlike the normal ones in the head. A normally positioned eye is shaped like the head of a mushroom, the outer surface bearing the facets corresponding to the top or convex surface of the mushroom. The transplanted eyes, however, were inside out; i.e., the facets, normally on the outside surface, were on the inside and the pigment cells were on the outer surface. Conveniently, then the color of the transplanted eyes could easily be assessed and compared to those in the head.

They wondered "Could we do it again? Following a systematic study, we demonstrated that we could."[49] Their excitement bubbled over into letters to Demerec and Sturtevant. Beadle asked Demerec about the possibility of presenting their preliminary data at an upcoming Genetics Society meeting in St. Louis.[50] Replying, Demerec invited Beadle and Ephrussi to submit an abstract for the meeting and a short paper for *Science* magazine with Beadle as first author as well as a picture of a three-eyed fly.[51] A month later, Beadle sent the abstract along with dollar bills to legitimize his and Ephrussi's membership in the Genetics Society.[52]

Pursuing their original goal, Beadle and Ephrussi excised eye buds from each of 26 different mutant larvae whose adult eye colors were clearly distinguishable from normal, and each was transplanted into both normal larvae and each of the 25 other mutant larvae. More than 600 separate transplantations were required, each set being repeated several times.[53] This prodigious effort showed that embryonic eye buds from 24 of the 26 mutants tested always developed "abdominal eyes" whose color was identical to that of the mutant from which the eye buds were taken. Thus, the mutations were autonomous and could not be overcome in the transplanted hosts. The two exceptions were eye buds from the mutants *vermilion* (*v*) and *cinnabar* (*cn*). When these were transplanted into normal or into most other mutant larvae, the transplanted eye buds produced normal-color eyes. Because the implant resided unattached to any of the larval internal organs, Beadle and Ephrussi surmised that the formation of normal eye pigmentation must result from the presence of some substance circulating in the fly's lymph fluid. They quickly realized that what was being supplied was needed for normal color development and could not be made in the mutant eye buds. The intriguing question was, What was the identity of the substance(s) that caused the *vermilion* and *cinnabar* eye buds to develop normal eye pigments?

The transplantation experiments were consistent with Sturtevant's earlier interpretation that *vermilion* eye cells are colored normally when surrounded by nonmutant eye cells, that is, the *vermilion* mutation is nonautonomous.[54] The new findings identified the *cinnabar* mutation as a second nonautonomous eye-color mutation. Believing at first that the *vermilion* and *cinnabar* mutants were deficient in the same process of eye pigment formation, Beadle and Ephrussi anticipated that transplanting *vermilion* eye buds into *cinnabar* larvae, and vice versa, would yield mutant color eyes in each case. To their surprise, however, *vermilion* eye buds developed into normal color eyes when transplanted into *cinnabar* larvae, but cinnabar eye buds produced only the abnormal cinnabar-colored eyes when introduced into *vermilion* larvae.

How could they explain this unanticipated finding? First, it was clear that the two mutations could not be affecting the same function, for then reciprocal transplants would have produced only mutant colored eyes. Furthermore, the two mutations were not affecting two unrelated substances, otherwise the reciprocal transplants would both have given wild-type eye color. They concluded that there must be two substances, each required for the production of the normal eye pigments; they dubbed them the v and cn substances. They reasoned further that the production of the v substance must be blocked in the *vermilion* mutant and that formation of the cn substance is blocked in the *cinnabar* mutant. Because the effect of the *vermilion* mutation was overcome by transplanting *vermilion* eye buds into *cinnabar* larvae, it seemed plausible that *cinnabar* larvae could make the v substance. In contrast, the *vermilion* larvae could make neither the v nor the cn substances.

Beadle and Ephrussi then made a big leap and concluded that the v substance is made before the cn substance during normal eye-color formation and is converted into the cn substance.[55] By this logic, they proposed that the function of the *vermilion* and *cinnabar* genes is to control two successive steps in the metabolism responsible for the production of the brown pigment.

$$\text{X substance} \xrightarrow{v} \text{v substance} \xrightarrow{cn} \text{cn substance} \rightarrow \rightarrow \rightarrow \text{brown pigment}$$

vermilion and *cinnabar* eye buds produced normal-colored eyes when transplanted into another eye-color mutant, *claret* (*ca*), but claret eye buds failed to develop normal color in either *vermilion* or *cinnabar* flies. They concluded that the three genes, *claret, vermilion*, and *cinnabar*, were responsible for the formation of ca, v, and cn substances, respectively. The ca substance, they believed, gives rise to the v substance, which in turn is converted to the cn substance, which ultimately is used in the formation of the brown pigment. Although Beadle and Ephrussi presumed that the ca, v, and cn substances were successive products in a reaction chain, they cautioned against concluding that this represented a direct chemical transformation of one substance into another. The *vermilion* and *cinnabar* eye buds produced normal-color eyes in many of the mutants that were unable to make the normal brown pigment. Thus, it seemed likely that these mutants were able to make and provide the v and cn substances but were unable to use them in the later steps.

Sturtevant urged them to publish their results quickly and to send a brief description of their experiments and their scheme for eye-color development for the upcoming genetics meeting at Cold Spring Harbor. Beadle and Ephrussi promptly submitted several short papers, some to the *Comptes Rendu Acad. Sci. Paris*, a French journal known for its rapid publication of important findings.[56] Notably, these reports and the more extensive manuscript published later in *Genetics*[57] were submitted within about six months of the time the experiments were initiated.

News of their success, especially the dramatic image of a third eye developing in the abdomen of the hatched flies, spread to America like wildfire. Sturtevant could even show off a "three-eyed" fly that Beadle had sent him as visible evidence of their result. He was especially gratified that the transplantation experiments confirmed his earlier conclusion that the *vermilion* mutation was nonautonomous. He also foresaw that the transplant technology paved the way for new approaches to developmental genetics. Both Sturtevant and Dobzhansky realized that embryonic transplants made it possible to create interspecies hybrids that were precluded by the inability to breed different *Drosophila* species. Sturtevant commented that "the attack on development can really go ahead...I can think of no technical advance that seems to me as influential in 'Drosopholistics [*sic*].' Even the salivary chromosomes are put in the shade." Dobzhansky reacted even more effusively by referring to their accomplishment as "an advance of extraordinary importance." Morgan, too, was especially pleased: He placed their discovery on a par with Muller's finding that X rays could induce mutations and Painter's discovery of polytene chromosomes.[58]

The French scientific establishment hailed and rewarded Ephrussi by promoting him to associate director of the tissue culture laboratory at which the transplantation work was done. A year later he was appointed maitre de recherche in the Centre Nationale de Recherche Scientifique as well as director of the new Laboratory of Genetics at the École des Hautes Études. Beadle also was pursued with attractive job offers. On his return to the United States, he presented a seminar at Cold Spring Harbor Laboratory. Demerec was convinced that the transplantation technology opened new opportunities for exploring *Drosophila* development. He worked to persuade George L.

Streeter, chairman of the Carnegie Institution of Washington's Department of Embryology, to offer Beadle a position as a research associate in the Department of Genetics at Cold Spring Harbor. The offer was quite generous with an annual salary of $3500, an additional $1500 for equipment, and $500 for a research assistant.[59] Beadle turned down the offer within a week saying that he would likely go to Harvard. But Demerec was not put off so easily, and he suggested a summer associate position between Harvard and Cold Spring Harbor. The chief of the Bureau of Plant Industry at Columbia, Missouri tried to recruit him to work on the cytology of wheat with Lewis J. Stadler, the codiscoverer of the mutagenic effects of X rays and ultraviolet radiation. Barbara McClintock's plan to move to the nearby University of Missouri may have tempted Beadle to accept, but in the end he declined all the offers.

Morgan hoped that Beadle could be persuaded to remain at Cal Tech as an assistant professor. To sweeten the offer, he approached the Rockefeller Foundation about providing an "elaborate microscope," so that Beadle could continue the work he had begun in Paris. But the Foundation refused his request.[60] Beadle may well have been tempted by Morgan's offer since he was grateful to him for making the trip to Paris possible. Moreover, he felt at home at Cal Tech and had many close friends there, but Beadle had been urged once before to move on from the place in which he had trained, advice he felt stood him in good stead and which he later passed on to others.[61] Perhaps wishing to strike out on his own was one of the reasons he changed his mind when Harvard renewed their offer. He accepted, intending to take up the appointment in the autumn of 1936.

With Beadle in Pasadena and Ephrussi in Paris, they established that the v substance, and in some instances the cn substance as well, could be obtained from different species of *Drosophila* and even from quite unrelated insects. When extracts of such insects were injected into *vermilion* or *cinnabar* larvae or pupae, normal eye color was restored in the emerging adult flies.[62] Beadle also discovered that several *Drosophila* tissues, notably Malpighian tubules (excretory tubes leading from the digestive tract in insects), produced abundant quantities of the v and cn substances.[63] This indicated that these substances were widespread in insects and therefore fundamental to normal eye development. During that time, Beadle and Ephrussi also demonstrated that several of the 26 other genes affecting eye color were in some way involved with the production of the v and cn substances, but exactly how was unclear.[64] The central problems remained: What was the chemical nature of the v and cn substances, when and where during development were they made, and how were they involved in determining the fly's eye color? The latter two questions were answered soon after Beadle returned to Cal Tech. Both labs confirmed that the v and cn substances were absent throughout the larval stage and appeared only during the middle to late stage of pupal formation. Furthermore, both substances affected eye-color development only during the latter phase of pupal development.[65]

Ephrussi missed Beadle and decided to return to Pasadena for a second year, thinking that Beadle would be there. He expected that they would collaborate in efforts to identify and characterize the v and cn substances. However, the Rockefeller Foundation grant he received to support his return to Cal Tech could not begin until the autumn of 1936, and Beadle was gone by then. While preparing to return to Cal Tech, Ephrussi persuaded Jacques Monod, a young French biologist working at the Pasteur Institute, to join him there. Beadle noted many years later that he had been impressed with Monod, but thought the Frenchman was conflicted about whether to be a musician or a biologist. On this occasion, science won out; initially reluctant, Monod decided that his career would benefit from learning some genetics and Cal Tech was an ideal setting for that purpose. He was well aware that France was lacking in opportunities for acquiring training in that field.[66] In accepting the Rockefeller Foundation fellowship that Ephrussi had obtained for him, Monod had to cancel the arrangements he had made to participate for a second time in a biological expedition to Greenland aboard the French research ship *Pourquoi Pas*. It was a fateful decision, since the *Pourquoi Pas* met with disaster off the coast of Greenland and all aboard were lost.[67] The year Monod spent at Cal Tech turned out to be scientifically unfruitful because he rarely spent any time in the laboratory. Instead, he organized a choral group under the auspices of a newly founded Bach Society and conducted vocal recitals throughout southern California. His reputation as a musician and social charmer grew to such an extent that he was offered an opportunity to remain in the United States as a musician.[68] Needless to say, Ephrussi was miffed by Monod's behavior and decried his having wasted a special opportunity to further his scientific career. Nevertheless, the two remained steadfast friends throughout their lives.

By the time Ephrussi and Monod arrived in Pasadena, Beadle and his family were already settled in Cambridge. They had agreed while in Paris to continue their collaboration and had settled on several important questions that still needed to be answered. They also agreed that it was best for each to pursue his respective research separately but to remain in close communication about their findings. When Ephrussi completed his second stay at Cal Tech, he returned to his new laboratories planning to continue the transoceanic collaboration with Beadle. Although the separation precluded the challenging discussions and exchanges they had enjoyed while together in Pasadena and Paris, they made substantial progress on answering some of the questions they had posed for themselves.

Looking back on those first attempts to explore whether and how an organism's genotype guides its embryological development, it is evident that Beadle and Ephrussi were ideally suited for that task with their complementary scientific experience and skills, their unusual self-assurance, and their consuming passion to crack an important problem in their science. Warnings of failure and, indeed, their own early failures might well have dissuaded others from continuing the effort, but they persisted. Their insightful interpretation of their results provided the first clear experimental evidence for the involvement of genes in embryonic development. As shown below, this breakthrough led Beadle to another, the discovery of the principal way in which genes perform that role.

CHAPTER 8

Becoming a Professor

Leaving their belongings and pets with Marion's mother in Claremont, Beadle, Marion, and David headed cross-country for Cambridge in their Model A Ford, arriving in time for the beginning of the 1936 autumn term. They settled in a small, pleasant apartment in a relatively unattractive neighborhood on Massachusetts Avenue, a location where few other professors lived.[1] The apartment remained sparsely furnished all year, perhaps because Marion was unhappy away from her native California surroundings and Beadle suspected his stay might be brief. Indeed, Sturtevant had predicted that Beadle would not like Harvard. David, however, could only think about getting to see snow and using the sled Beadle had bought for him before they left California.[2]

Once settled, Beadle also had to get his research program under way. He told F.B. Hanson, the program director of the Rockefeller Foundation's biological research grants, that he was worried about whether the research funds provided by the university were sufficient. Besides a salary of $4000 per year, Harvard provided funds to purchase equipment and $500 for technical assistance. Although the equipment funds were adequate, Beadle worried that the amount for assistants was insufficient, considering that engaging a graduate student required payment of $1 per hour, twice what he had paid at Cal Tech. Hanson's diary noted that "Beadle makes an exceptionally fine impression and is undoubtedly one of the most promising men of his age in biology—a man to be watched."[3] With that kind of assessment, the Foundation readily granted additional research funds. It was the first of a continuing flow of Rockefeller funds directly to Beadle.

Beadle wasted little time before plunging into experimental work. He soon found a friend and collaborator in another recent Harvard recruit, Kenneth V. Thimann, whom he had known at Cal Tech. Thimann was the first to characterize an auxin chemically as indole acetic acid, and this achievement won an offer for an assistant professorship at Harvard. Beadle persuaded Thimann to help him isolate and identify the v and cn substances. First, they needed a way to measure the amounts of these substances in various sources. They hit on the idea of creating "tester flies," flies with two mutations, one affecting the ability to make either the v or cn substance and another mutation (*bw*) that blocks the production of the red pigment. The eyes of such tester flies produced neither the red nor the brown pigments, and, therefore, were pale. By injecting a sample into such mutant larvae, the amounts of the v and cn substances could be assessed from the change in the adult's eye color.[4] Their chemical studies revealed that the v and cn substances, whatever the source, were unlikely to be proteins, enzymes, or fats. Instead, their properties resembled those of amino acids, the constituents of proteins, but not any of the known ones; thus, the identification of the two hypothetical substances eluded them.

While deeply immersed in the *Drosophila* eye-color work, Beadle found time to participate in some of the departmental social activities. Lunch at the faculty club was a time-honored tradition at Harvard, as it was at Cal Tech. He welcomed these occasions to meet colleagues and sample the institution's culture. Afternoon teas provided a relaxed setting for meeting faculty, students, and staff. As a sign of the times, it was uncommon for the faculty to mix with the few women students who attended the teas. Beadle, however, was not at all standoffish: A Radcliffe student, who later married Beadle's future collaborator and close friend, Norman Horowitz, recalled that he may have been the only biology professor who deigned to speak to women taking courses at Harvard.[5] Although he tried, Lloyd Law, a graduate student whom Beadle befriended, despaired of getting Beadle interested in sports or other university activities. The young professor had little time for such distractions.

Besides doing research, Beadle was expected to teach, and he was assigned an undergraduate genetics course and an advanced laboratory course that attracted only a single graduate student! Much to the consternation of some of his colleagues, the laboratory course consisted mainly of experiments he was doing in his own laboratory; namely, tissue transplantation in *Drosophila*.[6] Beadle's teaching skills left much to be desired, and the Harvard undergraduates who took his course let him know it. It was customary for the students to voice their approval or disapproval of a lecture by applauding or booing. As Beadle recalled it, he was booed quite frequently. Sensitive to these rebuffs, he cornered a student and asked why they were giving him such a hard time, particularly since they didn't boo at all after what he thought were poorer lectures by one of his colleagues. As he recalled it, the student said that Beadle's colleague wasn't worth the effort.[7]

Beadle's recollections of the year at Harvard were mostly negative. In his words, "Harvard was too formal for a Wahoo farm boy." Used to the open, frank, and collegial atmospheres at Cornell and Cal Tech, he railed at the biology division's "closed-door" culture that required an appointment to meet with a professor or even engage in a social discussion. Also, the morale among assistant professors, of which there were too many, was not high. They resented the lack of respect for junior faculty and the university's well-known reluctance to promote them. He was also annoyed by the lack of interest from his colleagues in *Drosophila* as a model for genetic research.[8]

About midway through the academic year, Beadle was asked to consider moving to Stanford University. The offer of a tenured full professorship in the division of biological sciences was attractive, but before accepting he sought the advice of the distinguished Harvard genetics professor, E.M. East, who, he recalled, had been Emerson's mentor 25 years earlier. "In his usual gruff and direct manner East said, 'well what do you want me to say about it?' I said I just want to know what you think about it. His response: 'Oh you want to know what I think about it? I'll tell you what I think. Stanford isn't any good, it never was any good, and it will never be any good. That's what I think about it." Despite a counter offer of promotion to tenure at Harvard, Beadle ignored East's advice and decided to move. Later, East said "Beadle I knew you would go to Stanford. I knew you didn't like Harvard." Beadle reminisced later, "In retrospect, my decision to leave Harvard at the end of the year was the right one for me."[9]

Beadle accepted the Stanford offer on April 5, 1937, and planned to arrive in Palo Alto, the university's home, in the upcoming fall. He was to be the first to carry out genetic research at Stanford. To entice him to Stanford, he had been promised sufficient freedom from teaching obligations for the first year to get his research program under way. C.V. Taylor, the chairman of the biology department, wrote immediately to Warren Weaver, the scientific director of the Rockefeller Foundation, outlining their plans for Beadle and requesting $3000–$3500 to support his research in the first year.[10] Hanson's high regard for Beadle assured approval by the board.

Marion probably welcomed the decision to return to California. She and David were frequently away from Cambridge during the winter and left for California by train sometime during the spring term. Beadle was left to dispose of their apartment and sparse furnishings and to drive to California without them. Instead of heading west immediately after Harvard's spring term ended, he visited Woods Hole where Sturtevant and his family spent the summer holiday working at the Marine Biological Laboratory. Law, the lone graduate student who had worked in Beadle's underattended laboratory course, accompanied him. They roomed in a house on a hill overlooking the harbor and Martha's Vineyard in the distance. Beadle relished the physical exercise and walked the mile or so down to the lab and back up to his room at the end of the day.[11] In summertime, Woods Hole is a gathering place for biologists of every stripe, primarily because of the steady stream of colleagues and the varied assortment of lectures and seminars. It was famous for its easy mix of intellectual activity with sand and sea, as well as for the opportunity to work undisturbed by the usual intrusive academic duties. Beadle also enjoyed the wide-ranging, informal discussions that he had missed at Harvard and that were so much in tune with his own preferences. It's doubtful that he conducted any experiments that summer. More likely, he and Sturtevant initiated the collaboration that led to their classic textbook *An Introduction to Genetics* that was published two years later.[12]

At summer's end, Beadle and Law set off for California in Beadle's newly acquired Plymouth. They planned to stop in Wahoo to visit with his father who still worked and his sister Ruth who came home to see him. The long trip west was made more tolerable by Beadle's suggestion that he and Law create a genetic crossword puzzle; Beadle composed and Law served as scribe for what remained an unfinished work. After several days at the farm, Beadle backtracked to Lincoln to visit with Professor Keim, who had steered him to a career in science. From there, the drive was nonstop until they arrived in Palo Alto, a bucolic community some 35 miles southeast of San Francisco. Nestled between San Francisco Bay and the golden, gently rolling hills leading down to the Pacific Ocean, Stanford is blessed with the kind of year-round mild climate to which he had become accustomed and which Marion welcomed.

Stanford University was established in 1885 by Leland Stanford, a founder of the Republican party in California who was credited with keeping the state in the Union during the Civil War. He had achieved national prominence for finishing the Union Pacific's first transcontinental railroad, which helped his election as governor of California (1862–1864) and subsequently to the United States Senate (1885–1893). The Governor and Mrs. Stanford founded the university to commemorate their son, Leland Jr., who had died of typhoid fever at the age of 16. The famed El Camino Real, the Royal Road that runs south from San Francisco to San Jose, separates the 8800-acre campus from the city of Palo Alto. The city derives its name from the tall spindly redwood tree that served as a landmark for the earliest Spanish settlers and still exists nearby. Because the campus had served as Governor Stanford's retreat for breeding and training prized trotting horses, the campus was and still is teasingly referred to as "The Farm." Beadle's passion for growing corn would find a home on The Farm.

The Stanfords insisted that the university stress research along with teaching in undergraduate and graduate courses. From its beginning in 1891, it was coeducational and had a strong tradition in biology, established and nourished by its first president, William Starr Jordan, a well-regarded ichthyologist. At the time Beadle arrived, Ray Lyman Wilbur, a physician by training, had presided over the university for nearly 22 years and would continue for another 6 years. Wilbur recruited a vigorous and forward-looking biology department faculty: the distinguished Dutch microbiologist, C.B. van Niel; the embryologists V.C. Twitty and D.M. Whitaker; L.R. Blinks, a marine electrophysiologist; and A.C. Giese, a cell physiologist. The department was well funded by a 10-year block grant from the Rockefeller Foundation, but Beadle's decision to come to Stanford was also influenced by the youth and vitality of the group.[13] Whether the president had a role in persuading Beadle to move from Harvard is unclear. It seems more likely that Taylor and Whitaker, T.H. Morgan's son-in-law, took the lead. By the time Beadle arrived, Palo Alto was a typical academic community with much of its business focused on university affairs and people.

Beadle's assigned research space was renovated and ready to move into when he arrived.[14] The laboratories were located in the basement "catacombs" of Jordan Hall, one of the buildings built at the turn of the century. They were hardly a match for the well-equipped labs of today. The windowless but well-lighted space, with the building's steam pipes running overhead, was divided into small separate working areas, allowing people to move freely from one activity to another. Beadle thought nothing of fixing things or painting the lab himself—what he called an "old-style do-it yourself" way of science.[15] These laboratories were home for Beadle and a diverse collection of graduate students, postdoctoral fellows, research associates, and visitors for the 9 years he was at Stanford. Following the first year, during which he was relieved from teaching, Beadle taught a spring quarter course in comparative genetics and in subsequent years a genetics laboratory as well. Later, he taught a more advanced course with laboratories in genetics in the winter quarter of alternate years.

By the time Beadle arrived, Marion had already rented a house in Palo Alto, on the corner of Lowell and Waverly streets, in what the locals referred to as "Professorville." The rustic one-floor rambling house, shaded by large old California oak trees, was situated at one end of a rather large yard. Soon after moving in, Beadle planted a garden where he grew vegetables and flowers, and he built a badminton court for David and his playmates. David was enrolled in the private Peninsula School, where Marion became an active participant. By the time he reached third grade, David had transferred to the public elementary school on the university campus, perhaps because Marion was becoming increasingly associated with other faculty wives and sought to be involved with campus activities.

Even before Beadle left for Stanford, probably while at Woods Hole, he concluded that if he was to make progress in establishing the chemical nature of the v and cn substances he would need to collaborate with a biochemist expert in chemical isolation and analysis. Beadle sought the advice of Professor W.H. Peterson and Dean E.B. Fred at the University of Wisconsin, both experts on the isolation of natural products. They recommended a former graduate student, Edward L. Tatum, who had been an undergraduate at the University of Chicago and earned M.S. and Ph.D. degrees at Wisconsin. His work on the nutrition and metabolism of bacteria, particularly lactobacilli, helped pave the way for the identification of vitamin B1. After completing his doctoral work and marrying June Alton, Tatum remained at Wisconsin for a year, during which his work with a variety of bacteria established that each species had diverse requirements for amino acids and vitamins. A Rockefeller Foundation fellowship provided him with an opportunity to work in Holland with Professor F. Kögel at the University of Utrecht, where his project to define the nutritional needs of hemolytic streptococci failed because of the complexity of the requirements. Although jobs were scarce because of the lingering economic depression, Tatum sensed that he was ready to move.

Beadle wrote to Tatum from Woods Hole describing his need for a chemist.[16] Despite Peterson's and Fred's advice that opportunities at Iowa State College and Wisconsin were preferable to the one at Stanford,[17] and his father's concerns about the nature of the position, Tatum accepted Beadle's offer. He was to begin as a research associate in the fall at a salary of $2000 per year.[18] Beadle assured Tatum that Marion would arrange for a suitable apartment for his family at a monthly rental of about $40–$50.[19] He was unconcerned that Tatum had very little background in genetics or experience with *Drosophila;* what he saw in him was someone with microbiological and chemical experience and skills that would be needed to isolate and identify the v and cn substances. Tatum seems not to have been put off by the prospect of working with an unfamiliar organism or in an area for which he had no preparation. The future proved that Beadle's selection of Tatum for the next phase of the work was foresighted, rivaling in its impact Morgan's choice of Sturtevant at the start of the *Drosophila* work.

Ephrussi also realized that he needed to team up with an experienced chemist if he was to make headway on identifying the v and cn substances. When he returned to Paris in the spring of 1937, he hired Madame Yvonne Khouvine, a biochemist, who had been working for some years with Professor R. Wurmser at the Institut de Biologie Physique-Chemie (IBPC). She joined his lab in mid-September, about the same time that Tatum arrived in Palo

Alto. As Ephrussi's principal research associate, Khouvine was to oversee the day-to-day efforts in isolating and identifying what Ephrussi had begun to refer to as the v and cn "hormones." He used the term hormones because they were produced in tissues where they can act in situ, diffuse, or be transported where they can act on other tissues.[20]

For the next 2 years, the Beadle–Tatum and Ephrussi–Khouvine teams engaged in a "competitive collaboration," keeping abreast of one another's experimental results by a remarkably frequent transatlantic exchange of letters and manuscript drafts. Discoveries and preliminary observations were communicated for the purpose of verification, and comments on interpretations and speculations were welcomed. Technical improvements in measuring quantities of the v and cn substances were rapidly incorporated into each other's experiments, and *Drosophila* stocks with particularly advantageous mutations affecting eye-color development were regularly shipped from California to Paris. Ephrussi encouraged Beadle to quote his unpublished results freely and to include mention of his results when distributing related reprints.[21]

Even while endorsing this free and generous exchange, they agreed to pursue their own leads and publish independently. From time to time, however, there were hints of concern about who was going to do what and how manuscripts would be coordinated to avoid preemption of priority. At one point Ephrussi wrote, "I agree with you on the non-desirability of 'pushing' for cooperation. So we have just to continue what we did before, in other words to publish as soon as something worth [*sic*] comes up. As before we will send you the manuscripts at the same time that we send them to the editors. I am sure also that this actually is a form of cooperation, and perhaps the most convenient under the present conditions."[22] However, more explicit concerns about competition surfaced in Paris soon after Tatum and Khouvine joined the efforts. Ephrussi asked Beadle to provide a "clearly defined" description of Tatum's plans for the isolation of the v and cn substances while assuring him that Khouvine was agreeable to cooperation.[23] A few weeks later he reminded Beadle that "I am anxiously waiting for your proposals for cooperation. As you know from our conversation at Cold Spring Harbor, I always was in favor of it if you offer a plausible scheme and am sure also it will be profitable to everybody,"[24] and again a month later, "why don't I hear from you about the cooperation project? Khouvine and I expected to hear about this some time ago but nothing came."[25] All the while, however, both laboratories made considerable progress. They clarified where and when the v and cn substances were produced in the developing fly and even obtained strong leads about their chemical nature.

Initially there was no clue to the chemical structure of the v and cn substances, and both laboratories had to rely on biological assays for their detection and quantification. Using the tester fly method that Beadle and Thimann had developed,[26] Tatum established a metric between the relative amount of v or cn and the intensity of the eye color by injecting extracts containing calibrated amounts of either the v or cn material into the corresponding double mutants. The real value of Tatum's "eye-color chart," however, relied on a key observation by Law that the tester fly's eyes responded well to the presence of v and cn in pupal extracts when the extracts were mixed with the fly food.[27] These improved methods were promptly made available to Ephrussi in Paris.

Using these double-mutant tester flies, the Paris group discovered that feeding *vermilion* larvae with yeast or partially digested proteins caused a change in the eye color to that closely resembling the wild type. Even more surprising, both Khouvine and Tatum found, independently, that normal eye color developed in *vermilion* flies if they were starved. They surmised that partially digested protein and something produced during the breakdown of some cellular constituents accompanying starvation led to the production of the v substance. This idea, however, did not bring them closer to a solution of the hormone's chemical nature.[28]

Tatum then focused his efforts on obtaining pure v and cn substances from hot water extracts of wild-type pupae, the only reproducible source of these substances. He began with tens of thousands of dried wild-type pupae and used chemical procedures to obtain a highly enriched preparation of the v substance, which he subjected to a variety of chemical and physical analyses. He learned that the v substance was a relatively small molecule (440–660 molecular weight), confirming that it resembled an amino acid.[29] But which one of the 20 amino acids found in proteins was it? When *vermilion* larvae were fed a mixture of amino acids, they matured into flies with nearly normal eye color. The data even hinted that tryptophan, one of the 20 amino acids in proteins, might be responsible, but the variability in the results deflected attention from its significance. One of Tatum's tryptophan samples did indeed impart normal eye color, but none of the others produced the same effect. The reason for this discrepancy, he discovered, was that the successful sample was contaminated by bacteria. Being an accomplished bacteriologist as well as a biochemist, Tatum surmised that the unidentified *Bacillus* species that he isolated from the contaminated material probably converted tryptophan into the active v substance, a deduction that proved to be correct.[30] Material isolated from cultures of the bacterial species growing with tryptophan was indistinguishable in its biological properties from the v substance isolated from *Drosophila* pupae. The biologically active material was crystallized, and this pure material was indistinguishable by biological tests from the v substance recovered from larvae.[31] However, the crystalline material was not tryptophan, but something resembling it.

All the while that Beadle was immersed in *Drosophila* research and organizing new courses he was to teach, he and Sturtevant were hard at work on the genetics textbook they had planned at Woods Hole in the summer of 1937. The book was completed in 1939, two years after Beadle arrived at Stanford,[32] and rapidly became a classic, serving as the magnet and primer for the next generation of geneticists. The term "introduction" belied its comprehensive and in-depth coverage of the fundamentals of genetics known at that time. It was, as they said in the preface, not an easy treatment of the subject, and the problems would require serious effort by students. Given his intense commitment in pursuit of the *Drosophila* eye-color puzzle, it's a testament to Beadle's ability that he could manage more than one project at a time.

The eye-color puzzle was coming to a head and the upcoming Seventh International Congress of Genetics scheduled for Edinburgh, Scotland, during August of 1939 provided the two groups with an opportunity to report on their progress. Ephrussi had been charged with organizing the program for the section on physiological genetics, and he invited Beadle to present one of the principal lectures; the others were to be given by J.B.S. Haldane, S. Wright, and N.W. Timofeeff-Rossovsky. Hoping to persuade Beadle to spend the summer at the Institut in Paris, Ephrussi offered up tempting recollections of scientific discussions at La Coupoulade, a favorite Montparnasse café-restaurant they had frequented during Beadle's previous stay. But as the time for the Congress drew nearer, the political tension in France had reached such a level that Ephrussi felt it impossible to work or even to maintain interest in the *Drosophila* eye-color work.[33] A month later he admitted that it was almost useless to plan experiments more than a fortnight ahead.[34] Perhaps Beadle had second thoughts about spending the summer in France under those circumstances, but in any case, he, Marion, and David planned to spend much of the summer in their cottage at Corona Del Mar. He was looking forward to a rare opportunity to be a real father to David.[35] As the summer drew to an end, he sailed from New York on the French liner Normandie on August 16, arriving in Edinburgh just as the Congress began.

Preparations leading up to the Congress had been troubled. At the time of the Sixth International Congress of Genetics in Ithaca during the summer of 1932, the International Genetics Society accepted a generous invitation from the presidium of the U.S.S.R. Academy of Sciences to hold the next Congress in Moscow in 1937. Professor N.I. Vavilov, Russia's most passionate supporter of classical genetics, was invited to serve as the Congress's president.

Vavilov saw this as an opportunity to showcase Russian genetics and to stem some of the derision generated by Lysenkoist practices. But as preparations for organizing the Congress were under way and commitments to attend were being received from scientists throughout the world, the Soviet committee without consultation or explanation canceled the event and proposed that it be held the next year. The decision was not Vavilov's; it had been ordered by V. Molotov, the Prime Minister of the Soviet Union, with the connivance of Lysenko. The reason for the cancellation was clear enough: Vavilov's status as the leading figure of Soviet biology would have been strengthened and the hoax that Lysenko was promoting would have been exposed.

Although the Soviets' stated intention was to hold the Congress in 1938, there was little evidence of any preparations. Accordingly, the International Genetics Society's executive committee canceled the event and offered the privilege of hosting the Seventh Congress to the British Genetics Society during the summer of 1939. That invitation was accepted, and the University of Edinburgh was chosen as the venue. Founded in 1583, this venerable institution's science faculty was among the best in the United Kingdom, particularly in genetics. Besides its scientific standing, Edinburgh's ancient battlements, regional charms, and relative isolation from the political storms enveloping England made it a desirable site for the Congress. To assuage their Soviet colleagues, the organizing committee again invited Vavilov to accept the presidency of the rescheduled Congress and Professor A.E. Crew from the University's Institute of Animal Genetics to serve as its general secretary. Preparations and registrations continued well into the summer, when less than a month before its opening, Vavilov informed the local organizers that neither he nor other Soviet geneticists would attend. The confrontation between Lysenko and Vavilov had come to a head with Vavilov the loser. He was regarded as a nuisance by the Soviet press and the country's ruling hierarchy; by 1941 he was imprisoned and 2 years later was dead.

H.J. Muller, by then among the most eminent geneticists in the world, had spent 3 years in Vavilov's Genetics Institute in Leningrad (now St. Petersburg) and was well acquainted with the devastating mischief that Lysenko's rise to prominence was having on Russian genetics and agriculture. Becoming more and more troubled by the rising chorus of Lysenkoist nonsense, Muller declared that any comparison between Mendelism–Morganism, the Russian derogatory reference to classical genetics, and the Lamarckian theories espoused by the Lysenkoists, was as absurd as one between "medicine and shamanism, chemistry and alchemy or astronomy and astrology."[36] Muller's increasingly outspoken criticism of Lysenko's unorthodox views and his political power had jeopardized the American's research activities and made him persona non grata in the Soviet Union. He found refuge in Edinburgh, where he remained until the middle of the war.

Muller and a group of largely American and European geneticists asked the Congress to support what they called a charter for the genetic rights of man. This "Geneticists Manifesto" labeled the Lamarckian doctrine for improvement of humankind as fallacious. Instead, the manifesto identified major sociologic and economic barriers to achieving a genetically improved humankind, the most important features of which were improved health, enhanced intelligence, and "those temperamental qualities which favor fellow-feeling and social behavior...which make for personal success." Unlike the more coercive proposals such as forced sterilization to reduce the propagation of undesirable traits, which characterized the more virulent forms of eugenics then prevalent, the manifesto called for a humanistic approach in applying genetic principles to human reproduction. Genetic improvement was viewed as being possible through "some kind of selection" on the part of persons "who had better genetic equipment, having produced more offspring on the whole, than the rest, either through conscious choice, or as an automatic result of the way in which they lived." This reflects Muller's early inclination to finding a means to human improvement. Such selection was viewed as unlikely in a civilized society unless it was encouraged by "some kind of conscious guidance," which in turn would require an appreciation of the value of that approach to genetic betterment. Because the manifesto seemed to be a subtle and gentle support of eugenics, it did engender considerable discussion. In the end, it was signed by 21 leading British and American geneticists, among whom were F.A.E. Crew, J.B.S. Haldane, J.S. Huxley, H.J. Muller, J. Needham, Th. Dobzhansky, R.A. Emerson, and J. Schultz.[37] Whether Beadle was asked and chose not to sign is not known.

Whatever the problems posed by the Soviets or the manifesto, the most serious threat to the Seventh International Congress was the war brewing in Europe. Ephrussi had frequently commented on the danger in his letters to Beadle. Hitler was becoming more strident and threatening. Austria and Czechoslovakia's Sudetenland were annexed under the guise of protecting their ethnic German populations. Emboldened by the British and French policies of appeasement, Hitler pressed his demands for all of remaining Czechoslovakia. France backed down on its pledge to provide assistance to the Czechs, and Neville Chamberlain, prime minister of Great Britain, seeking "peace in our time," acceded to Hitler's demands for the takeover of Czechoslovakia in the now infamous Munich pact on September 30, 1938. Much to Chamberlain's embarrassment, Hitler was not sated by his triumphs in Austria and Czechoslovakia, nor did he mean to stand by his pledge that he would not seek additional territory following the annexation of Czechoslovakia. Now he had designs on Poland, cloaking his ambitions by laying claim to Danzig, the territory ceded to Poland after Germany's defeat in World War I. In mid-August, just as the Congress participants were gathering in Edinburgh, the world was startled by the news that Germany and the Soviet Union had agreed on a mutual nonaggression pact. The purpose of their alliance became clear within a week. The Germans and Soviets simultaneously invaded Poland's western and eastern frontiers, respectively, with the aim of dividing the conquered territory between them. War was inevitable, as the British and French governments had reaffirmed their commitment to Poland's defense if the Germans attacked. That commitment led Britain and France to declare war on Germany three days after the invasion of Poland.

Remarkably, just before these catastrophic events unfolded, the first plenary session began on August 23. Professor O.L. Mohr, a Norwegian geneticist serving as chairman of the permanent international committee, reported the events that led to the postponement and relocation of the Congress. He cited "exaggerated and utterly incorrect statements pertaining to very serious differences on fundamental scientific principles" as having created an atmosphere that threatened an open and unbiased exchange of scientific ideas and opinions. Mohr made his feeling plain that the introduction of political influence in science was deplorable and that "the *conditio sine qua non* for scientific progress is freedom of thought and freedom of discussion" and that without those freedoms a truly international scientific spirit was impossible to achieve. He urged the meeting to "do its share in helping to increase respect for the basic scientific principles, to the greater welfare of mankind."[38]

Seemingly oblivious of the impending "gathering storm," the assembled scientists continued to meet, and Beadle presented his lecture the day after the Congress convened (see Chapter 9). As the week went on, the meetings were well attended and the discussions lively and profitable, in spite of the fact that the German and Italian governments ordered their scientists to return home, and the French and most other continental geneticists left Edinburgh soon after. Some of the British geneticists, including the Congress president F.A.E. Crew, left to join their military units, but most remained. It is unclear whether Ephrussi remained; he was to present his paper on the last scheduled day. Together with the Canadians, Australians, New Zealanders, and South Africans, the 125 or so American geneticists remained, most because they were unable to arrange passage home on such short notice. With about two-thirds of the registrants gone, the Congress was on the verge of disintegration, and it was decided to end the proceedings one day earlier than planned.

At the farewell party on August 29, "the rebellious rump was most unwilling to depart, and it was not until the early hours of the 30th that it was forced to accept the view that all good things must come to an end."[39] The next day the Americans learned that certain sailings had been canceled and many others were doubtful. The next few days saw fighting begin on the German–Polish border and the entry of Britain and France into the battle. World War II was on, ultimately to engage the United States and to continue until the spring of 1945. The final report of the Congress carried this doleful observation. "Had we been left undisturbed, this Congress would have been like many another, merely a pleasant affair of no great significance. But the external circumstances that attended it gave it a special quality. We had met as geneticists sharing the same interests and enthusiasms: suddenly we were required to behave as nationals with fiercely conflicting views. We found this demand difficult, irritating, saddening. Our memories of this Congress will make us even more fervent servants of peace and of scientific humanism."[40]

A stark reminder of the challenges faced by scientists with the onset of the conflict appeared in the preface to a volume containing a series of lectures by J.B.S. Haldane presented at the University of Groningen, Holland, in March 1940.[41] "At that time Holland was an island of peace and culture. It has now been swamped in the flood of war, and I do not know how many of my hosts are still alive, and how many of the survivors are in concentration camps. Alive or dead, I must thank them for their hospitality."[42] Called on to undertake intense research by the British government, Haldane apologized for not being able to give the book all the attention he wished but he wrote that "If the book is not published now it may never be published at all. I do not even know whether I shall enjoy life and liberty long enough to correct the proofs." He concluded "with the hope that genetical research may once again be possible in Europe in the near future."

Beadle's original plan to visit with Ephrussi in Paris after the Congress was no longer tenable. The *Normandie*, on which he planned to return to the United States, had sailed earlier than scheduled, and, given the menace of the German U-boats, she remained in the United States for the rest of the war. Some of the Americans gained passage on a British liner, the *S.S. Athenia*, which sailed from Liverpool on September 2 and made stops to pick up additional passengers in Glasgow and Belfast. Most of the American geneticists favored an American ship over one from a warring country and decided to wait for the U.S. government to arrange their passage. The group of scientists stranded in Glasgow, of which Beadle was one, set sail on a U.S. Maritime Commission ship, the *City of Flint*. She was bound for New York with a crew of 35 and a cargo of English wool and 30,000 cases of Scotch whiskey.

Although the ship was not equipped to carry passengers, makeshift arrangements were made to take 30 people: a group of Midwest and Texas college women and about a dozen American scientists.

The *City of Flint*, under the command of Captain Joseph L. Gainar, a fifty-ish New Englander who had served in the Merchant Marine Service and survived several encounters with torpedoes during World War I, left Glasgow the night of September 1. As a precaution, a United States flag painted on each side of the ship was illuminated. All was quiet until the night of September 3, when the captain ordered all hands to emergency stations to ready the lifeboats and gave instructions for all on board to sleep fully clothed with life jackets that night. He had already received a SOS from a ship in distress, and, hoping to rescue as many of the passengers as possible, he changed course heading for an area 200 miles west of the Hebrides. The *S.S. Athenia* had been hit by a German submarine torpedo and was sinking. In less than an hour, the *City of Flint* had been prepared for taking on survivors; professors and college students yanked out mattresses and began building stretchers and organizing first aid. When *Athenia* left the U.K. she carried about 1400 passengers and crew, well beyond the number usually carried on transatlantic voyages. On board were 212 Americans, 10 having been Congress participants.

When the *City of Flint* arrived at the disaster site in the early morning hours, the fairly heavy sea, illuminated only by moonlight, was filled with lifeboats struggling to stay afloat and escape from the sinking ship. The ship's crew and passengers plucked about 110 frightened passengers from the boats and the water. Another 120 or so were transferred from the *Southern Cross*, a Swedish yacht in the area. Suddenly, a ship that lacked adequate facilities to carry passengers had more than 260 people on board. The first hours were a nightmarish struggle in cramped quarters with limited supplies and facilities to deal with the more seriously injured among the chilled and battered men, women, and children. Altogether, 112 of the crew and passengers on the *Athenia* perished, 28 of them Americans.[43] The torpedo hit about the time of the second dinner sitting, but had it hit while the passengers were asleep, the death toll might have been much higher.

The *City of Flint* set course for Halifax in what became an arduous ten-day trip. A very rough sea battered the ship and took its toll on morale. But the selflessness of the officers, crew, and the original passengers bordered on heroic. Beadle and those who were able-bodied gave away most of their spare clothing and the costumes that resulted were amusing; slippers were constructed from braided rope and canvas normally used to cover the cargo. Meal service was a community enterprise, with passengers and able survivors serving and cleaning up during the eight shifts of three meals a day. Beadle gave up his cabin and slept on a coil of rope. His carpentry skills stood him in good stead when the crew and passengers were summoned to build tiers of bunks in the shelter deck by stretching long sheets of tarpaulin across the supports to form hammocklike bunks. News coming over the ship's radio was typed by those with stenographic skills and distributed to the passengers and survivors; in some cases it brought word of family and friends who survived and were picked up by other ships.

Huge crowds and the news media gathered to greet the *City of Flint* as she entered Halifax harbor. The captain was congratulated and praised, his response being "there's no such thing as a hero. You're either a man or a bum." Besides, he had been torpedoed three times before and "could not stand to see people kicking around in the water." Ironically, on a subsequent voyage to Europe, the *City of Flint* was captured by a German warship and taken to Murmansk.[44] After the survivors disembarked in Halifax, the *City of Flint* continued on to New York where Beadle immediately set off for California by train. He had given up all his clothes to the survivors and wore only a crewman's jacket and trousers and carried a cashmere wool blanket given to him by a woman who grabbed it as she left the ship.[45] While en route, Beadle wrote to his father apologizing for not being able to stop over for a day at home because he was already a week late getting back to work! His description of the ordeal was similar to the story in the *New York Times* of September 14; his own role in the rescue went unmentioned.[46]

CHAPTER 9

From Flies to Molds

Once Beadle and Tatum had crystallized the putative v substance, they were reasonably confident that they would soon solve its chemical structure. However, unknown to them when they went to Edinburgh, someone else had beaten them to it. Adolf Butenandt, an eminent German chemist working closely with Alfred Kühn on the meal moth (*Ephestia*) eye-color hormone, was probably aware of Tatum's report describing the production of the v substance from tryptophan by bacterial action[1] and tested a variety of known intermediates of tryptophan metabolism for v substance activity. One, kynurenine, which had previously been identified as the first product in the breakdown of tryptophan in mammals, restored normal eye color to both *Drosophila* and *Ephestia* eye-color mutants.[2] Tatum had missed identifying kynurenine because the crystalline material he obtained from the *Bacillus* cultures was kynurenine combined with sucrose. Removal of the sucrose did not affect the v substance activity, and he quickly confirmed that the resulting product was, indeed, authentic kynurenine as the German group had reported.[3] As little as one-hundredth of a microgram of isolated kynurenine injected into a *vermilion* pupa fully restored normal eye color in the emerging fly.

Beadle contacted Professor Clarence P. Berg at the University of Iowa for samples of authentic kynurenine. Berg had been involved in the compound's initial identification, and his kynurenine samples were chemically identical to the material Tatum had isolated. Ironically, it was not until Berg read about Beadle's fame in the *Wahoo Democrat* of January 9, 1947 that he learned that Beadle was the same person who grew up in Wahoo. Berg's family had settled in Wahoo in the 1930s, and his father, a harness maker, knew the Beadle farm and probably his father.[4] Isolation of kynurenine directly from pupal tissues of *cn* mutants, where it was presumed to accumulate, provided the final proof of the chemical nature of the v substance.[5] It wasn't until 1949 that the cn substance was identified as a previously unknown derivative of kynurenine, 3-hydroxykynurenine.[6]

Thus, the substances that Beadle and Ephrussi believed were *Drosophila* hormones are not unique to insects; rather, they are produced in bacteria and mammals during the normal breakdown of tryptophan. Today we know that the *vermilion* gene codes for an enzyme that converts tryptophan to *N*-formylkynurenine, after which another enzyme produces kynurenine. The *cinnabar* gene product is the enzyme responsible for the transformation of kynurenine into 3-hydroxykynurenine. The enzyme activities are absent from the respective *Drosophila* mutants.[7]

Butenandt's subsequent studies of the chemistry of 3-hydroxykynurenine confirmed that two molecules are used to form a product that polymerizes into the brown pigment, each step being performed by a genetically specified enzyme.[8] In contrast, virtually nothing is known about the series of reactions forming the red pigment, although there are numerous mutations that block that pathway. The *white* mutation, which produced Morgan's famous white-eyed fly, prevents the formation of both the red and brown pigments, most probably by blocking a step that is common to the formation of both pigments.

The failure of the Beadle and Ephrussi team's five-year effort to be first to determine the chemical identity of the v and cn substances was a disappointment. The frustration at having been scooped was galling, especially since the German group's identification of the v substance as kynurenine had relied on clues they had discovered. The Paris and Stanford groups both despaired that their efforts had been for naught. Beadle was disconsolate and considered Butenandt's action to be unseemly, even referring to it as "bitchy." Although he was clearly disturbed, it's unclear whether Beadle believed that Butenandt had been unethical in using their findings as a stepping stone to the identification of the v substance. But Butenandt was already engaged in chemical studies of tryptophan-like metabolites and through his association with the Kühn group was not only aware of Tatum's clue but also in a position to test his compounds straight away. One must wonder why the Beadle–Ephrussi groups failed to test a variety of chemical derivatives of tryptophan. They knew that tryptophan itself was not the v substance and that it was converted to active material by bacteria. Perhaps, they failed to consider that possibility or were unaware of compounds metabolically derived from tryptophan or lacked access to them.

It's uncertain how Ephrussi learned about Butenandt's discovery or what his reaction was. At the time of the German's publication, Ephrussi was more concerned about finding a way for his family to escape from France than he was about science. Ephrussi was mobilized almost immediately on his return to Paris from Edinburgh. His family stayed in Orléans while he was stationed at a laboratory in the neighboring countryside where he was assigned work that was quite unrelated to *Drosophila*. While in Orléans, he urged Beadle to send reprints and other scientific news to his laboratory address, planning to retrieve them on occasional undercover trips into Paris.[9] During the next half-year, he managed to go to Paris once a week, but all he found were empty laboratories. Khouvine and Mme. Auge, another of Ephrussi's assistants, stowed his flies, although he doubted that he would see them again.[10] Just as in most of Europe, genetic research in France ground to a halt. Many of the scientists were leaders of the resistance operating underground. Early in February 1940, Beadle and Marion offered to have Irène Ephrussi, then only a child, come to live with them to ensure her safety, but being relatively safe at the time, Ephrussi and Raja decided they should all stay together. Ephrussi responded: "I will nevertheless keep in mind your proposal and consider it as valid in the future. And we are grateful for having thought of it, anyway, [*sic*] to both of you."[11]

Sometime near the end of 1940, Ephrussi determined that it was becoming increasingly precarious for a Jewish family to remain in the German-occupied region of France. He communicated with Demerec and others about the prospects for a position or financial support should they be able to escape. During January of 1941, Ephrussi was warned that the Gestapo was planning to arrest him and his family. Early the next day, in complete secrecy and without any of their belongings, the Ephrussis fled, accompanied by Boris's father, Mrs. Louis Rapkine, and her daughter.[12] They traveled by night train from Paris to Bayonne, hoping that no one would ask for their papers, which identified them as Jewish. On their arrival in Bayonne, one of Ephrussi's colleagues put him in touch with a man who guided their trek from occupied into unoccupied France. From Clermont-Ferrand, they made their way to Juan-les-Pins on the Riviera where, after several months and considerable uncertainty, Louis Rapkine arranged for them to receive visas to enter and stay in the United States. The party traveled legally from Marseille to Barcelona and through Spain to Lisbon, where they sailed for New York on the steamer *Excalibur*. After a brief stay with the Dobzhanskys for part of the summer at Cold Spring Harbor, Ephrussi settled in Baltimore. The Rockefeller Foundation had arranged a two-year fellowship for him in the department of genetics at Johns Hopkins University beginning September 1, 1941. In the 1942–1943 academic year, Ephrussi was appointed associate professor of genetics.[13]

Ephrussi remained at Johns Hopkins for three years, focusing on the composition, properties, and measurements of the amount of the *Drosophila* eye pigments rather than the genetics of their development. The experience in Baltimore served as a "holding operation" until he could leave for London to join the Free French forces where he helped evaluate the results of allied bombing operations during the invasion of Normandy. When the war ended, he returned to Paris and was appointed the first professor of genetics in

France. Soon thereafter, Ephrussi and Raja were divorced and he married Harriet Taylor, an American geneticist. Years later, he lamented how the war and the disruptions it caused had "robbed" him of precious time to pursue the research he and Beadle had begun.

Beadle's plenary lecture at the Edinburgh Genetics Congress summarized the progress that he and Tatum, and Ephrussi and Khouvine, had made in revealing the genetic control of *Drosophila* eye-color development. The lecture[14] began by citing the work of others, notably R. Scott-Moncrieff, J.B.S. Haldane, and their collaborators at the John Innes Horticultural Institution, on the formation of plant pigments that account for the various colors of ornamental flowering plants,[15] and A.E. Garrod's analysis of alcaptonuria, an inborn inability to metabolize the amino acids phenylalanine and tyrosine beyond homogentisic acid.[16] Beadle inferred that "the defect or deficiency in the enzyme systems concerned with these processes...can be attributed to a change resulting from a single gene substitution." Beadle was evidently aware of the gene–enzyme connection in the systems he cited, but he seemed not to regard them as sufficiently robust for pursuing that connection.[17]

Regarding *Drosophila* eye-color development, Beadle concluded "that the process proceeds by a system of reactions occurring in series and in parallel, each of which is gene controlled." At the end of the lecture, he acknowledged the tentativeness of the scheme and the likelihood that it would be modified as more information became available, but he thought that it was essentially correct. His and Tatum's proposal that "genes acting through the intermediation of enzymes" were responsible for the formation of *Drosophila* eye pigments was, he admitted, a "purely gratuitous assumption" for which there was no direct knowledge of the enzyme systems involved. Nevertheless, in defense of the proposition, he added that "we know," that "in any such system of biological reactions, enzymes must be concerned in the catalysis of the various steps, and since we are convinced by the accumulating evidence that the specificity of genes is of approximately the same order as that of enzymes, we are strongly biased in favor of the assumption." Speculating further, Beadle was more explicit in stating "that the immediate products of many genes may be enzymes or their protein components." But, he added, "at the present time the facts at our disposal probably do not justify the elaboration of hypotheses based on this assumption."[18] That speculation is important, he concluded "not because it tells us much about what genes do but rather because it may indicate a method of attack that we may hope will become increasingly useful to both geneticists and biochemists." Earlier, he had suggested such an approach: "There are many ways in which genetics can aid biochemistry and perhaps even more ways in which the geneticist can profit from co-operation with the biochemist."[19] Although he lacked a formal background in biochemistry, Beadle was nevertheless convinced that biochemical and genetical approaches were complementary, a view he reiterated on many later occasions. Perhaps the lesson of having missed out on the identification of the v substance as kynurenine, a known metabolic product of tryptophan, had persuaded him of the importance of biochemistry.

The lasting achievement of the Beadle–Ephrussi collaboration was the realization that genes controlled individual steps in a metabolic pathway. Having missed out on the identification of the v substance only reinforced Beadle's belief that neither the *Drosophila* system nor any other system being studied anywhere was optimal for exploring the broader problem of the physiology of gene action, or, as he referred to it, biochemical genetics. Beadle recognized that a more amenable biological system and experimental strategy were needed, one in which mutations would affect easily recognizable, already known, physiological functions. Logically, it seemed, it would be considerably easier to assess the physiologic function of a gene if mutations affected the production of known compounds rather than unknown ones, as was the case with the *Drosophila* eye-color pigments.

While he and Tatum were struggling with how best to pursue the gene–enzyme paradigm experimentally, Beadle was struck by something Tatum said during one of his lectures in a course on comparative biochemistry given during the winter quarter of the 1940–1941 academic year. Many years later, he recalled the circumstances: "Sitting in on one of Tatum's lectures...and observing him writing sequences of reactions on the blackboard, I suddenly realized how stupid we had been all these years. Here were all those enzymatic reactions already worked out by competent biochemists. If our gene–enzyme concepts were correct, then we ought to be able to identify the genes immediately responsible for specifically known enzyme-catalyzed reactions. So why not reverse the approach? Instead of looking for reactions by enzymes controlled by known genes, why not look for genes that control already known chemical reactions and thus make the chemistry far easier? How? It should be easy. Start with an organism capable of carrying out reaction chains for a number of known end products. Then find genes that control the specific steps in the chains of reactions." Beadle's logic was elegant in its clarity and simplicity. But the idea had to be translated into an explicit experimental approach: "Since essentially all processes in an organism are biochemical, each should be resolvable into a series of specific reactions.... If all such syntheses are gene controlled, it should be possible to block the production of specific substances in the cell by inducing mutations in the genes controlling these syntheses. We might then expect to find mutations...characterized by an inability of the organism to synthesize essential diffusible substances such as vitamins, amino acids and other building blocks of the cell's protoplasm."[20] Beadle was confident that this approach would succeed.

What exactly in Tatum's lectures triggered Beadle's epiphany? Fortunately, the contents of the lectures were preserved in a set of handwritten notes by Carleton Schwerdt, a young faculty member in the department of medical bacteriology.[21] Throughout January and well into February 1941, Tatum presented a comprehensive review of what was known at the time about the intermediary metabolism of microbes, plants, and animals. Tatum implied that the wide variety of nutritional requirements of microorganisms for vitamins, amino acids, purines, pyrimidines, etc. reflected the organism's inability to make these metabolites from simpler precursors. More specifically, he proposed that complex nutritional requirements reflected the loss of one or more genes responsible for the production of a required nutrient, a view that had been developed by others.[22]

As the course neared its end, Tatum asked rhetorically "What do genes do?" He concluded that genes control metabolic reactions and that mutations accumulated over evolutionary time cause a loss of function of some particular metabolic function. He provided numerous examples in which microbes differ in their ability to synthesize amino acids, vitamins, etc.; even humans, he speculated, were unable to make a variety of amino acids and vitamins because reactions involved in their synthesis were blocked probably by the loss of the corresponding genes. But Tatum failed to provide any insight as to how genes control metabolic reactions or, for that matter, any property that is determined by a gene.

Beadle recognized immediately that if mutations were found which caused known nutritional deficiencies, they could link gene action to metabolic processes. However, to succeed they needed an organism that had as few nutritional requirements as possible. The organism would also have to be mutable by radiation or other mutagenic agents. Additionally, it should be amenable to rapid genetic analysis so that any nutritional deficiency could be attributed to a single gene, preferably one that could be mapped to a particular chromosomal location. Tatum scouted around in the nearby medical bacteriology department for a microorganism that might meet their needs but got little help from that group.[23] One candidate was the ascomycete, *Neurospora*, a fungus whose properties and genetics Beadle was already familiar with.

The organism had attracted the attention of the famous Dutch botanist, F.A.F.C. Went, during his travels in Java, and he cataloged many different species. The mold is ubiquitous in nature, its most obvious feature being a bright salmon-orange-colored "fuzz" covering the surface of various forms of vegetation. The year 1927 marked the birth of the genus *Neurospora* and the beginning of its use in genetic studies. What made *Neurospora* attractive to Beadle was that it could be propagated asexually and could also be induced to undergo a sexual cycle. Like *Drosophila, Neurospora* exists in two mating types, and when these mate the products of crossing-over during the meiotic divisions are preserved in the resulting spores. In a practical sense, just as with *Drosophila*, the segregation pattern of genetic markers in the spores provides a way to determine the order of genes in the chromosome. For Beadle's purposes, however, the important aspect of *Neurospora*'s sexuality was that mutations causing a change in nutritional requirements could be verified as being single gene changes and assigned a specific chromosomal location.

Beadle first heard about *Neurospora* as a graduate student when B.O. Dodge, a noted plant pathologist at the New York Botanical Gardens, had visited Cornell and presented a seminar on the mold's cytogenetics.[24] More than any other individual, Dodge laid the foundation for *Neurospora* genetics by working out much of its life cycle and mating behavior, particularly details of the meiotic events leading to spore formation. His serendipitous discovery that spores produced in the sexual phase of the life cycle could be stored and then germinated after heating facilitated the fungus's genetic analysis.

Dodge did not graduate from high school until age 20 because he needed to help out on his father's Wisconsin farm. After 15 or so years teaching high school and attending the University of Wisconsin intermittently, he obtained a doctorate in mycology at Columbia University. Only after he was appointed principal pathologist at the New York Botanical Garden at age 56 did Dodge do his most notable research with *Neurospora.*[25] Dodge believed that *Neurospora* deserved to share the genetics limelight with *Drosophila* and convinced T.H. Morgan, then a close friend, to take some *Neurospora* cultures with him to Pasadena. But Morgan's attempts to work with *Neurospora* in the new laboratories resulted only in getting them contaminated with bacteria and *Penicillium*, the common bread mold, both of which were frequent inhabitants of the *Drosophila* culture medium used in the lab. When Carl C. Lindegren joined Cal Tech's new biology division as a Ph.D. student, Morgan suggested that he work with the fungus for his thesis. With considerable help and advice, principally from Sturtevant and Bridges, he advanced the basic genetics of *Neurospora;* new genes were found, methods were developed for examining the progeny of genetic crosses, and a good start was made toward mapping the chromosomes.[26] Beadle was well aware of the mold's potential usefulness because Lindegren was still working on *Neurospora* in 1931 when he arrived at Cal Tech, and he and Sturtevant had reviewed its genetics in their textbook.[27]

About the time Tatum's lectures were addressing the question of the function of genes, Beadle wrote to tell Dodge that "Dr. Tatum and I are interested in doing some work on the nutrition of *Neurospora* with the eventual aim of determining whether the requirements might be dependent on the genetic constitution." He asked Dodge for different isolates of the fungus so that "if preliminary experiments prove to be encouraging we will be interested in trying out the available *Neurospora* species and various 'wild-types.'"[28] Two weeks later, Beadle wrote Dodge that Tatum's survey of the nutritional requirements of three different strains of *Neurospora* from different sources revealed that none of them would grow without the newly discovered B vitamin, biotin.[29] As it happened, Tatum was familiar with biotin; he had been in Utrecht with Professor F. Kögl at the time the vitamin was isolated and identified as a required growth factor for yeast and fungi.[30] Coincidentally, Nils Fries, a Swedish research fellow who shared Tatum's lab in Utrecht, was exploring the requirements of a wide range of fungi for biotin. In the process, Fries developed a very simple diet for growing *Neurospora:* It consisted of a sugar (several different ones could serve), a source of nitrogen (either ammonia or a nitrate salt), phosphorus (in the form of a phosphate salt), several minerals, and biotin. This diet, named for Fries, is still used in *Neurospora* research. Although biotin was somewhat scarce at the time, Tatum's past association with the Kögl lab assured him a ready supply.[31]

Vegetative growth

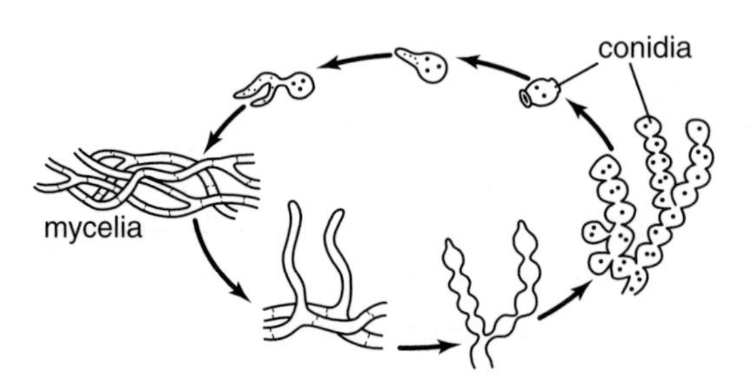

Neurospora's growth behavior and its elementary genetic properties were central to Beadle and Tatum's plan for obtaining mutants with readily identifiable nutritional deficiencies. On a solid surface containing all the essential nutrients, *Neurospora crassa*, the species they adopted, propagates asexually as a tangled mat of highly branched, tube-like filaments referred to as mycelia or hyphae. Septa divide the filaments into irregular-sized segments, within which nuclei are distributed among the other intracellular cytoplasmic organelles. Growth occurs by extension at the tips of the filaments; the rate at which the tip extends is a measure of the fungus's growth, a feature that Beadle used to good advantage. Occasionally, specialized aerial or surface filaments differentiate into spherical salmon-colored structures called conidia that are easily spread by mild air currents. Under suitable conditions, the dispersed conidia germinate, initiating new sites of filament growth. In this phase, referred to as vegetative growth, the nuclei of both filaments and conidia are haploid; they contain one copy of each of *Neurospora*'s seven chromosomes.

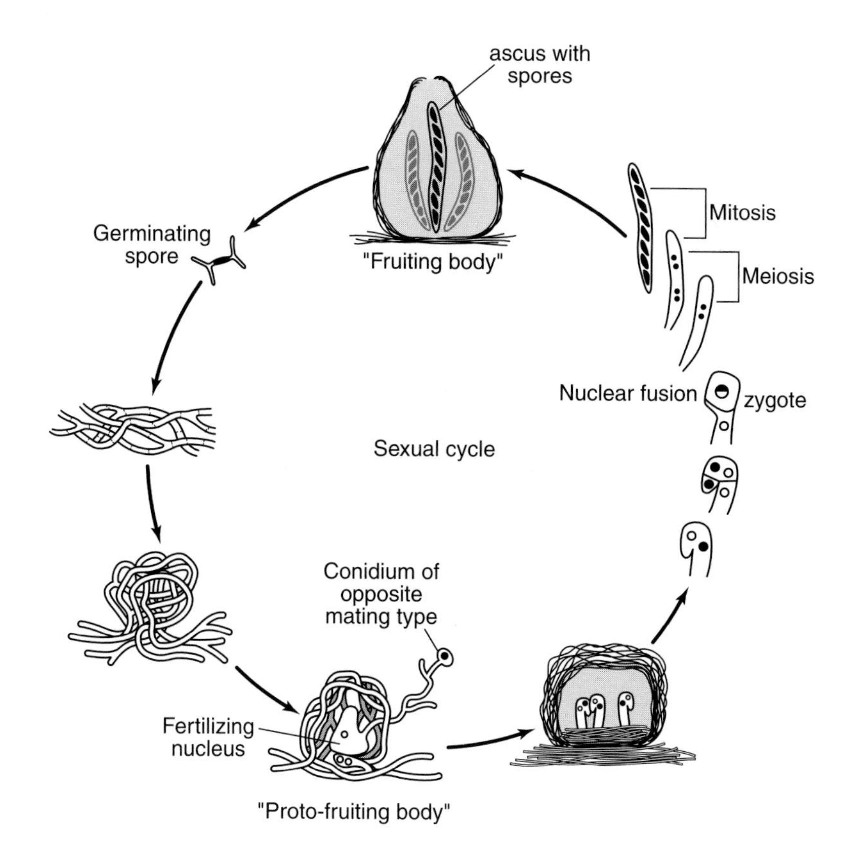

Neurospora was especially attractive for Beadle because of its ability to enter a sexual phase. Haploid cells exist in either of two mating types, referred to as **A** or **a** types. When starved, portions of the mycelium differentiate into a "fruiting body" within which haploid nuclei of **A**- and **a**-type conidia fuse to form cells containing the two types of nuclei. After the nuclei fuse, the cells elongate to form asci, sac-like structures within which the diploid nuclei undergo two successive meiotic divisions. In the process, four haploid spores are formed, and each undergoes a single mitotic division to yield eight or four pairs of spores. At this point, walls are laid down around each of the spores.

The two spores produced after the first meiotic division lie side by side, and when each divides in the second meiotic division, the resulting four spores are arranged as side-by-side pairs. Each of the four spores then undergoes a single mitotic division yielding eight spores. Consequently, the four spores in the top originate from one of the two spores formed at the first meiotic division and the four in the bottom half of the ascus originate from the other spore produced at that stage. Most important, the products of the successive divisions occupy a specific spatial arrangement in the asci, specifically, in the linear order of the four pairs of spores. To analyze the spores without disrupting their order, *Neurospora* geneticists have over the years adapted dissecting needles to pry open individual asci and remove the spores in order, one by one.

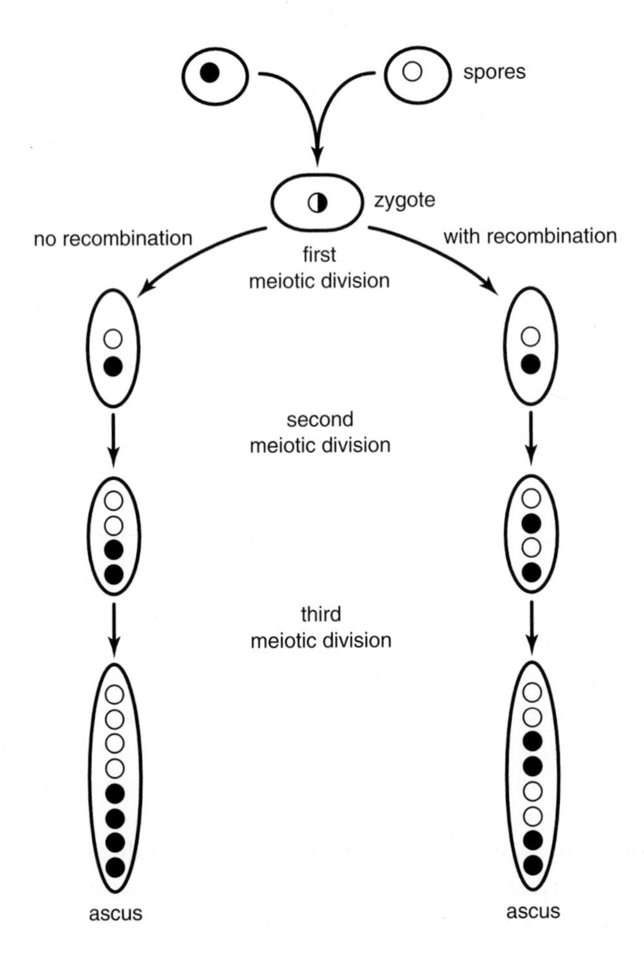

Genetic analysis of the four pairs of spores in each ascus allows inferences about the genotypes of the two haploid conidia that initiated the process as well as what transpired during the meiotic divisions. For example, because the difference between the **A**- and **a**-type mating forms depends on alternative alleles of a single gene, four spores in the asci will be of the **A** type and four of the **a** type. Similarly, when a wild type and a nutritional mutant are mated, four spores will have the wild-type allele and four the mutant allele. If more than one mutation is responsible for the altered phenotype, the arrangement and types of spores will deviate from the above outcome.

The arrangement of spores in the ascus also reflects whether recombination occurred during meiosis. If recombination does not occur, the ascus will contain four spores of one type adjacent to four spores of the other type (4-4). However, if recombination occurs during meiosis, the spores are arranged in pairs, one pair of one type adjacent to a pair of another type (2-2). The fraction of asci with the 2-2 spore arrangement relative to those with the 4-4 array is a measure of the frequency of recombination during meiosis. The genotypes and phenotypes of the four pairs of spores in multiple asci can confirm that the two mating forms differ by a single genetic difference. For example, Beadle could determine whether a nutritionally deficient isolate resulted from a single mutation by examining the spore pattern in a mating with the wild type. If he obtained four mutant spores and four wild-type spores in such a mating, he could be certain that the nutritional requirement was the consequence of a single mutation. Two isolates with an identical nutritional requirement would be considered to be alleles in the same gene if they produced the same ratios of 4-4 and 2-2 spore arrangements after mating with the wild type. If two mutants with the same nutritional requirement produced distinctly different ratios of the two spore arrangements in different asci, they were said to be affected in different genes.

Another property of *Neurospora* that proved invaluable is the ability of hyphae of the same mating type to fuse. In this case, the two haploid nuclei remain separate in the cytoplasm. If the resulting hyphae contain nuclei that have different genotypes, they are referred to as heterokaryons. In a genetic sense, heterokaryons are the equivalent of diploid animal or plant cells that contain two different haploid genomes in the same nucleus. If heterokaryons formed from a mutant and a wild-type strain have properties of the wild type, the mutation is recessive. If, however, the heterokaryons have a mutant phenotype, the mutation is dominant. Heterokaryons formed from strains with mutations in different genes generally have a wild-type phenotype because each haploid nucleus is able to provide the wild-type allele missing in the other. Heterokaryons formed with strains whose muta-

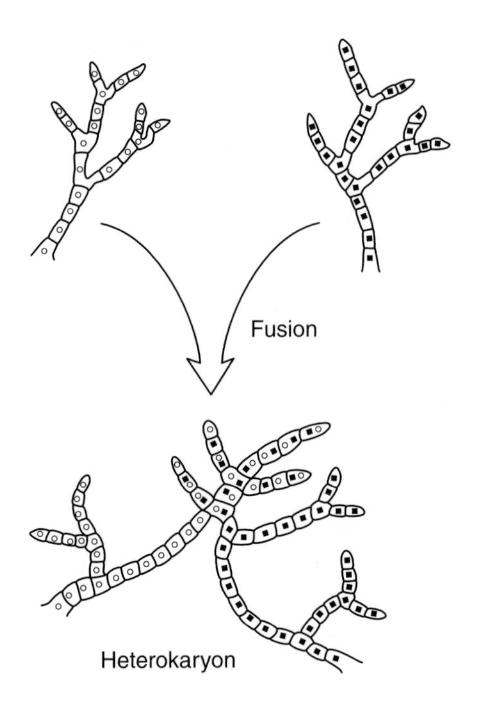

tions are in the same gene have the mutant phenotype because no normal allele is available.

Beadle and Tatum could, therefore, characterize their mutants in two ways: by the arrangement of ascospores when a mutant was mated to the wild-type parent, and by the properties of heterokaryons resulting from fusions between mutant and wild type or between two mutants. They could readily determine whether independent isolates of mutants displaying the same nutritional requirement reflected a change in the same gene or in different genes. They could also localize mutations to a particular chromosomal location.

Neurospora was, therefore, perfectly suited for discovering and characterizing mutations that create identifiable nutritional requirements by determining what each mutant required to restore its ability to grow in Fries' simple culture medium. Beadle and Tatum's confidence that this strategy would work was bolstered by their belief that single mutations affecting *Drosophila* eye color blocked what appeared to be single reactions. Most geneticists, however, if they were aware of the experiments that were in the offing, would have been skeptical, because a direct connection between a gene and a single biochemical function seemed unlikely. This view was predicated on the notion that multiple genes were required to determine any one trait or biochemical function. There seemed little reason to doubt that individual genes had multiple functions, and therefore mutations were likely to have more substantial consequences, probably causing multiple deficiencies. That view had been expressed by Muller more than 10 years earlier: "When we find, for example, that a certain gene difference results in the presence or absence of a particular enzyme, we have not proved that the gene directly produced the enzyme; it may merely have caused, through a series of intermediate processes, the production of an acid that inactivated or destroyed that enzyme, the acid having in turn been produced by another enzyme, and that activated by a coenzyme, and that produced by a protein—when the latter was ionized by the gene! Who can tell, in this house that Jack built?"[32] Others were even more pessimistic, believing that an understanding of gene action was beyond experimental dissection. Even 5 years after these experiments were initiated, Muller remained vague about the nature of the gene–enzyme relationship. But Beadle and Tatum had little doubt that they were on the right track. Their lone uncertainty was whether the frequency of mutations would be too low for them to find the mutants they sought.

They generated mutations in *Neurospora* by irradiating spores with various intensities of X rays. The physics department provided a homemade X-ray machine that was surrounded by hammered-down lead plates because "it leaked radiation like crazy."[33] Exposing spores to ultraviolet light also produced mutants. They collected only one spore from each irradiated culture to ensure that each mutant arose by an independent event.[34]

To avoid being needlessly discouraged, Beadle and Tatum collected and stored 5000 irradiated spores before systematically searching for mutations. The irradiated spores were germinated, and portions of each culture were deposited on agar surfaces containing a rich mix of every known component of proteins, nucleic acids, sugars, and all of the vitamins. Both mutant and wild-type *Neurospora* grew on such a rich medium and produced the characteristic filamentous mycelia. Samples of each of the mycelia from the rich medium were then placed on separate agar surfaces containing the simple Fries' diet. Wild-type *Neurospora* grew out normally on the simple diet, but any mutant that required a missing nutrient failed to grow. It remained to determine that required nutrient.

Substances known to be present in the rich nutrient mixture were added back to Fries' simple mixture, first in combinations and then singly, to determine which one restored a mutant's ability to grow. For example, mutants were tested for their ability to grow on the simple mixture to which all the amino acids or all the known vitamins were added. If the simple mixture containing all the vitamins supported growth, the mutant would be tested with

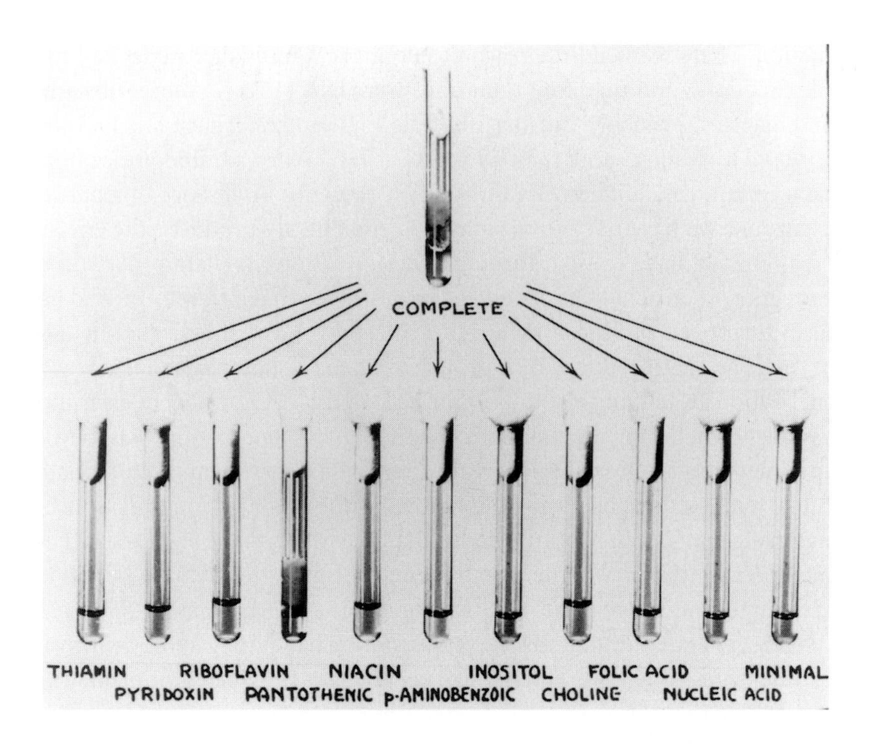
COMPLETE
THIAMIN
PYRIDOXIN
RIBOFLAVIN
PANTOTHENIC
NIACIN
p-AMINOBENZOIC
INOSITOL
CHOLINE
FOLIC ACID
NUCLEIC ACID
MINIMAL

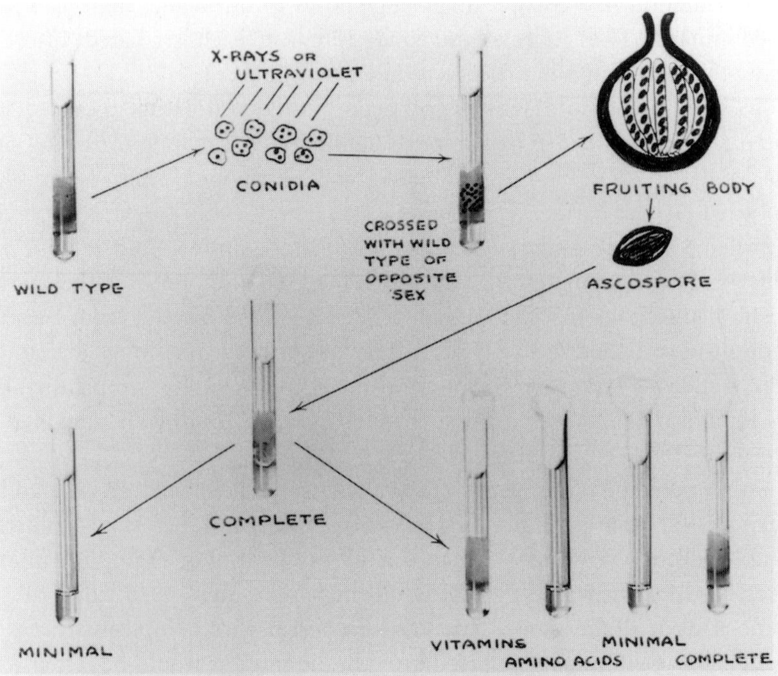
X-RAYS OR
ULTRAVIOLET
CONIDIA
WILD TYPE
CROSSED
WITH WILD
TYPE OF
OPPOSITE
SEX
FRUITING BODY
ASCOSPORE
COMPLETE
MINIMAL
VITAMINS
AMINO ACIDS
MINIMAL
COMPLETE

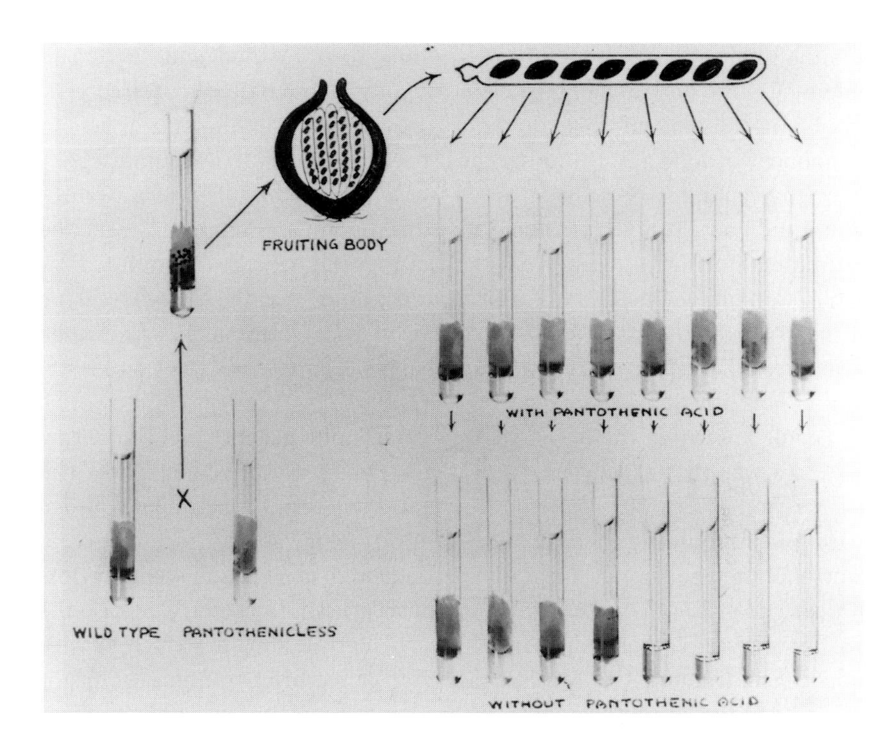

groups of vitamins at a time and, where growth occurred, the vitamins were tested individually. In this way, mutants could be identified as requiring a single vitamin, a single amino acid, or even a metabolic precursor to one of these. When cultures grew in the rich medium but failed to grow when any of the defined supplements was added to the simple medium, the mutants were presumed to be unable to make some unknown but essential compound. Because the unknown compound was contained in the rich mixture, it could be isolated and identified. Candidate mutants were then mated with the wild type of the opposite mating type to determine whether they had a single gene difference. Thus, if crosses between the putative mutant and the wild type of opposite mating type yielded four haploid spores that grew on the simple medium and four that did not, the mutation was judged to result from a single genetic change.

Beadle and Tatum found their first mutant after analyzing only 299 spores. Within about three months after receiving the *Neurospora* cultures and deducing their requirements for growth, Beadle reported to Lindegren that he had "rather good luck with the *Neurosporas*," adding that he "always knew they were fine bugs to work with but never appreciated all their advantages"[35]; the same news was reported to Dodge two weeks later.[36] Within two months, they published a paper describing their method and success in obtaining three nutritionally deficient mutants.[37] One mutant required vitamin B6 (pyridoxin), another could not grow without *p*-aminobenzoic acid (a component of folic acid), and the third needed vitamin B1 (thiamin). They knew that vitamin B1 is made up of two parts—a thiazole and a pyrimidine—and had already determined that this mutant was unable to make the pyrimidine portion. Judging from the fact that the paper was submitted for publication on October 8, 1941, and appeared in print in the November issue of the *Proceedings of the National Academy of Sciences*, the manuscript must have received accelerated treatment, probably because it was viewed as a scientific "bombshell."

Beadle knew that the new discovery could unite genetics with biochemistry in a way that had defied them in the *Drosophila* eye-color work. The unique properties of the *Neurospora* system provided an experimental wedge to explore not only the genetic control of morphological features, but also the control of metabolism; in short, the mechanism of gene action seemed within their grasp. To pursue that goal vigorously, he would need to increase the size of the research group substantially, particularly adding more people with biochemical training. Cal Tech seemed a fertile recruiting ground for bright young biologists, and no sooner had the paper been submitted for publication than Beadle went there looking for people to join him at Stanford.

Those who attended the seminar he gave during his visit were stunned. The title, if there was one, gave no clue about what was to come. As far as any of the attendees knew, it was the usual general biology seminar held in Kerckhoff. One vivid description of the occasion recalled that "the talk lasted only half an hour, and when it was suddenly over, the room was silent. The silence was a form of tribute. The audience was thinking nobody with such a discovery could stop speaking after just 30 minutes—there must be more. Superimposed on this thought was the realization that something historic had happened. Each one of us, I suspect, was surveying, as best he could, the consequences of the revolution that had just taken place. Finally, when it became clear that Beadle had actually finished speaking, Professor Fritz Went, whose father had carried out the first nutritional studies on *Neurospora*, got to his feet and, with characteristic enthusiasm, addressed the graduate students in the room. This lecture proved, said Went, that biology is not a finished subject—there are still great discoveries to be made!"[38] As usual, there were questions, many of which Beadle had anticipated. For some he had answers and for others he said that experiments were under way. It was a "tour de force" both scientifically and by way of showmanship. Lacking any description of the lecture other than what is quoted above, it's uncertain whether Morgan's criticism of his earlier presentations had improved Beadle's style. It seems more likely that the substance of his electrifying presentation overcame any stylistic shortcomings.

As word spread of the Beadle–Tatum discovery, even Stanford's first- and second-year medical students were thrilled by its implications.[39] Ephrussi, who by this point was working at Johns Hopkins University on the *Drosophila* eye-color problem, wrote in his usual effusive style, "I want to congratulate both you and Tatum. I believe that these first results leave no doubt that you are entering an unexplored field of most promising possibilities."[40] In their ensuing correspondence, Ephrussi expressed some concern about their collaboration on the *Drosophila* eye-color project, but Beadle's replies spoke mostly of the latest successes in obtaining additional *Neurospora* nutritional mutants. His response to Ephrussi's question about some data and the disposition of several fly cultures was "unless I can find a student to interest in it, i.e. *Drosophila*, I'm not sure I'll ever get back to it. The *Neurospora* work is constantly accelerating and I find I'm more and more tied to it. This state of affairs seems to apply to all the 'fly lab' workers so *Drosophila* seems to be at least temporarily out of luck at Stanford."[41] Beadle never returned to *Drosophila*, although Ephrussi continued to characterize the eye pigments for a while and then he, too, moved on to other areas.

At the time, Beadle's research was supported from the 10-year $200,000 grant to Stanford from the Rockefeller Foundation. His share of those funds was, however, insufficient to support the increased pace of activity and the personnel he knew had to be added. The American Philosophical Society's Penrose Fund, which had a history of providing limited amounts of money to support innovative research, had provided incremental funding for Beadle and Tatum's initial experiments, but only the Rockefeller Foundation could provide a substantial increase in support. A week or two after the initial paper appeared in print, Beadle wrote to Weaver at the Rockefeller Foundation, advising him of the breakthrough and including a detailed description of the principle and methods underlying their strategy. By then, he and Tatum had accumulated a dozen or so confirmed mutants. Beadle emphasized that "we have an approach not only useful in biochemical genetics but of great potential value"; he foresaw that simple and reliable vitamin and amino acid assay methods would be a by-product of the work.[42] This was the first of many times that Beadle stressed the advantages of their mutants for assaying vitamins and amino acids more specifically and rapidly than possible with existing methods.

Hoping that Rockefeller would increase its support for his research, Beadle offered to fly to New York in mid-December, 1941, to discuss that possibility.[43] Characteristically, he was not deterred from his planned meeting with Weaver by Japan's attack on Pearl Harbor the week before and the ensuing declaration of war. Barring blackouts, the possibility of Japanese invasions, and such, Beadle was intent on making it to New York. There is no hint from the communications between Beadle and the Rockefeller Foundation in the months immediately following the entry of the U.S. into the war that either of them had changed their priorities. Nevertheless, both recognized that the war effort would impose certain limitations on basic research in the life sciences. Beadle was determined, however, to drive the new research forward even as he realized that ways had to be found to juggle basic and applied investigations to keep his research on track.

Before he met with Weaver, officials from Merck indicated that they were prepared to fund his entire program in return for exclusive patents on any discoveries his group made that they deemed important. Merck believed that the *Neurospora* work could facilitate their own efforts to identify new growth factors and that they could enhance Beadle's work by providing him with raw materials and considerable expertise in the isolation of natural products. Although Merck was prepared to provide substantial funding for his operation, Beadle sought Weaver's advice about the advisability of entering into a collaborative agreement with Merck. Stanford President Wilbur and his department chairman, Taylor, encouraged Beadle to explore the Merck connection. But Beadle was concerned that such an arrangement, while assisting the research, might bring with it problems concerning manufacturing and patent rights. He was also well aware that Stanford University policy was opposed to its faculty seeking patents on research. In that event, Merck officials suggested the possibility of some sort of collaboration, one that precluded patents but assured Merck scientists an exclusive "early look" at the laboratory's findings.

During his meeting with the Rockefeller staff, Beadle told Hanson that he had "no interest in patents nor any personal profit for himself but, on the other hand, must find outside assistance to push this work rapidly." He also told Hanson that besides the interest expressed by Merck, he had also received an offer of $10,000 from the Research Corporation of America (RCA) to fund his research program. RCA was a nonprofit organization that held university-generated patents for the purpose of using their royalties to support promising research. However, Beadle confessed to Hanson that he was leery of both these propositions because of potential complications. In his report of their meeting, Hanson noted that Beadle's first preference "would be a grant from the Rockefeller Foundation that would free him of all obligations other than to work hard and publish freely his results. His second choice would be the Research Corporation and third, Merck."[44]

Hanson had long regarded Beadle as "one of the most promising men of his age in biology."[45] He recognized the enormous significance of the *Neurospora* breakthrough and agreed to increase Beadle's annual support. At the time, Beadle was receiving $20,000 per year from the biology department's Rockefeller Foundation grant. Hanson notified him that the Foundation would provide a supplement of $7500 in 1942 and the same amount for each of the 2 succeeding years.[46] Beadle asked that the Rockefeller grant be reviewed on a yearly basis so that his needs could be periodically assessed; he was concerned about a possible reduction of his funding needs should he lose some of his group to the draft. Although the Research Corporation never funded his research, Beadle was reassured by their promise to do so should the funding from Rockefeller prove to be inadequate.[47]

Soon after the Pasadena seminar, and with funding in hand, Beadle invited two recent Cal Tech Ph.D.s, Norman H. Horowitz and David M. Bonner, to join the *Neurospora* project. Years before, Beadle had put in a "good word" for Horowitz when, in reviewing his application to Cal Tech for graduate work, he learned that Horowitz had already published work in transplantation as an undergraduate at the University of Pittsburgh. A native of Pittsburgh, Horowitz arrived at Cal Tech in the fall of 1936 expecting to earn a Ph.D. in genetics, but Morgan assigned him to work with Albert Tyler on the embryology of marine organisms. His thesis research explored the nature and temporal aspects of respiration as they pertain to morphological development in the eggs of two marine invertebrates. Horowitz received his degree in 1939, after which he was awarded a National Research Council fellowship to study abroad. But the war in Europe forced him to change his plans, and he went to Stanford to continue his research with Professor Douglas Whitaker in embryology. While there, he renewed his connection with Beadle, who was then deeply involved in the final stages of the *Drosophila* eye-pigment work. At the end of the year, in need of a job, Horowitz was persuaded by Professor Henry Borsook to return to Cal Tech as a postdoctoral fellow to work on the enzymology of tooth formation, a project that was being funded by a local dentist. Not long afterward, Horowitz attended Beadle's electrifying lecture on the initial *Neurospora* findings and was thrilled when Beadle offered him a postdoctoral position in his group at Stanford. Accepting without hesitation, Horowitz delayed his arrival at Stanford until mid-1942, when he and his family rented a house at the edge of the campus, an easy ten-minute bicycle ride or brisk walk to the lab. Cars were still a luxury, especially with wartime gasoline rationing.

David Bonner had also just completed his Ph.D. research when he heard Beadle's seminar and he, too, eagerly accepted the invitation to work on the

Neurospora project. Bonner had grown up in a large Mormon family in Salt Lake City, Utah, where his father was a professor of chemistry at the university. After completing an undergraduate chemistry degree in 1937, he came to Cal Tech, where the oldest of his four brothers, James, was already one of the biology division faculty. At the time, the hot topic in plant physiology was how various plant growth factors, auxins, stimulated growth of only certain parts of the plant; e.g., roots, shoots, or flowers. Working initially with F.H. Went, and then with A. Haagen-Smit, Bonner identified growth factors that specifically affected leaf development. Subsequently, he examined the relationship of the leaf auxin to those that affect other plant tissues. Because of his chemical experience and interest in metabolism, Bonner seemed well suited to characterize the metabolic deficiencies caused by individual *Neurospora* mutations. He and Horowitz arrived at Stanford at about the same time and worked in the part of the basement adjoining Tatum's labs.

Herschel K. Mitchell turned up at Stanford uninvited after he heard about the *Neurospora* experiments and pleaded with Beadle to hire him. A native Californian and chemistry graduate from Pomona College, Mitchell had already had two notable scientific successes. While still a master's degree student at the University of Oregon, he had worked out the structure of the vitamin pantothenic acid and then a few years later, as part of his Ph.D. thesis research with R.J. Williams at the University of Texas, he identified folic acid. Mitchell's experience with enzymes won Beadle over, although funds to pay him were not yet in hand. He never regretted the decision. Mary Houlahan, whose husband had just taken a position in San Francisco, was another "walk-on." She brought with her strong recommendations from Cold Spring Harbor for her work in radiation biology. Told that she was a wonderful worker and that he would not be sorry if he hired her, Beadle was persuaded to add her to the team. Several years later, she and Mitchell divorced their respective spouses and married.

Besides the comparatively well-trained postdocs, there were many graduate students who wanted to be part of the new venture. But Beadle had to find money for their support. Fortunately, realizing that *Neurospora* mutants with defined and specific nutritional deficiencies could be used for bioassays, the Nutrition Foundation, with the blessing of the Rockefeller Foundation, agreed to fund several graduate fellowships for work along those lines. Over the ensuing years these were held by Adrian M. Srb, August H. (Gus) Doermann, Frank C. Hungate, Taine T. Bell, Verna Coonradt, and David Regnery.

Like Beadle, Srb had grown up on a small Nebraska farm, attended the university in Lincoln, majored in agronomy, and been introduced to research by Professor Keim, Beadle's undergraduate mentor and friend. Sent by Keim to study with Beadle, Srb arrived in 1941 expecting to work on the *Drosophila* project, but "the bulk of workers in the laboratory were manipulating bits of 'orange fuzz,' an organism not yet recognized in the Agronomy Department at Nebraska"[48] or for that matter hardly anywhere else. Doermann, an Illinois native, had studied biology at Wabash College and the University of Illinois, and arrived at Stanford just as the *Neurospora* work was picking up steam. Regnery had been a student in Beadle's undergraduate genetics class, but not being enthusiastic about continuing to work on the *Drosophila* eye-color problem as a graduate student, he went to Cal Tech. He too was overwhelmed by Beadle's seminar on the *Neurospora* work and returned to Stanford to join the *Neurospora* group. Soon after, he was called for military service and did not return to complete his graduate research until 1945.

Beadle believed strongly that students should generate their own research problems. It was the way he was raised at Cornell and at Cal Tech. He had been expected to identify and chart an experimental program of his own design while being assured of help along the way. But the "mutant-hunting operation," manned primarily by undergraduates and technicians, was turning up mutants at a great rate, and they all needed to be studied. Graduate students, generally working with the more senior postdoctoral fellows, selected groups of mutants whose nutritional requirements needed to be identified. They were to characterize the mutants they selected, establish whether the deficiencies were due to single mutations, pinpoint their chromosomal locations, and identify the affected biosynthetic reaction.

Beadle and Tatum opened the way for a new assault on the problems that had eluded geneticists from the beginning: What exactly do genes do and how do they do it? They believed they had developed an experimental approach that could lead to an answer to the first question. Their experiments pointed to a role for genes in specifying the production of enzymes that catalyze the myriad reactions that comprise the organism's metabolic and developmental capabilities. How genes and enzymes were related remained to be determined, and the mechanism by which genes "guide" the production of enzymes, not at all indicated by these early experiments, remained a matter for speculation and debate. Nevertheless, together with the very talented and devoted efforts of the team that Beadle assembled so quickly, and the wisdom of the Rockefeller Foundation in providing the needed resources, he and Tatum were able to build on their breakthrough on the gene–enzyme connection. Although they had provided a way to explore the functional relationship between genes and enzymes, an understanding of how genes were involved was not achieved for another 20 years.

CHAPTER 10

One Gene–One Protein

The brilliant scientific achievements during the summer of 1941 coincided with the need for Beadle to resettle his family in a new house. Soon after Stanford University was established, senior professors were encouraged to build homes on sites leased from the university. Over time, as older faculty retired or moved away, existing homes came up for sale and the Beadles were attracted by the prospect of living on the campus. Marion and David spent the summer with friends near Los Alamos and Beadle stayed with the Tatums while negotiating the purchase of the new house. Located on a secluded cul-de-sac within easy walking distance to his lab, the adobe-style house with a sloped red-tiled roof was characteristic of many of the Spanish-style homes in the area. From the house, as David recalled it, he could easily see the protecting blimps hovering over Moffett Field, a Navy air base about ten miles away. The Tatums, too, moved onto campus not far from the Beadle house, but the two families rarely socialized.

One of Beadle's first projects was to install a substantial flower and vegetable garden that extended from the front of the house around to the side as well. Creating and tending a garden seemed a necessity wherever he settled, perhaps because it rooted him in one of the favorite activities of his youth. In this case, the psychological lift also served another purpose. Over the next few years, the garden's ready supply of produce compensated for the wartime shortages. Welcoming friends and colleagues to the house and its garden gave the Beadles a great deal of pleasure. On many weekends, the lab people and their families were invited to come to share steaks and roast sweet corn that he had grown on a campus plot set aside for his farming. He never lost the Nebraska farmer's love of growing and eating corn; indeed, it was to be a constant throughout his life. Students and people from the lab often showed up in his garden to help with the "farming," thereby earning the right to carry away the fruits of their labors. Marion enjoyed her role as hostess and cook, both of which she did well.[1] She particularly enjoyed her role as "house mother" and welcomed the opportunities for David to get to know the lab people.

Everyday life at the laboratory was busy and sociable, largely because the graduate students and postdoctoral fellows were a lively and congenial group.

Undergraduates, who did most of the tedious work connected with the experiments, were viewed as part of the lab's community. Beadle understood, however, that days of all work and no play were bound to erode the lab efforts. Sometimes, he gathered a group together to hike up into the hills early in the morning where they built a fire and made fried egg sandwiches.[2] Occasionally, he declared a partial lab "holiday" for a party at the beach and led a demanding ten-mile bicycle trip over the coastal range, on one occasion carrying a watermelon on the bike's handlebars.[3] Marion brought the noncyclists and lunch in their Ford roadster. She was a gracious, yet reserved, hostess at their annual Halloween apple dunking and frequent barbecue parties. The Friday lab teas, complemented by Marion's cakes, functioned as research meetings very much in the spirit of Morgan's at Cal Tech in the early 1930s. Beadle even continued Morgan's custom of reading interesting letters to the group.

Beadle and Adrian Srb developed a special relationship, more like the one he enjoyed with Emerson at Cornell. Aside from their mentor–student bond, they enjoyed a spirited competition in tennis, broad jumping, and growing the best corn. It was not long before Srb recognized, as did others who challenged Beadle to a competition, that he was not likely to give any quarter to his opponents. Winning, whatever the game, was important, but it had to be honestly and fairly done.[4] Competing and being first were probably important drivers of his scientific ambitions, but honesty and magnanimity tempered those occasions when he failed to come out on top.

The Srbs lived in a small cabin located on the biology department's experimental corn field near the laboratory. Beadle was comfortable enough with them to wander into their cabin early in the morning when they were still in bed and gather up their months-old daughter into a pack on his back and take her out to inspect the corn.[5] Marion, too, welcomed Jo, who was a graduate student in social sciences, and on occasion went with her on fishing trips at a nearby Pacific Ocean beach.

With the outbreak of war and a need to help out in activities for which men were usually employed, Marion volunteered to serve as driver for the local Motor Corps. In addition, she was increasingly engaged with activities at the university and neighboring community. Being attractive, intelligent, and outgoing, she had little difficulty making friends with the wives of many Stanford faculty. It was largely through Marion's friendship with Connie Merritt that Beadle established a close personal relationship with Walter Merritt, a professor of English and Stanford's authority on Old English. Their friendship flourished because Merritt was in awe of Beadle's ability to talk about genes and enzymes, and Beadle marveled at Merritt's ability to recite Beowulf from memory.[6] He may have been the only close friend Beadle made outside of the biology division during his tenure at Stanford. However, Marion's demanding and dominating personality finally alienated Connie, and the two couples' relationship cooled and ended. A close relationship between the Beadles and Carleton and Kit Lewis, a local couple, also foundered because of Marion's quirky personality. Her domineering manner repeatedly soured friendships just as it was affecting her relationships with Beadle and David.[7] In time, his and Marion's relationship grew more strained and his withdrawal from family life grew more pronounced. It was to grow worse as time passed.[8]

David appeared, at least to adults, to be shy and unhappy. He recalls that his mother's overbearing manner kept him from establishing a closer relationship with his father. His difficulty in school dismayed Beadle so much that he was sent away to boarding school at Mecca in the desert near Los Angeles when he was about 12 years old. About this time, Marion began taking long trips with obscure purposes in Mexico. Only when she was away could David visit with his father in the lab and feel free to invite his friends home for barbecues. On these occasional one-on-one situations, Beadle and David "hit it off" even to the point of collaborating on David's ambition to earn a Boy Scout merit badge in plumbing. David recalled visiting his Dad's lab and peering down a microscope to see a third eye growing in the "belly" of a fly. David learned early in his childhood that although Beadle was generally even tempered, he would not tolerate abuse of animals. On one occasion, his Dad's temper "really blew" when he cruelly taunted an animal.[9]

The Japanese attack on Pearl Harbor and the United States' entry into World War II transformed the university as strikingly as it affected Beadle's research program. In keeping with the need for an accelerated war effort, Stanford changed its academic schedule of three terms per year plus a free summer to a year-round schedule with classes beginning at 7:30 a.m. and continuing until 11:30 p.m. Several new courses were instituted and some existing research efforts were modified to contribute more directly to the war effort. Prominent faculty members disappeared from campus to take up positions with newly created government agencies in Washington. In response to the threat of air raids, Beadle took on air raid warden duties and had to leave the campus briefly for training. Within half a year, Stanford was transformed into a military ground. It was not uncommon for troops and motorcades to parade by the back of the Beadles' garden.[10] Some of the frills of college life such as fraternities and intercollegiate sports were discontinued. By the time Beadle's newly recruited students and fellows showed up at Stanford in mid-1942, the student body was dominated by the presence of army, navy, and marine trainees.

The number of young men plummeted as high school graduates were volunteering or being called into military service. David recalled that Encina Hall, the university's administration building, which had been taken over by the military, was the favorite hangout for kids because the military post exchange there was the place to get free Hershey bars.[11] Along with about 4500 army, navy, marines, and a contingent of the Woman's Army Corps, there were 2100 civilian students, 1400 of whom were women, including the student body president and newspaper editor. The university's financial situation, which had been basically sound before the war-caused disruptions, began to generate budget deficits that hampered activities for years after the war ended. Genetics was not high on the male students' agenda, and those taking Beadle's course were largely women and the few men preparing for medical school.

The grim war news seemed not to diminish the excitement that the Beadle–Tatum breakthrough generated within the genetics community. Weeks after his seminar at Cal Tech, Beadle presented his and Tatum's discoveries at an American Association for the Advance of Science (AAAS) meeting in Dallas, Texas. Returnees from the meeting reported tremendous enthusiasm about the new approach. Muller referred to it as "the one of two most important happenings at Dallas," and Sewall Wright, one of the pioneers of genetics, "qualified it as the most important discovery in genetics since the finding of *Drosophila* salivary polytene chromosomes." Ephrussi wrote to say how disappointed he was that Beadle was not awarded the $1000 prize for the meeting's best paper.[12] Nevertheless, there were skeptics of both the experimental approach and the inferences drawn from the experimental findings. Beadle recalled later that during the discussion of the paper, a fellow scientist commented that either he and Tatum were incredibly industrious or he would have to be skeptical, hinting, perhaps, that he did not believe the results. Beadle was certain that their hypothesis seemed much too simple to him.[13]

Meanwhile, science in the lab was booming. The flood of new mutants that poured out of the mutant-hunting room astonished everyone and inspired the same degree of awe and wonder as had the pace of new mutant discoveries emanating from Morgan's Fly Room almost 30 years earlier. Beadle commented to Sterling Emerson that "mutations in *Neurospora* are popping up so fast we can't keep up with them."[14] Beadle appreciated that the procedures had to be efficiently organized in much the same way as the country was organizing the mass production of tanks and warplanes for the war effort. He hired and trained students, mostly women, since they comprised the great majority of students on campus at the time, to screen putative mutants for their nutritional requirements and conduct the needed tests to determine whether the mutations had occurred at unique sites.

From the outset, Beadle was confident that mutant *Neurospora* could be used to assay for substances the mutants required. In the initial publication describing the isolation of the mutant that could not synthesize pyridoxine (vitamin B6), Beadle and Tatum noted that there was only a marginal increase in the weight of the mycelial mass after six days in a medium lacking pyridoxine. With increasing amounts of pyridoxine added to the medium, there was a nearly proportional increase in the weight of the mycelium after the same amount of time. Clearly then, the amount of pyridoxine in a particular sample could be deduced from the amount of mycelial growth it supported. Where only one compound among many similar or related compounds could replace a mutant's particular requirement, the assay for that substance was quite accurate even if the samples being tested were rather impure. Where the mutant could utilize substances metabolically or chemically related to the natural substance for growth, the assay's specificity was, of course, compromised. Horowitz and Beadle established the utility of this approach for measuring choline, a component of cellular lipids, in various foodstuffs. In doing so, they discovered that their choline-requiring mutant would also grow normally if the amino acid methionine replaced choline; in *Neurospora*, as in many other organisms including humans, choline and methionine are metabolically related. Fortunately, choline and methionine can be readily separated from one another, thereby allowing accurate estimates of the choline level to be made in a wide variety of foods.[15] Beadle recognized that being able to assess the nutritional quality of the nation's food supply was of considerable commercial value in wartime. The food and drug industries particularly were eager to adopt these new methods for their own research programs, and Beadle encouraged their interest, even welcoming them to his laboratory to gain firsthand experience.

Characteristically, Beadle was impatient with a procedure that relied on weighing mycelium to determine a particular mutant's need for a nutrient. Seeking a faster way, he reasoned that because *Neurospora* grows by extension of its mycelial tips, growth could be followed by measuring the rate at which the leading edge of the mycelial mass advanced along a horizontal glass culture tube half filled with an agar medium. Proud of his glassblowing skills, he made the lab's supply of what came to be called "race tubes." In practice, he made glass tubes of about a half-inch inside diameter and 15–20 inches long, with their turned-up ends plugged with cotton. After inoculating the agar-containing medium at one end of the tube with a bit of mycelia, he recorded the positions of the advancing tips at convenient intervals. The advancing front was remarkably sharp, and its position could be determined within a millimeter. When either wild-type or mutant *Neurospora* was inoculated into tubes with the agar-containing complete medium, the tips advanced at a constant and identical rate. However, when, for example, the pyridoxineless mutant was deposited in a medium lacking pyridoxine, the mycelial tips failed to advance. With increasing amounts of pyridoxine in the medium, the tips advanced along the tube at rates governed by the amount of pyridoxine in the medium. The same procedure could be used to measure the amount of virtually any substance required by a *Neurospora* mutant.[16] Although Beadle had devised this improvement in the assay, it was Francis J. Ryan, a visiting National Research Council fellow, who sorted out the variables and conditions of the race tube method so that the results were reproducible and reliable.[17] Ryan had completed his Ph.D. dissertation on amphibian development at Columbia University in New York and had come to spend a year at Stanford in Whitaker's lab, studying the embryological development of marine invertebrates. Ryan was alert to the implications of even the earliest *Neurospora* experiments and decided to switch labs. Moreover, having become enamored of the use of *Neurospora*, Ryan gave up on his first love, embryology, and began to generate his own collection of *Neurospora* nutritional mutants soon after returning to Columbia. By the time he was settled in his new laboratory, he was astonished and envious at how quickly new mutants were being generated at the Stanford lab.[18] Beadle too was amazed at how rapidly their search for nutritional mutants was proceeding. He boasted to Ephrussi, "We now have about 10 people doing *Neurospora* in one way or another. The mutants are turning up so fast we haven't had time to digest them yet."[19] One of the mutants that Ryan chose to study required lysine, the same mutant A.H. (Gus) Doermann was studying for his Ph.D. thesis. Beadle, who was intent on protecting his student's lead, issued a mild rebuke to Ryan, insisting that he keep Doermann informed about his work and explore the possibility of publishing their findings at the same time.[20]

Horowitz characterized the Stanford years during 1942–1946 as a "scientific paradise," the most exciting years of his life and, he imagined, of everyone else's in the lab as well. "Before the *Neurospora* breakthrough, the idea of uniting genetics and biochemistry had been only a dream with a few scattered observations. Now, biochemical genetics was a real science and it was entirely new. Incredibly, we privileged few had it all to ourselves. Every day brought unexpected new results, new mutants, new phenomena. It was a time when one went to work in the morning wondering what new excitement the day would bring."[21]

Beadle and his colleagues were well aware that a new area of genetics had been revealed, for now it was possible to relate genes with enzymes and the metabolic reactions they catalyze more directly. Genetics was no longer to be preoccupied solely with experiments that examined the mechanics and mode of transmission of genes from one generation to the next, a focus that began and was emphasized by the "Morgan school." It's unlikely that Beadle, or anyone else at the time, foresaw that his experiments had nudged genetics from its abstract phase to its molecular beginnings.

What were these new mutants and new results, and how did they hasten the "marriage" of genetics with biochemistry? Most significantly, the number and variety of mutants with only a single altered gene and a single nutritional requirement were increasing explosively. A preliminary but logical inference was that each *Neurospora* gene controlled a single metabolic reaction. But there were also instances for which a set of genetically distinguishable mutants had the identical nutritional requirement. To explore this apparent anomaly, Srb and Horowitz examined one such case in depth.[22] They used fifteen mutant strains whose growth defect was overcome by arginine, one of the protein amino acids. Seven of these proved to be genetically distinct; the remaining eight represented recurrent mutations of genes included in the first seven. They established this by examining the progeny of pairwise matings among the fifteen strains, by matings of each to the wild-type strain, and by the pairwise fusion of mutant hyphae to form heterokaryons. Each of these methods was an independent test of whether the mutations were in the same or different genes. The growth response to compounds structurally related to arginine was especially revealing. One mutant responded to arginine only, two grew normally with either citrulline or arginine, while the remaining four grew equally well with ornithine, citrulline, or arginine. Presumably, those mutants were blocked in separate reactions leading to the production of ornithine. Srb and Horowitz inferred correctly from information about the metabolic breakdown of arginine in animals that ornithine and citrulline were most likely precursors for the synthesis of arginine.

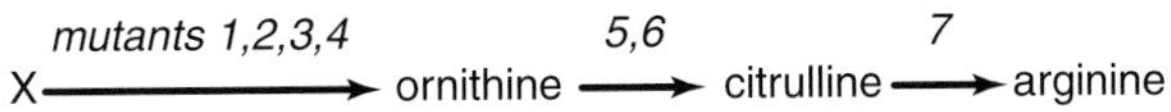

$$X \xrightarrow{\textit{mutants 1,2,3,4}} \text{ornithine} \xrightarrow{\textit{5,6}} \text{citrulline} \xrightarrow{\textit{7}} \text{arginine}$$

This interpretation was strengthened when they discovered that citrulline accumulated in the culture medium when the mutants that responded solely to arginine were grown with limited amounts of arginine; thus the blocked reaction in mutant 7 caused the accumulation of the immediate precursor of arginine. Similarly, when either of the mutants that could grow with citrulline exhausted that compound from the medium, it excreted ornithine, indicating that mutants 5 and 6 could not convert ornithine to citrulline. Because Srb and Horowitz had four different kinds of mutants that could grow with ornithine in place of arginine, they concluded that there were probably at least four discrete reactions involved in the formation of ornithine. They inferred that *Neurospora* makes its arginine from ornithine via citrulline, a pathway that mammals use for the same purpose.

Carrying Srb's and Horowitz's reasoning further, Beadle realized "There is no inherent reason why the course of biosynthesis can not be determined by the method of biochemical genetics all the way back to the inorganic starting materials."[23] Although true in principle, this rarely turns out to be practical. The mutants blocked in reactions leading to the production of ornithine, for example, do not accumulate measurable amounts of the intermediates that precede ornithine in the pathway. Moreover, the cells are sometimes unable to take up the intermediates from the medium, and such intermediates are therefore not likely to be detected by their ability to support a mutant's growth. Now, the formation of ornithine is known to involve five reactions, four of which are specified by the genes that Srb and Horowitz identified.

The mutant-generating "machine" turned up about 30 mutants requiring tryptophan for growth. Bonner accepted the challenge of characterizing them. Despite extensive testing, he found only two genetically distinguishable classes. Besides being able to grow with tryptophan, one class also grew normally with indole but not with any of a large collection of compounds chemically related to indole. When the indole-requiring mutants were grown with limited amounts of indole, they accumulated a compound that was able to replace indole or tryptophan for the growth of some other mutants; that compound was identified as anthranilic acid. That suggested that anthranilic acid was very likely an intermediate in the pathway for tryptophan formation. Because some of these mutants were able to grow on indole, but not on anthranilic acid, it seemed plausible that indole and tryptophan were derived from anthranilic acid.

$$X \longrightarrow \text{anthranilic acid} \xrightarrow{\textit{mut 1}} \text{indole} \xrightarrow{\textit{mut 2}} \text{tryptophan}$$

How is indole converted to tryptophan? Hints from studies of tryptophan synthesis in bacteria led Tatum and Bonner to test whether *Neurospora* could convert indole and serine into tryptophan. Sure enough, indole disappeared rapidly in the presence of serine, and an almost equivalent amount of tryptophan was produced.[24] We now know that the reaction of indole with serine to form tryptophan occurs in a single reaction. So it is surprising that mutants blocked in this reaction were not detected in the mutant search. Equally surprising is that only one genetic class of mutants was obtained that could grow with indole, yet several discrete reactions are involved in the formation of anthranilic acid and its conversion to indole. Today we know that some of the metabolites prior to indole do not diffuse outside the cells and would not have been detected by their experiments.

By early 1944, virtually all the results were consistent with the view that single mutations affect single primary reactions: one gene–one metabolic reaction. Occasionally, the group encountered mutants whose properties or behavior was anomalous or at least inconsistent with that simple view. One such anomaly was the discovery of a mutated gene that caused a requirement for two supplements to grow; neither alone was adequate. This mutant, which differed from the parent strain by only a single gene, required both valine and isoleucine to grow.[25] Because valine and isoleucine are very closely related in structure, Bonner thought that the mutation might affect a reaction that was common to the synthesis of both amino acids, but the explanation turned out to be more complex.[26] This particular mutation blocked the last step in the formation of isoleucine, which is consistent with the requirement of isoleucine for growth. But why was there a requirement for valine as well? As it turned out, the block in isoleucine formation caused the accumulation of the compound that is normally converted to isoleucine. This compound, in turn, inhibited the nearly identical reaction that leads to the formation of valine. Thus, the double growth factor requirement associated with a single mutation was accounted for by the inhibition of one reaction by an intermediate that accumulates as a result of a genetic block in another. There were also other anomalous findings that appeared to challenge the one gene–one biochemical reaction generalization. One was the double nutritional requirement for two amino acids, threonine and methionine, by a mutation that affected only a single gene. Here, the explanation turned out to be that the mutation blocked the synthesis of homoserine, the common precursor for threonine and methionine.

Some of the mutants were unable to make vitamins rather than amino acids. One that appeared to be blocked in the formation of niacin involved intermediates that Beadle had previously encountered in his work on *Drosophila* with Ephrussi. Three genetically distinct classes of mutations were found to require niacin for growth; for purposes of illustration these are designated as mutants 1, 2, and 3. Bonner tested a wide variety of compounds "ordinarily found in the cupboard of a chemical library" but none of them replaced niacin for the growth of any of the three mutants. However, when grown with limiting niacin, mutant 2 accumulated a compound that supported the growth of mutant 1 but not mutant 3. This result indicated that a compound B is formed beyond the block created by mutation 1 but before the block caused by mutant 3.

$$\text{tryptophan} \xrightarrow{mut\ 1} \text{A} \xrightarrow{mut\ 2} \text{B} \xrightarrow{mut\ 3} \text{C} \longrightarrow \text{niacin}$$

This was the same logic used by Ephrussi and Beadle when they ordered the reactions responsible for the formation of the v and cn substances in *Drosophila.* Bonner failed to identify compound B while at Stanford, but several years later, after moving to Yale, he discovered that the complete pathway involved more than three reactions. Quite unexpectedly, tryptophan was the metabolic source of niacin and is converted sequentially to *N*-formyl-kynurenine, kynurenine, and 3-hydroxykynurenine, the latter two being the v and cn substances involved in *Drosophila* eye pigment formation. Additional reactions, each one specified by a different gene, convert 3-hydroxykynurenine to niacin.

The group also identified mutations that blocked various reactions required for the formation of nucleic acid constituents and some affected fat and carbohydrate metabolism. "Indeed, the Beadle and Tatum team had produced more knowledge of biosynthetic pathways than had been accumulated by two or three generations of biochemists using traditional methods."[27] As the pace of the efforts in the lab increased between mid-1942 and late 1944, there was a greater need to organize an efficient infrastructure for generating, verifying, and characterizing the mutants. Large numbers of spore cultures had to be irradiated and tested for mutations. The innumerable crosses and production of heterokaryons were tedious and time-consuming. "Lots and lots of technicians" were needed for this routine work... "It was kind of a factory, with a kind of competition as to who could find the most interesting mutants."[28] Beadle and the people working with him on the genetics were at one end of the labyrinthine lab, and the people sorting out the phenotypic effects of the mutations were localized in Tatum's group at the other end. Although the lab's activities were physically separated, the group had long accepted that "If you find something exciting, you tell everybody."[29] Steeped in the ways of the "Morgan Fly Room" and the "Hole" at Cornell, where new discoveries were promptly shared, Beadle was well aware of the value of having everyone sharing in what was happening throughout.

One of the devices Beadle adopted for keeping himself and others up to date in what was happening in the lab was the daily tea. Everyone was expected to attend with few excuses accepted. Depending on how things were going, Beadle was eager to hear "new stuff." Sometimes, it was the amount of new mutants that had been generated and by whom, at other times it was a presentation and discussion of a new result, or occasionally it was a seminar by one of the students on a topic of his choice. Quite often, Beadle invited the group to discuss one of the scientific issues about which he was frequently being contacted. The daily tea was also an occasion for him to "read the riot act" to the group for their transgressions in what he felt was irresponsible laboratory practice. Contamination of the group supply of biotin with *Neurospora* spores, keeping food alongside mold cultures in the refrigerators, and general sloppiness in the crowded labs were sure to raise his ire. His anger, when it flared up, was justified; he was strict but fair and moved on quickly once he had let people know about a problem.[30] Generally, people in the lab were not devastated because they knew that even if they were at fault he did not carry a grudge. One thing the students could count on was that Beadle expressed his feelings straightaway; "if he was angry, he damned us to hell and after making his point the matter was dropped."[31] It was a characteristic his son David had seen when he had transgressed on something his father felt strongly about.

At times, Beadle could be a hard taskmaster, insisting before going off on a trip that everyone was going to have to work hard to have something new when he returned. Students who were not producing soon found that he became inattentive to their work. He had his favorites, but those who were at the top of his list could easily find themselves near the bottom if their efforts flagged. Driven by the excitement of discovery, Beadle could not understand those who did not share his sense of urgency. Throughout his life he had been excited by big challenges, and he expected others would be equally driven by the urge to succeed. Nevertheless, overall, he was much admired for the way he ran the group's research activities even as he spent less time in the laboratory and more time traveling and raising money.

Throughout the *Neurospora* work, Beadle continued to teach the same courses he had given since arriving at Stanford: a general genetics course of two lectures a week that used his and Sturtevant's textbook and five hours of laboratory work with *Drosophila* as the principal experimental material. Every other year, he offered an advanced course on special topics and an opportunity to do original laboratory investigation. If he was tough on those in the research lab, he seemed more relaxed with the students taking his courses. It was not uncommon, at the end of class, for Beadle and the students to congregate outside the central quadrangle to smoke and chat (smoking was forbidden in the quad). Relishing competition, and possibly believing that doing so relaxed the formality between professor and students, Beadle would occasionally challenge them to a standing broad-jumping contest. There is no record of how well he fared against his considerably younger students.[32]

As the research and large group consumed more and more of his attention, and his travel increased in frequency and duration, his relationship with Marion began to unravel. His life was so wrapped up in his work that he could go back to the lab after finishing Christmas dinner without appreciating the effect it would have on Marion and David. Oftentimes, he would arrive at the lab very early in the morning without breakfast and stay late well past dinnertime.[33] From early in their marriage, Marion's keen interest in music, art, and literature and Beadle's disinterest in these and other intellectual pursuits had been a contentious issue. She often belittled his lack of interest in artistic and literary matters. Although the rift between Beadle and Marion widened, neither one seemed inclined to discuss their problems with others.[34] Beadle was a compulsive achiever, a workaholic in today's terminology, while Marion was a chronic underachiever, often falling short of her own goals.[35] They had little in common except for raising David and that, too, was often the subject of contentious exchanges, sometimes in the lab in the presence of his students.[36] By staying long hours in the lab, Beadle could avoid being confronted by Marion's perplexing emotional reactions and the ensuing shouting matches.[37] David, away at boarding school, no longer served as the day-to-day glue for the family. Remembering those days decades later, David believed that his father understood neither his mother's nor his own emotions. Whether willed, learned, or instinctive, Beadle's emotions were held in check. Reason, not emotion, seemed to predominate in his personal dealings with his family, and he often behaved as though his and other's emotions had minimal influence on what he was pursuing. At the lab, he didn't need to confront Marion's or David's difficulties, and in that setting there was no question that he was in charge.

None of the core members of the lab group had been lost to the war effort, but after the summer of 1942, Beadle was uncertain about how much longer that would be so. Indeed, that was why he requested that the Rockefeller Foundation's supplemental grant be reviewed annually. He obtained occupational deferments for the students supported by the Nutrition Foundation on the grounds that training and skill in the field of nutrition were likely to contribute more to the war effort than military service. Bonner was deferred after Beadle wrote to his draft board saying that his work for the "war effort and civilian welfare" was irreplaceable.[38] However, Doermann, one of the promising graduate students, was notified by his local draft board that Beadle's request for a deferment was denied and his induction was imminent. Seeking some help to head off this loss, Beadle appealed to Hanson at the Rockefeller Foundation to contact Doermann's local board and the state appeal agency and attest to the fact that the value of his work was that it benefited the war effort.[39] Hanson promptly informed him that the Foundation had adopted a policy of not intervening in the actions of draft boards.[40] Beadle acknowledged the Foundation's "wisdom" and foresaw that the lab would have to increase the emphasis on applied work if it was to continue. Although he hoped to avoid the inflexibility of government contracts, the squeeze on his personnel and supply requirements was making the situation precarious. Accordingly, he informed Hanson that he was seeking a contract for the further development of vitamin and amino acid assays.[41]

Instead of a contract, Beadle got an even more important and unanticipated prize from the Committee on Medical Research (CMR), an offshoot of the war-created Office of Scientific Research and Development (OSRD). Dr. A.N. Richards, the Committee's chairman, informed Beadle of CMR's conclusion "that your research work is of sufficient fundamental importance and potential practical usefulness that it should not be interrupted in favor of other research which may seem to have more immediate practical utility to the war effort." The Committee could not guarantee deferments for the lab staff in the absence of a government contract for continuation of the work, but he assured Beadle "we will endeavor to give such requests the full influence of the OSRD in the case of any of your investigators whom you certify as essential and irreplaceable. Similarly in the case of critical materials for which high priorities are needed, we will do everything in our power to assist you."[42] The problems plaguing Beadle had been solved and he could proceed with the lab's basic research program and publish the results freely.

Ever on the lookout for additional financial support, Beadle met with A.D. Welch, director of research at Sharp and Dohme Laboratories in Philadelphia. Welch suggested that some of the existing *Neurospora* mutants and possibly newly created ones should be examined for their ability to produce penicillin. Although the likelihood of this being the case was small, Beadle thought it was worth a try. A memo to Hanson outlined the theoretical considerations, which presumed that penicillin was a normal metabolic intermediate in some reaction sequence in *Neurospora* and that it might accumulate if that pathway was blocked by mutation. It was considerably less likely that a mutation would add a new property, i.e., the production of penicillin. In agreeing to consider the project, Beadle assured Hanson that it would not impede the main line of their efforts and could be done during a few months with a single technician.[43] A few weeks later, Sharp and Dohme accepted Beadle's proposal and provided $600 to support one technician for a few months. In the meanwhile, however, Beadle was unexpectedly asked by the War Production Board to determine whether penicillin production could be increased by genetically modifying *Penicillium notatum*, the mold that normally makes penicillin.[44] Penicillin was proving to be very effective for treating infections caused by war wounds and, with the upcoming invasion of northern Europe, there was an urgent need to increase its production. Both Welch and Hanson approved of Beadle's decision to accept the War Production Board's assignment: *P. notatum* seemed a better choice than *Neurospora* for making penicillin. Although Bonner, who was given the responsibility for guiding the work, was concerned primarily with looking for mutants that produced higher levels of penicillin than the parent strain, he also recovered nutritional mutants of the type found with *Neurospora*.[45] He did succeed in getting mutants that produced moderately elevated levels of penicillin, but the project was abandoned because more efficient strains were produced elsewhere, the best being a mutant of *P. chrysogenum* generated by Milislav Demerec at Cold Spring Harbor.[46] The penicillin project was in fact the only direct war-related research carried out by the Beadle lab.

As a graduate student at Cornell and then as a postdoctoral fellow at Cal Tech, Beadle had been concerned with mutants affecting meiosis in maize and *Drosophila*. As the number of *Neurospora* mutants accumulated and their genetic locations were assigned to "linkage groups," it became urgent to relate these groups to *Neurospora* chromosomes, as had been done with the two organisms from his earlier studies. Beadle believed that there was no better cytogeneticist in the world than Barbara McClintock and so, early in 1944, he invited his old friend and colleague to spend ten weeks at Stanford to study *Neurospora* chromosomes. The cytology of *Neurospora* was not entirely new to McClintock: She had examined the meiotic figures in developing asci some years before. Her long experience and broad knowledge of the technology for visualizing meiotic chromosomes allowed her to follow the chromosomal mechanics from zygote formation through the two meiotic and single mitotic divisions. Besides confirming that meiosis in the developing ascus is typical of other eukaryotic organisms, she established unequivocally that *N. crassa* has seven chromosomes. That number conformed to the number of linkage groups that had been estimated from genetic crosses. In addition to her description of chromosomal behavior during ascus development, McClintock established that the anomalous behavior of several mutants in genetic crosses was explained by their having chromosomal rearrangements. The ten weeks McClintock spent at Stanford, although only a minor triumph in her illustrious career, yielded a major advance for fungal cytogenetics.[47]

As the work with *Neurospora* progressed, Tatum decided to determine whether nutritional deficiencies could also be induced in bacteria. His bacteriological training and instincts probably led him to realize on his own that the approach he and Beadle had used with *Neurospora* could be applied to bacteria. At the time there was considerable uncertainty as to whether bacteria even had genes.[48] Beadle encouraged Tatum to have his "own special responsibility" and enhance his chances of finding suitable employment.[49] Tatum selected two strains of bacteria from the Stanford bacteriology department collection because they could grow on a simple medium much like the one used to grow *Neurospora*. One of the strains, *Escherichia coli* K12, later became a standard for study of bacterial genetics. He subjected the bacteria to X irradiation and obtained mutants that required single nutritional supplements, such as amino acids, vitamins, or nucleic acid precursors. Tatum concluded "that biosyntheses in bacteria are controlled by specific genes," although it was not possible at the time to determine whether each nutritional requirement resulted from a mutation in a single gene.[50] Later that year, Tatum created double and triple mutants of *E. coli* by successively irradiating mutants with single requirements and recovering mutants with two or three requirements for growth. It was clear, then, that specific "growth-factor requirements result from heritable changes analogous to true gene mutations."[51]

From the time that Tatum joined Beadle in 1937, there was some concern about his future employment opportunities. One day, Tatum's father, then a professor of pharmacology at the University of Wisconsin, told Beadle that he was concerned about the professional future of his son. "Here you have him in a position in which he is neither a pure biochemist nor a bona fide geneticist. I'm very much afraid he will find no appropriate opportunity in either area." Beadle reassured Professor Tatum not to worry, that "it is going to be all right."[52] In 1940, after working together for three years, Beadle urged that the biology department consider Tatum for a faculty appointment. He also alerted Hanson about Tatum's availability for an academic position. Hanson had admired Tatum's contributions to Beadle's program and recommended him for a position to R.J. Williams at the University of Texas. Stanford's department of biology held Tatum in high esteem but had no place for a biochemist. Nevertheless, Tatum was appointed an assistant professor in 1941, about the time of the *Neurospora* breakthrough.

Although there is no outward evidence of friction or ill feeling between Beadle and Tatum during the years they collaborated at Stanford, there seems never to have been a close personal or social relationship.[53] Tatum's family rarely participated in Beadle's garden parties or outings.[54] By 1944, Tatum was ready to go out on his own. He had clearly acquired a strong genetics background to go with his biochemical skills. At the urging of Cornelis B. van Niel, with whom Tatum had taught a course on the use of microorganisms for vitamin assays, Tatum was offered a position as an associate professor at Yale University. He would have preferred to remain at Stanford, but the biology department, much to van Niel's consternation, turned him down again presumably because of faculty disputes that had little to do with his qualifications. Beadle's role, if any, in the department's deliberations and decision is unknown. Tatum accepted the Yale offer and left Stanford in early 1945. There he continued collecting nutritional mutants of *E. coli,* but more importantly, he and a student, Joshua Lederberg, exploited the properties of the double and triple mutants to discover that *E. coli* can participate in a sexual exchange of genes. This paved the way for a molecular genetics of these microbes.[55] Beadle congratulated Tatum: "The sex life of bacteria seemed darned interesting," Beadle wrote, and added, "It looks to me like it is the most important advance in bacteriology in the last 100 years."[56] Nominating Tatum for election to the National Academy of Sciences (1952) was one measure of Beadle's high regard for Tatum's accomplishments. Indeed, they maintained an intermittent but warm correspondence and occasionally visited with each other on their cross-country trips.

Ironically, three years later, after Beadle moved to Cal Tech and Stanford had a new president and biology department leadership, Tatum received and accepted an offer to return to Stanford as a full professor. Nearly ten years later, when Stanford planned to create a new department of biochemistry in the medical school, Tatum was selected to be its first chairman. However, marital problems leading to divorce caused him to move to Rockefeller University, where he remained until he died in 1975.

The Beadle–Tatum collaboration marked a turning point in the evolution of genetics. Until then, the principal focus of geneticists was on the genotypic properties of genes and chromosomes, namely, their mode of segregation in succeeding generations of offspring. Beadle was quick to acknowledge, perhaps under Tatum's tutelage, that gene action had become a biochemical problem and the forerunner of what became molecular biology. Before the *Neurospora* breakthrough, the idea of uniting genetics and biochemistry had been only a dream with a few scattered observations. But the "marriage" between the two made the new field of biochemical genetics a real science, posing new and challenging problems. Genes were clearly the determinants of an organism's metabolic and biosynthetic capacities and the Beadle–Tatum experiments provided a new way to examine the molecular details of that connection. Indeed, their findings over the course of just a few years had contributed more knowledge of amino acid biosynthetic pathways than had been accumulated during decades of traditional study.

Linking gene action to the production of enzymes and Beadle's speculations about how a gene determined the specificity of a protein forced renewed attention on the structure of the gene. Although the complexity of protein structures was not fully appreciated until nearly ten years after the *Neurospora* experiment's revelations, Beadle's speculations that proteins were somehow assembled on a gene serving as a template were provocative. They also provided skeptics with a target for their disbelief. Solving the molecular nature of the gene was now high on the agenda for understanding how genes were implicated in the synthesis of proteins.

Photo Gallery

Highway sign on entering Wahoo, Nebraska. *(Photo by Paul Berg.)*

Beadle house in Wahoo, circa 1900; figures are presumed to be GWB's father, mother, and older brother Alexander.

Hattie Albro Beadle, GWB's mother. *(Courtesy of the California Institute of Technology Archives.)*

Ruth Beadle, GWB's sister, during World War II. *(Courtesy of Ruth Beadle.)*

Beadle home interior circa 1901; GWB's father (Chauncey), mother (Hattie), and brother (Alexander); paintings and piano were Hattie's. *(Courtesy of Ruth Beadle.)*

GWB, 2 to 3 years old. *(Courtesy of Ruth Beadle.)*

Family portrait; Ruth Beadle, Chauncey Beadle, and GWB, circa 1915. *(Courtesy of Ruth Beadle.)*

Bess McDonald, circa 1921. *(Courtesy of Ruth Beadle.)*

GWB and his dog Buddy, circa 1921. *(Courtesy of the California Institute of Technology Archives.)*

GWB and friend at University of Nebraska, Lincoln, circa 1922–1926. *(Courtesy of Ruth Beadle.)*

Charles Burnham, Marcus Rhoades, Rollins Emerson, Barbara McClintock (standing, left to right) and GWB (kneeling) at "The Hole," Cornell University, circa 1929. *(Courtesy of the California Institute of Technology Archives.)*

Cal Tech Biology Division staff, 1931. Seated from left: Borsook, Dolk, Simms, Sturtevant, Emerson, Huffman, and Morgan. Standing from left: Schott, Burnham, Lammerts, Lindstrom-Lang, Ellis, Keighly, Bonner, Tyler, Beadle, Schultz. *(Courtesy of the California Institute of Technology Archives.)*

Thomas Hunt Morgan and Rollins Emerson, 1932. Meeting at the Sixth International Genetics Congress in Ithaca, New York. (Photo by Trevor Teele, Ithaca, New York.) *(Courtesy of the California Institute of Technology Archives.)*

Boris Ephrussi and GWB discussing *Drosophila* work. *(Courtesy of the California Institute of Technology Archives.)*

Marion and David Beadle in Palo Alto. *(Courtesy of Ruth Beadle.)*

Cal Tech Biology Division Faculty, March 1947. Standing, from left: Keighly, Sturtevant, Went, Haagen-Smit, Wildman, Beadle, Lewis, Wiersma, Mitchell, Van Harreveld, Alles, Anderson. Seated from upper left: Borsook, Emerson and lower left: Dubnoff, Bonner, Tyler, Horowitz. *(Courtesy of the California Institute of Technology Archives.)*

GWB and Gunnar Bergmann on the trip to climb Mt. Donnerak, Alaska, 1952. *(Time photo by Check. Courtesy of the California Institute of Technology Archives.)*

Left to right: Lee DuBridge, George Beadle, Harry Earhart, K.V. Thimann, Frits Went, and Robert Millikan, Cal Tech, 1949.Taken on the occasion of the dedication of the Earhart Plant Laboratory of Biology. *(Courtesy of the California Institute of Technology Archives.)*

GWB and Ernest Anderson inspecting corn mutants grown from seeds exposed to atom bomb test, The Farm, 1950. *(Courtesy of the California Institute of Technology Archives.)*

Muriel and GWB at their wedding, August 12, 1953. *(Courtesy of Ruth Beadle.)*

GWB and Alfred Sturtevant, 1951. *(Courtesy of the California Institute of Technology Archives.)*

GWB and James D. Watson at the University of Chicago, 1963.

GWB and Edward Tatum, Nobel Prize Ceremony, Stockholm, Sweden, 1958. *(Courtesy of the California Institute of Technology Archives.)*

Linus Pauling and GWB at Church Laboratory Dedication, 1955. *(Courtesy of the California Institute of Technology Archives.)*

Max Delbrück and GWB, 1978. *(Courtesy of the California Institute of Technology Archives.)*

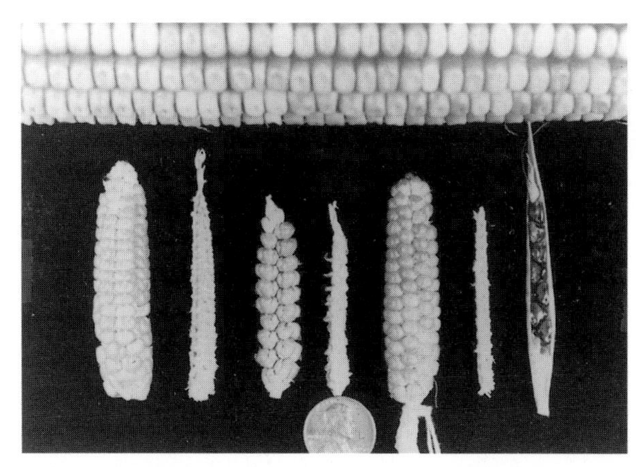

Left to right: A primitive type of corn ear like 7000-year-old specimens; cob of previous; F1 hybrid of primitive corn and teosinte; cob of previous; corn-like ear from F2 generation; cob of previous; teosinte ear (spike). (Reprinted, with permission, from GWB. 1978. Teosinte and the origin of maize. In *Maize breeding and genetics* (ed. D.B. Walden), pp. 113–128. John Wiley, New York.)

GWB grinding teosinte with primitive pre-Columbian grinding stones. (Photo by Ed Jarecki. Published in The Mystery of Maize, *Field Museum Bulletin* (Chicago). 43: 10, November 1972. *(Courtesy of the California Institute of Technology Archives.)*

Left to right: Ramana Tantravahi, H. Garrison Wilkes, Paul Mangelsdorf, William Davis, GWB, Umesh Banerjee, Elso Barghoorn, Walton Galinat at the conference at Harvard University, June, 1972. (*Photo by Hugh H. Iltis.*)

CHAPTER 11

Confronting the Skeptics

By early 1945, Beadle's day-to-day contacts with the lab's experimental efforts had diminished. The search for mutants and their biochemical characterization had become nearly routine. The number of challenging problems that could still be investigated with *Neurospora* were being taken up by the excellent graduate and postdoctoral students he had assembled three years earlier. Indeed, most of them—Norman Horowitz, Herschel Mitchell, David Bonner, Gus Doermann, and Adrian Srb—would soon become leaders in "biochemical genetics" in their own right. They formed a tight-knit, highly interactive, and remarkable group. He was as fortunate in selecting them as his team as Morgan had been in assembling the fly group. Few would contest that Beadle was the driving force behind the effort, but there is little doubt that their intellectual inputs were as important as the experiments they performed. Some credit Horowitz with being the midwife in conceiving the name one gene–one enzyme that was used to refer to the hypothesis that emerged from the *Neurospora* experiments. Evidence of Beadle's scientific stature was his election to the U.S. National Academy of Sciences in 1944, and in the following year the Genetics Society of America chose him to succeed Barbara McClintock as president. His increasing reputation also drew him to leadership responsibilities in several scientific organizations and the National Research Council. All of this, in addition to a heavy travel and lecture schedule, made it difficult to follow up on Sturtevant's suggestion that they bring their 1939 textbook of genetics up to date.[1]

Nevertheless, although the *Neurospora* work was viewed as a major breakthrough, there was a problem. Many in the scientific community were skeptical of the one gene–one enzyme hypothesis, questioning whether the inferences drawn from the *Neurospora* work were relevant to gene function in other organisms, e.g., animals and mammals. Characteristically, Beadle met skepticism and criticisms head on. In a comprehensive review entitled "Biochemical Genetics," he compiled and described all of the characteristics of animals and plants known to be controlled by genes; these included metabolism, pigment formation, morphogenesis, chromosome behavior, disease resistance, and psychological traits.[2] In a second review, appearing the same year, Beadle enumerated, one by one, the many examples from their *Neurospora* studies in which mutations in a single gene resulted in the loss of a single metabolic function. The inferences, he argued, were compelling; because metabolic reactions were known to be catalyzed by enzymes, one enzyme per reaction, and because the more complex characters mentioned above are presumably also the product of enzymatic action, genes must control the activity of enzymes directly or indirectly.[3]

In addition to the review articles, Beadle spent considerable time on the road giving lectures that argued his case. One set, under the auspices of the Sigma Xi Society, was given at about two dozen institutions throughout the country over the course of about 12–18 months.[4] In each, he laid out in considerable detail, as he had done in the review articles, the many examples where the loss of a single enzymatic reaction was attributable to a single gene mutation. Despite the acclaim by audiences that heard him speak, the skeptics were in the majority. At most of the institutions, the audiences "were sure gene action could not be generally described in the simple way we had postulated."[5]

In spite of, or perhaps because of, the widespread disbelief of the one gene–one enzyme hypothesis, Beadle was invited by the Harvey Society to deliver the prestigious annual lecture at the New York Academy of Sciences in February 1945. The inconveniences of midwinter and wartime travel did not deter him from the opportunity to present his views before the scientific elite. Traditionally, before the lecture, the speaker and a select number of friends and guests, decked out in black tie for the occasion, are feted with a sumptuous dinner accompanied by excellent wines. But the highlight of the evening is the lecture itself before an audience of the nation's leaders in science and medicine. After reviewing the historical precedents favoring a role for genes in the elaboration of proteins, Beadle recounted the evidence from his lab's work. He reiterated his conclusion that "every enzymatically catalyzed reaction that goes on in an organism depends directly on the gene responsible for the specificity of the enzyme concerned." He seemed certain that "for reasons of economy of the evolutionary process, one might expect with few exceptions the final specificity of a particular enzyme would be imposed by only one gene."[6]

Enlarging on that same point elsewhere, Beadle emphasized that although a given enzyme might derive its structure and specificity from more than one gene, each gene specified one and only one enzymatic "function."[7] He could see no way around the conclusion that the gene determined whether the protein it controlled was active or inactive; consequently, he believed that "every enzyme whose specificity depends on a protein should be subject to modification or inactivation through mutation." Of course this meant that the reaction normally catalyzed by the enzyme in question would either be blocked completely or have its rate modified by mutation. On this view, "the final specificity of an enzyme molecule is set by a particular gene." But because proteins are synthesized from component parts, he foresaw that these parts themselves must be "synthesized by enzymatic reactions that in turn depend on the functioning of many genes." Given the existence of such a hierarchy, the need for it to be regulated seemed obvious.[8]

Convinced that genetics and biochemistry were sharing common ground, he assailed the barriers that existed between geneticists and biochemists: "Too often in the past," he observed, "[genes] have been regarded as the exclusive property of the geneticist." But "the biochemist cannot understand what goes on chemically in the organism without considering genes any more than a geneticist can fully appreciate the gene without taking into account what it is and what it does." It is most unfortunate, he charged, that in "our institutions of higher learning, investigators tend to be forced into laboratories with such labels as 'biochemistry' or 'genetics.'" Because "the gene does not recognize the distinction—we should at least minimize it."[9] The same plea had been made 25 years earlier by Sewall Wright: "by constant comparison of the deductions of the geneticist with the findings of the biochemist, it should be possible in the end to establish a very pretty correlation of results."[10] Continuing on this theme, Beadle commented that "some students in the university enter a laboratory on the door of which is printed 'Genetics Laboratory'; other students enter another door marked 'Biochemistry Laboratory.' But in the future Genetics and Biochemistry will be in the same laboratory and students will enter through a single door." He could not have known at the time that he would have the chance within the coming year to create a laboratory labeled only "Biology" housing both geneticists and biochemists. Whether he created any converts to his view is not known, but Hanson, his longtime Rockefeller Foundation champion, was there and recalled that Beadle received the greatest ovation ever given a Harvey lecturer.[11]

For Beadle, the gene was a reality, not an artifact as implied by Richard Goldschmidt, a prominent physiological geneticist.[12] "The gene as a unit," Goldschmidt conceded, "is a concept derived from the existence of a thing called the mutant gene. There is the possibility," he argued, "that a condition exists at a definite locus in the chromosome that we call a mutant gene but that no corresponding plus condition exists at that site as a separate unit." Instead he offered a somewhat metaphysical speculation in which the wild-type condition reflected a property of the entire chromosome organization and that what geneticists called mutations were perturbations of that organization causing the function of that region to be altered.[13] Goldschmidt seemed to reject the confidence that Beadle held, namely, that a gene was a discrete site in the chromosome, specifying a particular physiological purpose, and that it could exist in either wild-type or mutant states.[14] Sewall Wright also expressed some uncertainty about whether genes exist as discrete entities in either of two allelic forms, but he was less adamant about it.[15]

Near the end of the Harvey lecture, Beadle raised a fundamental question, one that Schultz had considered years before: "If we knew enough about what a gene does we might be able to deduce what a gene is,"[16] but because the lecture was necessarily brief he was unable to address it then. However, he was well aware that without knowing more about the nature of the genetic material, the debate on the nature of the gene and whether it could act in the way he proposed could not be resolved.[17] Biochemical genetics, he wrote repeatedly, had to be concerned with the gene's chemical structure and its function.[18]

Arguments over whether genes were localized to the chromosomes' deoxyribosenucleic acid (the term DNA was not widely used until the 1950s) or protein or both had been going on for years. The staining of cells with dyes specific for DNA, in particular using the Feulgen reaction, clearly showed that DNA existed in the cell nuclei of all plants and animals and even in bacteria. Attempts had been made to study chromosomes by digesting away either the proteins or the DNA with appropriate enzymes, but the resulting preparations were, as Beadle recognized, too impure to give definitive results. In any case, DNA seemed an unlikely candidate for the genetic material, primarily because of its apparently simple composition. The prevailing view was one proposed in 1921 by P.A. Levene of the Rockefeller Institute: DNA was composed of repeating tetranucleotide units, each one containing the four bases adenine, guanine, cytosine, and thymine.[19] Even though the justification for that structure was problematic, it seemed almost incomprehensible that the extent of genetic diversity and the complex properties of a gene could be accounted for by a substance of such limited information content.[20] Instead, together with many biologists, he favored proteins as the genic material.

Beadle's preference for genes being proteins was influenced by work on viruses. In 1935, Wendell Stanley had crystallized the tobacco mosaic virus, and everything he knew about the virus was consistent with its being a protein. Soon after, however, other workers realized that Stanley's crystals also contained ribosenucleic acid (RNA) and the virus was a nucleoprotein. It would take two more decades to prove that the gene-like properties of the virus resided in the small amount of RNA rather than the large amount of protein. Meanwhile, the view that genes were proteins also appeared to be consistent with work on bacterial viruses, bacteriophage. Modern research with bacteriophage was initiated by Emory L. Ellis and Max Delbrück while both were visitors at Cal Tech during the late 1930s. They believed that the viruses were composed of protein and "possess the property of multiplying within living organisms."[21] By the mid- to late 1940s, bacteriophage were shown to contain DNA, and work by Delbrück, Alfred D.. Hershey, and Salvatore E. Luria established that even bacteriophage have genes. Beadle was convinced, and admitted that "all living systems do contain genes [sic] as indispensible units."[22] But, as he did in the Harvey lecture and *Chemical Review* article, he still preferred to think of genes as nucleoproteins, with proteins as the primary source of the genetic information.[23] Having committed to proteins as the genes, he was forced to ponder how the protein portions of nucleoproteins could act as templates to direct the formation of functional copies of themselves. He was well aware that "there is no general theory of how genes duplicate themselves or control protein specificities that has any support in experimental observation." But he favored a widely held view that the gene was a "master molecule or template in directing the final conformation of the protein molecule as it is put together from its component parts."[24] This conception was clearly modeled after Linus Pauling's proposal that antigens direct the folding of antibody proteins by serving as a template. However, many found the notion of a protein gene being able to replicate itself and make the cellular enzymes implausible. It was a weakness that opponents of the one gene–one enzyme hypothesis found easy to criticize.

It was not until after 1945, when Erwin Chargaff at Columbia obtained valid measurements of the four bases from carefully prepared DNA samples from many organisms, that the tetranucleotide hypothesis was finally buried. Chargaff's data plainly revealed that the four bases were not always present in equal amounts. Indeed, the relative amounts of the four nucleotides differed from one species to another. The only constant Chargaff found in the base compositions among the DNA samples was that the amount of adenine always equaled the amount of thymine and that of guanine always equaled that of cytosine.[25] A few years later these data would be key to determining the structure of the DNA double helix, but in 1945, Beadle was stuck with the tetranucleotide hypothesis, which seemed to rule DNA out as the genetic substance in spite of two other observations that suggested otherwise.

Beadle knew that earlier in the 1940s his old colleague Lewis Stadler and, independently, Alexander Hollaender, had demonstrated that "ultraviolet radiation produces gene mutations and its efficiency per unit energy varies with the wavelength in a manner similar to its absorption by nucleic acid."[26] He was also aware,[27] but skeptical, of the recent work "on the nucleic acid of Pneumococci bacteria suggesting that this component played a part in determining the specificities of individual genes."[28] He was not alone in his skepticism. Nucleic acid experts thought that Oswald Avery's experiments were still open to question, and many were critical of the purity of the DNA that was purported to be behaving as genes.[29] Although Beadle accepted Avery's data, he saw no way to differentiate between two possible interpretations: Either the result reflected a specific mutagenic effect by Avery's DNA preparations, or the "nucleic acid itself is the gene."[30] Even a few years later, directed mutation was still invoked as the explanation for DNA-mediated genetic transformations, thereby avoiding any implication that DNA itself was the genetic carrier. Beadle also explained away the increased frequency of mutations caused by UV light at wavelengths at which nucleic acids absorbed most light by suggesting that the nucleic acid acted as a conduit for passing the energy of the radiation along to a protein.[31]

Another reason for Beadle's skepticism was that Avery's experiments were with bacteria. Like other bacteria, pneumococci reproduce by asexual division. Like most classical geneticists, Beadle believed that the existence of genes relies on their following known genetic rules. These depend on having sexually reproducing species carrying two different alleles of a gene, one mutant and one wild type. For him, therefore, "The genetic definition of a gene implies sexual reproduction."[32] Beadle's thinking seemed to be constrained by the abstract gene of the Mendel–Morgan context. If one could not do genetic crosses, it meant that there were no genes. Interestingly, Beadle's reservations were in spite of the fact that Tatum had already obtained stable bacterial mutants with single nutritional deficiencies[33] and that S.E. Luria and M. Delbrück had reported bacterial mutations that affected their sensitivity to virus infection.[34] But it is also possible that like Morgan and Emerson, he required direct evidence, not speculative interpretations offered in the Avery, Tatum, and Delbrück works.

Eventually, Beadle conceded that there was substantial experimental support for the existence of gene-like units in bacteria. In his review of a book on bacterial cytology,[35] Beadle acknowledged that bacteria have genes.[36] Beside Tatum's demonstration that X rays induced mutations in bacteria just as they had done with *Neurospora,*[37] he was persuaded that sexual-like processes were detected in *E. coli*. Using genetically distinct strains of *E. coli*, each with different nutritional requirements, Joshua Lederberg discovered that merely mixing these strains produced strains that lacked any of the parental nutritional requirements.[38] Beadle could understand the implications of this result because it was formally the same as ones he obtained when he crossed two yeast strains differing in nutritional requirements and obtained strains that grew without any nutritional supplementation.

But the one gene–one enzyme proposal encountered more substantive objections. These concerned the experimental methodology and the failure to consider alternative interpretations. Following Bonner's presentation of the Beadle lab's findings at the 1946 Cold Spring Harbor Symposium, the meeting ground of the genetics elite, Max Delbrück raised serious objections to the procedure for obtaining the nutritional mutants.[39] Delbrück conceded "that the evidence accumulated in the *Neurospora* work is compatible with the assumption that there exists a one-to-one correlation between genes and the various species of enzymes found in the cell." But he wondered "whether the evidence obtained actually supports the thesis, beyond the mere fact that it is compatible with it." Delbrück argued that "in order to make a fair appraisal of...a one-to-one correlation between genes and species of enzymes, it is necessary to begin with a discussion of methods by which the thesis can be disproved." Otherwise, he stated, "the mass of compatible evidence carries no weight whatsoever." Delbruck argued persuasively that the value of any scientific hypothesis or proof is that it embodies ways in which it can be disproved, and he could see no way to meet that ideal. He pointed out that the experimental procedure was unlikely to turn up *Neurospora* with properties other than those in which only one function was affected. For example, "the thesis might be untrue in two ways: First, if an organism had suffered a mutation in a gene that affects multiple enzymatic reactions, each of the deficiencies caused by the mutation would have to be overcome for it to grow. If, however, one or more of the defects can not be overcome by the 'complete' medium, the mutation is lethal and will not be recovered." Of additional concern was the lack of attempts to falsify the hypothesis by, for example, searching for nonallelic mutations that affect the same enzymatic reaction.

The best Bonner could do at the time was concede the possibility that some genes had multiple functions, but he insisted that in each of the cases studied, the mutation affected only one metabolic reaction. He also emphasized that there was no instance where nonallelic mutations caused a block of the same enzymatic reaction.[40] Bonner also acknowledged that the "one–one" concept is not intended as a law. Rather it is a useful working hypothesis, and as such it serves an important function. "Exceptions to the 'one–one' concept may well exist but it is difficult to devise an experiment that would yield an alternative explanation. Since the data now available are accounted for on a 'one–one' basis, it would be of little value to devise a more complex thesis of the relation of genes to enzymes."

Along with Delbrück's criticism that the Stanford group missed mutations that cause multiple defects, there was also Joshua Lederberg's outspoken skepticism of the Beadle hypothesis. He focused his objections on the unproven supposition that the "gene works as a unique template for stamping the specificity of an enzyme." Without knowing how proteins were assembled, he felt it was presumptuous to assume that a change in a gene was directly translated into a defective enzyme. Also unproven was whether "a given gene is the primary seat of specificity of an enzyme, or whether [the gene] has an indirect influence... to specify which of several latent potentialities will be realized. The main difficulty with the one-to-one theory," Lederberg concluded, was "that it is experimentally indefeasible, that there is no experimental test that can exclude it short of the knowledge of how the genes work." He added that ad hoc arguments rather than experimental evidence were being used to explain away anomalies or inconsistencies that failed to fit the model.[41]

The criticisms at the Cold Spring Harbor meeting and afterward dropped acceptance of the one gene–one enzyme concept to an all-time low. Beadle had the impression that "the number of people whose faith remained steadfast could be counted on the fingers of one hand—with a couple of fingers left over."[42]

At a symposium to celebrate the 50th anniversary of the rediscovery of Mendel's work, Beadle reflected on the status of what he now preferred to call the one gene–one (primary) function hypothesis rather than the one gene–one enzyme model, because the production of substances other than proteins might also be governed by genes. He reiterated a point that he had made in the Harvey lecture five years earlier that "the hypothesis does not require that the entire specificity of a given enzyme or other molecule be determined solely by a single gene, although this is a possibility, and is often erroneously thought of as a necessary consequence." Rather, "the hypothesis requires that a given gene be concerned in a primary way with only a single enzyme (protein)." At this point, Beadle turned conservative in his evaluation of the hypothesis, a trait that was often missing when he was promoting it during the five preceding years. At the time of this statement, Delbruck's refutation of the hypothesis still cast a pall over the model, and Horowitz's demolition of that criticism was still to come. Without being sure about the nature of the gene, Beadle was at a loss to determine how genes affected enzyme formation. He expressed his intuition, "I feel that in a great many cases in *Neurospora,* the reaction under primary gene control has been identified, but I confess I know no way of proving it in even a single instance." Summarizing, "it can be said that the one gene–one function hypothesis has without doubt served a useful purpose and while there are no compelling reasons for abandoning it at this time, neither should it be accepted without reservation. Even if it should prove correct in principle, like many useful working hypotheses in science, it may well be found to err in the direction of oversimplification."[43]

It took a while to figure out a way to deal with Delbrück's critique. Horowitz quipped that if it could not be answered "the one gene–one enzyme idea must be banished to the purgatory of untestable hypotheses, along with the proposition that a blue unicorn lives on the other side of the moon."[44] Delbrück's criticism presumed that there were many genes with multiple functions, and, if any one of the defects caused by mutation was irreparable, they would be lethal and therefore not recoverable by any form of nutritional supplementation. To test for that possibility, Horowitz compared the frequency of mutations that cause reparable and irreparable defects. The existence of mutations whose phenotype is dependent on the temperature at which the cells are grown provided the way.[45]

Assuming that both reparable and irreparable mutations were equally likely to have a thermosensitive phenotype, two kinds of measurements were made. One determined the frequency of mutants that grew at low temperature but not at high temperature in the rich medium; these are mutants for which even a full menu of "goodies" could not overcome the effect of the mutation. Such defects, Horowitz reasoned, were by Delbrück's contention lethal because they could not be overcome by nutritional supplementation, at least at the elevated temperature. The second class consisted of mutants that prospered in a rich medium at both high and low temperatures and failed to grow in the minimal medium at high temperature but grew well at the low temperature. This second class of mutations, which was the type with which they had been dealing all along, causes a deficiency that can be overcome by the rich medium but not by the unsupplemented minimal medium at the elevated temperature. Comparing the relative frequencies of the two types of mutants, Horowitz estimated that about 70% of genes affect only a single function. These were minimal estimates because, aside from the possibility that a gene specified multiple functions, there were numerous other reasons that a mutant's growth defect could not be rescued by any nutritional supplementation. A similar kind of analysis with the bacteria *E. coli* gave an estimate of 77% of genes specifying a single function.[46] The conclusion was clear: The genes Beadle's group had analyzed were not a rare class, rather they were representative of the majority, genes that specified only a single function. Delbrück's criticism was laid to rest without a whimper. Later at a symposium to honor him, Delbrück conceded that one of the factors involved in the production of any particular enzyme "is a genetic control" in which "a mutation of a single gene can be responsible for whether or not a cell can manufacture a given enzyme."[47]

In time, answers to all the objections emerged and more and more evidence accumulated to substantiate Beadle's proposal. Much of that evidence came from Horowitz and Mitchell after they moved to Cal Tech, and from Bonner who had settled at Yale. One set of experiments showed that the enzyme activity that condenses indole and serine to form tryptophan was indeed missing in one of the mutants that required tryptophan for growth.[48] But it was still unclear whether the enzyme was present, albeit in an inactive form, or the mutant was unable to make the enzyme at all. Independent discoveries favored the former explanation. One relied on studies of a human genetic disease. Sickle-cell disease, a serious affliction among individuals of African, Southeast Asian, and Middle Eastern origins, is the result of a single mutation that imparts an additional negative charge to hemoglobin. This alteration in the hemoglobin causes red cells to assume an abnormal shape during the delivery of oxygen to the tissues.[49] This qualitative change in the protein molecule supported a direct relationship between the gene and the structure of the protein. Some years later, the change in the hemoglobin molecule was identified as a substitution of a negatively charged amino acid for a neutral one in one of the two hemoglobin chains.[50]

Another demonstration that a mutation caused the production of an inactive altered enzyme used antibodies directed against an active normal enzyme. For example, among the mutants lacking the enzyme activity for converting indole and serine to tryptophan, some failed to produce any detectable immunoreactive protein while others made proteins that were detectable by antibodies to the functional protein.[51] Moreover, the inactive, antibody-reactive proteins produced by different mutants were physically distinguishable, suggesting that different mutations affected the protein differently. This strongly suggested that the gene had a direct role in determining the structure of the protein it controlled. Two additional studies supported that view. They relied on mutations that prevent an organism from growing at an elevated temperature while allowing normal growth at a lower temperature. In each of these cases, the isolated enzyme functioned normally at the lower temperature but was inactive at the elevated temperature, thereby mimicking the behavior of the mutant organisms.[52]

Experiments done years later showed that the structure of genes and proteins they control are colinear; that is, mutations affecting the beginning of the gene affected the amino acid sequence at the beginning of the protein.[53] While this work was in progress, efforts were under way to establish the precise nature of the genetic code and the correspondence between the order of bases in a gene's DNA and the amino acid sequence in the encoded protein. In one of the triumphs of molecular biology, the "genetic code" was solved in 1964.[54]

Years later, in a letter to Horowitz, Lederberg admitted that "in recent years, I have had a chance to reflect back on the noise I used to make about the one-gene one-enzyme theory, and I now see that I was not only factually wrong in opposing it, even as an intellectual exercise, but showed rather poor judgement in failing to defend it. Perhaps I was reacting to the idea that no one else *ever* had that it was the *ultimate* truth; what in science ever is!"[55] Beadle acknowledged "that when all our best friends had deserted the ship, Josh is the only one I knew of who has had the grace to admit a change of heart."[56]

Why was it so difficult for the biological community to accept the idea that genes act through their ability to control the formation of proteins that catalyze particular enzymatic reactions? Beadle's own studies with the *Drosophila* eye-color mutants had already planted the notion that genes worked at the level of metabolic reactions and that each chemical reaction is under the influence of one gene. Although he never cited it, the mutations he identified as affecting the pathway of germ cell formation in corn may also have been part of his thinking, even if connecting it with enzymes would have been a stretch at the time. There were also observations and explicit claims that genes acted through their ability to influence or control the specificity of enzymes long before Beadle's announcement of the one gene–one enzyme hypothesis. He reviewed many of these in his presentation five years earlier at the Seventh International Congress of Genetics in Edinburgh,[57] and in his major articles in 1945.[58] Others wrote about them as well.[59] The history of that early work has since been reviewed.[60]

One of the impediments to accepting the Beadle–Tatum model was the widespread view among geneticists that each gene had multiple functions and possibly that every gene is concerned with the production of every character.[61] There was also considerable skepticism that complex biochemical, morphological, and behavioral phenotypes could be accounted for solely by gene-directed specific enzymatic and structural proteins. There were some biologists who had difficulty accepting the notion that what was true about *Neurospora* could be generalized to include more complex organisms, like mammals. But these doubts belied the existence of reports and speculations from as far back as the turn of the century that hinted at a physiological connection between genes and enzymes. These entailed the inheritance of coat color and pattern in small mammals.[62] The conclusion that Mendelian-like genes controlled the nature, intensity, and distribution of color patches was clear. Furthermore, the nature of the pigments, chemical reactions, and role of enzymes was elucidated, and, in some cases, a mutant's lack of a particular coloring was correlated with the absence of an enzyme needed to produce the pigment.[63]

On another front, Archibald Garrod, as early as 1902–1908, had concluded on the basis of his observations of several inherited metabolic disorders that the enzyme catalyzing the normal metabolic reaction was absent in affected individuals. Indeed, Beadle believed and frequently stated that Garrod should be acknowledged as "the father of chemical genetics," for "the relation gene–enzyme–chemical reaction was certainly clearly in his mind."[64]

Who was Archibald Garrod and how did he make the connection between enzymes and genes?[65] Garrod (1857–1936) chose to follow a career in medicine, earned a bachelor of medicine and surgery degree from the University of Oxford, and qualified for membership in the Royal College of Surgeons. Inspired by the rapid scientific advances in medicine that were sweeping Europe, Garrod elected to pursue both the clinical practice of medicine and chemical pathology, a specialty his father had pioneered. Fascinated by the extent to which chemical changes in urine provided clues to underlying pathologies, he sought an explanation for a rare trait named alcaptonuria or "black urine disease."[66] Individuals afflicted with alcaptonuria are not ill in the classical sense, but their urine, initially normal in color, turns black on exposure to air. The blackening was attributed to the presence of a substance initially referred to as alcapton but identified as homogentisic acid, a metabolic product of the amino acids phenylalanine and tyrosine. Because there was often more than one affected child in a family, Garrod considered it likely that alcaptonuria represented an inherited inability to metabolize homogentisic acid. But he was puzzled because the parents of alcaptonuric children were invariably unaffected and therefore his explanation was inconsistent with the then conventional understanding of human inheritance.

This apparent inconsistency was resolved when the mother of one of his patients, who had normally colored urine, gave birth to a child whose diapers began turning black. Garrod surmised that the characteristic of black urine was inherited rather than acquired. In reviewing the family histories of his patient group, Garrod found a clue to support his theory of a familial susceptibility: Alcaptonuric individuals were most often the offspring of marriages between first cousins. Unaware of the rediscovery of Mendel's principles of inheritance a few years earlier, he could not yet make use of this clue to explain the inheritance pattern of alcaptonuria. William Bateson, one of Mendel's most aggressive advocates, learned about Garrod's work and contacted him with an explanation.[67] He explained that alcaptonuria behaved like a recessive trait; each of the parents contain one of the responsible alleles and have normal-colored urine but, on average, one in four of the children born of such a marriage would be affected. Because the factor causing alcaptonuria is rare, marriages between unrelated individuals are very unlikely to produce an affected child.

Garrod was the first physician to appreciate the significance of Mendelism for human disease. At a meeting in 1908, he suggested correctly that alcaptonuria appeared to fall into the category of a recessive character and might result from the absence of an enzyme that normally metabolizes homogentisic acid.[68] Meanwhile, Garrod was exploring other diseases: albinism, where the affected individuals fail to produce the normal melanin pigment in the skin and hair, and cystinuria, where abnormally high levels of the amino acid cystine appear in the urine.[69] These, too, showed up most frequently in the offspring of consanguineous marriages for the same reasons, and he concluded that these metabolic abnormalities were also probably the result of recessive mutations.

In the ensuing years, Garrod became increasingly intrigued by the notion that individual genetic variation affected many metabolic reactions, thereby influencing the phenotype in profound ways, particularly regarding susceptibility to disease. Garrod was fast becoming an authority on inherited diseases although he was far from educated in or taken with the emerging science of genetics. Garrod's classic book *The Inborn Factors in Disease* explored the ramifications of his view that chemical individuality and predispositions to disease were inherited.[70] Throughout the treatise, Garrod hinted that variations in the proteins originated in the chromosomes, and he came close to arguing that an individual's chemical uniqueness results from qualitative and quantitative variation in proteins. He might have been pleased to learn that the enzyme capable of metabolizing homogentisic acid, present in normal human plasma, was absent in patients with alcaptonuria.[71] Some 20 years after Garrod's death, that enzyme was isolated[72] and the gene encoding the enzyme has been localized to the long arm of human chromosome 3.[73]

Did Garrod anticipate Beadle and Tatum's conception of the one gene–one enzyme linkage? If so, it was in spite of a skimpy knowledge of genetics. He continued to refer to Mendel's factors, even 20 years after they had been supplanted by the word "gene." During the entire period Garrod was formulating his ideas on hereditary predispositions, he never mentioned Morgan's discoveries with *Drosophila* and how they revolutionized biology. Nor did he acknowledge the contemporaneous research on the genetic control of flower color ongoing in England. He was aware that enzymes were very likely proteins, but he never made the connection between genes and protein enzymes.

Nevertheless, Beadle acknowledged Garrod's priority in his Harvey lecture, and his Nobel lecture was even more effusive in crediting Garrod's insights: "In this long, roundabout way, first in *Drosophila* and then in *Neurospora*, we had rediscovered what Garrod had seen so clearly so many years before. By now we were aware that we had added little if anything new in principle... Thus we were able to demonstrate that what Garrod had shown for a few genes and a few chemical reactions in man was true for many genes and many reactions in *Neurospora*."[74] Was Beadle's tribute accurate or was he reading more into Garrod's insights than was there? J.B.S. Haldane, the distinguished British biologist and enzymologist, gave Garrod unambiguous credit for anticipating the one gene–one enzyme hypothesis.[75] There is little doubt that Garrod sensed a functional relationship between heredity and physiology in humans, but nowhere in Garrod's writing is there an explicit or even implied statement of a one-to-one relationship between a hereditary factor and a specific metabolic reaction or biological property. Garrod's biographer, A.G. Bearn, sought to place Garrod's contribution in context: "It was not that the fruits of Garrods research did not support Beadle and Tatum's one gene–one enzyme hypothesis. They did. Moreover, Beadle and Tatum's hypothesis was specifically and narrowly concerned with the one-to-one relationship between genes and enzymes. On the other hand, Garrod's central insight was not simply the absence of enzymes in recessively inherited disease. For him, these rare diseases were merely one set of examples that demonstrated the fundamental principles of biochemical individuality."[76] Lederberg's assessment of Garrod is that he failed to view the defect as the loss of a normal gene and the consequential loss of an enzyme rather than the acquisition of a gene that caused a pathological condition: "The normal gene sort of loses its identity and it's only the abnormal gene that he focuses his attention on."[77]

When Beadle lauded Garrod's acumen, there was considerable puzzlement as to why Garrod's discoveries were not recognized earlier. First and foremost, they were not prized for their biological significance. "Both Continental and British biochemists of Garrod's day were not very excited by the hereditary aspects of inborn errors, nor did they recognize that these metabolic disorders offered hints to the ways compounds are broken down in the body."[78] The relative obscurity and minor impact of alcaptonuria, cystinuria, and albinism on morbidity blunted any medical interest in their origin. Given the prevalence and more devastating consequences of infectious disease, inherited metabolic defects drew little or no attention. At the time, geneticists were aware of Garrod's findings, but only a few were concerned with the biochemical oddities he described and even fewer were inclined to think of hereditary traits in chemical terms. Also, geneticists tended to think of genes as having manifold effects, and the notion that a gene would control one biochemical reaction was too simple for its time.[79] Biochemists, even those as influential as F.G. Hopkins, were not drawn to investigate the inherited diseases that were noted in textbooks as metabolic anomalies. Ironically, Beadle was unaware that when he was a fellow at Cal Tech, Borsook covered Garrod's work on inborn errors of metabolism in his biochemistry course to undergraduates. Also, references to Garrod appear in Bodansky's widely used textbook of biochemistry.[80] Neither did he recall that the fly group never referred to Bateson's book where Garrod's work on alcaptonuria was mentioned.[81] No mention is made of the Garrod work in Goldschmidt's treatise on physiological genetics[82] or in the genetics book Beadle and Sturtevant had written.[83]

Until Beadle and Tatum's experiments with *Neurospora*, mutations were used primarily as genetic markers to study the mechanisms of inheritance. Geneticists relied on spontaneous, random events that altered an observable or measurable phenotypic property of an organism to establish the existence of a gene; the normal or wild-type form of the gene was more an inference than a measurable entity. Also, there was no need to know the function of the mutated gene or its corresponding normal allele in order to map it to a specific chromosomal locus, to determine its linkage to other genes, or to follow its inheritance from one generation to the next. The one gene–one enzyme paradigm proposed that the normal form of a gene directed the formation of the functional form of a single enzyme (polypeptide chain). The mutant form of the gene could then be understood as directing the formation of an impaired or inactive form of that enzyme. Geneticists were therefore obliged to view genes from three perspectives: as units of mutation, units of recombination (crossing-over), and units of function.[84] It wasn't until genes were known to be segments of DNA that these three descriptions of genes could be reconciled as fundamental features of particular sequences of bases in DNA. Thus, mutations result from one or more changes in the DNA sequence, genetic recombinations result from exchange of segments between two DNA chains, and gene function results from translation of a base sequence into a specific protein molecule.

Besides assigning a physiological role to genes, the Beadle–Tatum experiments introduced a powerful experimental approach for the analysis of biological systems. In essence, that strategy is to identify mutants that are affected in one or another of the steps in a particular biological process. The presumption is that every process is the consequence of many consecutive reactions, each catalyzed, facilitated, or structurally dependent on one or more proteins. For example, the biosynthesis of the cellular constituents such as amino acids, purines, pyrimidines, vitamins, sugars, and fats results from sequential enzymatically catalyzed chemical reactions. The assembly of amino acids into proteins, of purines and pyrimidines into DNA and RNA, of single sugars into complex polymeric carbohydrates, and the regulation of all of these processes are also dependent on the action of multiple proteins and, therefore, specific genes.

Almost immediately following Beadle and Tatum's reports with *Neurospora*, mutational analysis became the predominant approach to the analysis of metabolic pathways and the mechanisms organisms use to derive energy. The same strategy is now the dominant experimental approach to analyzing such complex processes as memory, learning, vision, smell, and even how the shape of an organism is patterned during embryonic and fetal development, and, more recently, the process of aging. Mutations that alter the normal course of events are, today, the starting point for identifying discrete steps in these complex processes. Generally, many genetically different mutants are isolated affecting the same process, and then the challenge is to determine the order of events in which the normal gene products act. Once a gene has been identified as being associated with a particular physiological function, the effort focuses on isolating the affected gene, examining its entire base sequence, identifying the particular protein that that gene specifies, and attempting to deduce the function it performs. The 11 discrete enzymes involved in the fermentation of sugar to alcohol or the 10 used to convert glucose to lactic acid during muscle contraction were identified over nearly 40 years of painstaking protein fractionation to separate and characterize each of the enzymes. Today, the identification of the enzyme proteins involved in yeast fermentation could be achieved more efficiently by employing Beadle and Tatum's mutational approach. Mutants blocked in each of the individual metabolic reactions would be the first step. The wild-type DNA sequence of the gene in which the mutation occurred can be converted through the genetic code to provide the identity of the proteins in the particular affected reaction.

CHAPTER 12

In Morgan's Footsteps

As 1945 began, Beadle's laboratory at Stanford was thriving and his position there and in the scientific community was secure and rewarding. Everyone anticipated that the war would be over before the year was out and the country and scientific research would return to normalcy. He had been working with *Neurospora* for about 4 years. Although there was some skepticism about the generality of the one gene–one enzyme concept, as far as he was concerned, it was established. Tatum was extending the approach they had taken with *Neurospora* to bacteria, which promised to be even simpler to work with and turned out to be enormously productive. Beadle's brilliant students and postdoctoral fellows were continuing to reveal essential information about metabolic pathways, and his own reputation and efforts assured a steady flow of funds.[1] But Beadle was restless.

The most challenging question facing geneticists, the chemical nature of genes, was much on his mind. Beadle referred to it repeatedly in the long review articles he published in 1945. However, neither the *Neurospora* experiments nor the many other known biochemical effects of genes led to an obvious strategy for revealing the structure of genes or the way that genes yield the corresponding enzymes. Convinced that "genes are as much a part of biochemistry as chemistry is of inheritance, development, and function,"[2] he could not think of ways to pursue the connection experimentally. The Rockefeller Foundation agreed that the next major advances would come at the intersection of these fields, and its future support was likely to depend on having the two disciplines joined in the same department. Beadle tried to promote interactions between the biochemists and geneticists at Stanford, but that turned out to be an uphill battle. Hubert Loring, the senior biochemist at Stanford, looked toward chemistry and was uninterested in biochemical genetics.[3] It seemed that Beadle would have to go elsewhere if he was to pursue this new direction. By the spring of 1945 he was considering an offer to become director of the Wistar Institute in Philadelphia.[4] Then, only a few days after Germany surrendered in May, Sturtevant invited him to join the Cal Tech faculty.[5]

Morgan had been reluctant to step down as chairman of the Cal Tech biology division and did not retire until 1942 when he was 76 years old. He had not made any recommendations about a successor, and the critical need for new leadership remained unresolved during the war years. Millikan, who was partial to executive committees because of the success that he, Hale, and Noyes had enjoyed, appointed a committee to run the biology division in Morgan's place. It did not function well. Sturtevant was unable or unwilling to exercise strong leadership, although Sterling Emerson and Albert Tyler tried repeatedly and unsuccessfully to convince him to take things in hand. The others, Henry Borsook, Frits Went, and Arie Haagen-Smit, used the situation "to feather their own nests."[6] Eventually, James Bonner accepted a position at the University of Chicago, and Emerson, who was sufficiently frustrated to complain to Linus Pauling, chair of the chemistry division, thought about leaving.[7] Importantly, the biology division no longer had direct representation on the Cal Tech executive committee and had to rely on Pauling to represent its interests within the university. It was essential that something be done about the deteriorating situation in biology. Beadle seemed an obvious solution.

The Pasadena biologists likely realized that they had missed hiring Beadle several times in the past, including in the early 1930s when his fellowship ended (Chapter 6). Then, after Beadle and Ephrussi's success with *Drosophila* in Paris, Sturtevant and Morgan realized too late that Beadle could help sustain Cal Tech's eminence; Beadle had already committed to go to Harvard. Morgan woke up to his mistake when word went around that Beadle was unhappy at Harvard. He told Weaver that he needed the Foundation's help if he was to get Beadle back to Pasadena, but Weaver would only consider an institutional grant, not salary money for an individual.[8] Without that help, Morgan decided he could not compete with Stanford and gave up. But 8 years later in 1945, the Cal Tech biologists thought they might succeed in a competition with Wistar.

Sturtevant's letter inviting Beadle to return to Cal Tech was curiously ambivalent.[9] He seemed more concerned about putting his friend in an uncomfortable spot than convincing him to come to Pasadena. The proposed salary and setup funds were only $7000 and $7500, respectively, and operating expenses were to be determined later. Sturtevant made it clear that he pulled these numbers out of a hat and anticipated that they were too low. Could Beadle tell him what would be realistic? He reported that the division staff was enthusiastic and unanimously approving but that he was keeping them in the dark about the financial aspects because the salary offer was more than any of them received, with the exception of himself. Although Millikan supported the invitation, he would not commit himself to raise additional money. If Sturtevant had been a conniving person, he might have written just such a letter in an effort to change Millikan's mind about raising funds.

Beadle responded promptly, indicating a willingness to talk. A week after Sturtevant's letter, Sterling Emerson, who learned of Beadle's interest, wrote to his old friend to say that Borsook, Went, Tyler, and "Haagy" were enthusiastic about the prospect of Beadle's appointment and the expected rebuilding of genetics. Of all of them, "I stand to gain the most from you if we can get you here."[10] Emerson also knew well that the real excitement in genetics was coming from Beadle's group and elsewhere, not from Cal Tech.[11] However, neither he nor Sturtevant gave any indication about critical issues such as how much space would be freed up for Beadle's large research group or how the required lab renovations would be funded.

Beadle's visit to Pasadena did little to encourage his interest. Afterward, both Emerson and Sturtevant regretted that they had not pressured him more to "do what we would like."[12] Instead of talking about the positive things, they had stressed the problems. Beadle's realistic financial requests, for example, were considerably more than any of the biology faculty had for themselves and more than Millikan had agreed to provide. Emerson was optimistic that this issue might be resolved by the expected appointment of a replacement for Millikan, who was now 77 years old. The trustees had already limited his activities. At the end of June, Sturtevant was able to send, at the trustees' request and with Millikan's approval, an improved offer for $8,000 salary, $10,000 for setup, and $8,000 toward underwriting his annual operating expenses in case no outside support materialized.[13] Beadle's teaching duties for undergraduate and graduate biology were to be organized by mutual agreement. But Sturtevant avoided a critical question: How much of a role would Beadle have in shaping a vision for the rebuilding of the division?

Beadle must have concluded that Cal Tech would not provide the opportunity he sought and decided to remain in Palo Alto.[14] Stanford had come up with more research support and he was hopeful that the Rockefeller Foundation's help would be sustained and even increase.[15] All Sterling Emerson said when he learned of the decision was "It would have been nice if you could have joined us."[16] Sturtevant, too, wrote a weak response: "I'm even prepared to grant that you may have made the right decision. I'm disappointed, but not too greatly surprised."[17] As far as they were concerned, the matter was finished. They had just lost out on a great scientist, one with whom they shared 15 years of common history, strong intellectual ties, even affectionate friendships. Yet all they did was regret his decision. Luckily for them and the Institute, there was another, more aggressive, more visionary person at Cal Tech who was not ready to give up so easily.

At the age of 44, Linus Pauling had long been the world's leading chemist. Cal Tech had been his academic home since he arrived as a graduate student. As director of the chemistry division, he was already steering it in the direction of his own growing interest in biology. He wanted a strong and complementary biology division and he wanted Beadle to lead it. He also knew that a Pauling–Beadle combination would improve Cal Tech's chances for continuing, major support from the Rockefeller Foundation.[18] The chemist had become a favorite of Warren Weaver's soon after Weaver joined the Foundation in 1933. Weaver, who always looked at the intersections of standard disciplines for exciting science, supported Pauling's first Rockefeller grant of $10,000 because the proposed research brought chemistry, physics, and mathematics together. Even that early, Weaver had begun to tell the Rockefeller trustees that the "time is ripe for application to basic problems in the biological and medical sciences of techniques and procedures developed in physical sciences."[19] Beside Pauling, Morgan had been one of the first to reap the benefit of Weaver's views and Beadle, too, had been well supported by Rockefeller. Pauling probably knew that the Stanford biochemists were not congenially disposed to Beadle's interest in collaboration in spite of the financial help that would have followed, and he likely played a part in Sturtevant's letter to Beadle in May. Perhaps he anticipated Weaver's belief that the connection between chemistry and biology at Cal Tech would work because of just two words: Beadle and Pauling.[20]

Shortly after the end of the war in the Pacific on August 7, Pauling visited with Weaver to propose that the Rockefeller Foundation make a huge grant to Cal Tech for work in biochemistry and protein structure—what he called chemical biology. His ambition encompassed $2 million toward two new buildings and $15 million for operating funds for 15 years. He surmised that if Cal Tech was to benefit from Weaver's great enthusiasm for what the Foundation official was already calling "molecular biology," he would have to team up with a biologist who shared that vision and had the stature and skill to help pull it off. Pauling was convinced that Beadle was the best possible candidate. As a member of the newly constituted executive committee of trustees and five faculty members who ran Cal Tech after Millikan relinquished his leadership in September, the chemist was in a good position to push for the appointment.[21]

Pauling wasted no time and quickly convinced Sturtevant to write to Beadle again, this time without the knowledge of the rest of the biology faculty, and offer "to make you chairman of the Division." Sturtevant added that he himself is "*completely* [sic] in favor of that" but he was less enthusiastic about Pauling's other ideas. "Pauling thinks that, if you accept, he can get a very large sum from the Rockefeller...such a sum would be very nice to have, but it doesn't seem to me that it is really necessary." His ambivalence showed in his description of the problems in the department: "don't feel that I've gotten myself in a jam and want you to get me out of it. The department *is* [sic] in sort of a bad fire, but I don't want you to feel that I'm hollering for help—we can manage the situation somehow."[22] Sturtevant didn't realize that Beadle probably shared Pauling's ambitious vision. "Somehow" was not the way either Pauling or Beadle usually managed situations. The younger men also realized that science would be a different endeavor now that the war was over, whereas Sturtevant, although only 54, seemed to look backward rather than forward to an expansive future. Besides his outlook, he also retained the kindly manners of an earlier time, ending the letter with a warning: "And don't let Pauling talk you into anything you wouldn't want to do."

Sturtevant was right about Pauling being aggressive. The chemist rushed to Stanford to talk with Beadle about what they could accomplish together.[23] Beadle knew that unlike Stanford and Loring, Cal Tech and Pauling were eager to take advantage of Weaver's willingness to support an interdisciplinary effort that drew biochemistry and genetics together. C.V. Taylor at Stanford quickly recognized that the game was lost[24]; there was nothing he could do to match the Cal Tech offer or the promise of a collaboration with Pauling. Beadle told James R. Page, chairman of the Cal Tech board of trustees, that he would move to Pasadena by July 1, 1946.[25] He reported his decision to the Rockefeller Foundation along with a request for additional funding to tide him over for the first six months of 1946.[26] Weaver was relieved and delighted. He and Hanson had not liked the idea of Beadle going to Wistar.[27] The Foundation quickly awarded Beadle the requested $5,000.[28]

The members of the Cal Tech biology division had mixed reactions to Beadle's appointment as chairman. Sterling Emerson and James Bonner were pleased and Bonner decided against going to Chicago.[29] Others, like Borsook, were peeved about having been kept in the dark once Pauling had gotten to work.[30] Went was concerned about his own activities, and with reason. Almost immediately after Beadle accepted, Millikan urged him to look into whether Went's work warranted an investment of $300,000 for new green houses and $6,000 a year for their maintenance.[31]

Neither Page nor Pauling was about to wait seven months to engage Beadle in Cal Tech affairs. They expected him to start work immediately on recruitments, appointments, and the proposal to the Rockefeller Foundation.[32] Sturtevant, too, assumed that from then on Beadle would be making the decisions.[33] In addition, Pauling expected his new colleague would promptly share in his active engagement with the national issues engendered by a rapidly changing scientific scene.

Wartime experience had established the effectiveness of federal support of scientific research in the nation's universities. In 1944, President Franklin Roosevelt asked Vannevar Bush, who was managing a huge scientific research program in support of the war effort, to prepare a peacetime plan. After Roosevelt's death, Bush proposed to President Harry Truman establishment of what would eventually be called the National Science Foundation.[34] Two opposing bills were promptly introduced in the Senate and shaped a debate that postponed establishment of the foundation until 1950. The Magnuson bill closely followed Bush's report in giving a board of private citizens, mainly scientists, authority over the proposed foundation. In contrast, the Kilgore-Johnson-Pepper bill gave more authority to the president and insisted that the sponsored research support social purposes, thereby raising the specter of political influence on the research programs. Pauling wanted Beadle to organize support for the Magnuson bill in the San Francisco Bay region as he was doing in southern California.[35] More generally, scientists were excited about the idea of the foundation and the promise of continued, even expanded, federal support for research. McClintock, in a letter telling Ephrussi about Beadle's move, wrote "The upswing in science, though, is sensational. If it lasts, science in the United States will receive much support... Big ideas are hatching and they certainly look good on paper. The genetics groups are feeling it too."[36] But she was quick to empathize with Ephrussi, his family, and colleagues who were grappling with much more basic issues than science funding in that first winter of peace in Paris.

Besides all the Cal Tech affairs, Beadle had to plan for moving his family, laboratory, and research group to Pasadena. Happily, Marion "developed real enthusiasm for getting back to Pasadena" where they would be closer to David, who was in school in southern California.[37]

Pauling learned from Weaver that their proposal ought to reach Rockefeller by the first of the year. Beadle immediately began to draft the biology section and Pauling the chemistry portion of a proposal which was meant to be an imaginative and comprehensive plan for a decade or more of work.[38] Weaver "managed to keep himself from dropping the telephone" when Pauling told him they were thinking of asking for about $6 million for a 15-year period.[39] Beadle himself thought the amount was "staggering" but that he could "get used to the idea." He cautioned Sturtevant not to advertise the sum, to avoid disappointment in case it didn't succeed, and asked for Sturtevant's views regarding the proposal.[40] Sturtevant himself had previously offered only "vague" and "broad" ideas about microbiology including phage, antibiotics, yeast, and bacterial genetics.[41]

Beadle's initial draft proposal began by saying, "Developments in the last decade or so in nuclear physics, chemistry, and various phases of biology have put the biological sciences in a most favorable position for future advances." He pointed out the special opportunities inherent in the development of isotope tracer techniques and chemical genetics, especially if the two are combined. "Unfortunately," he continued, "there were two reasons why few institutions were able to take advantage of these opportunities: (1) Conservatism and tradition that so often retard science in general have kept biology and the physical sciences separated in most institutions and (2) the methods and techniques necessary for maximum progress involve expenditures of funds that are out of the question for most colleges and universities." He then explained why Cal Tech was well situated to make such a synthesis and emphasized the need to strengthen enzymology and microbiology, including virus research, and to obtain the needed facilities and instrumentation. On one additional page he listed the division's current programs and personnel and those he would want to recruit. The estimated annual budget was $265,000, of which $170,000 was for the ongoing programs, his own included.[42]

Pauling was not satisfied with Beadle's plan. "The Rockefeller Foundation would be much more interested in a program involving new departments than in the heavy subsidy of all the members of the present staff," he commented.[43] He especially emphasized, from Beadle's list, nucleic acids and nucleoproteins. By early December Pauling had revised Beadle's concise proposal, incorporated it into a hefty, detailed "draft" plan, and sent it off to Weaver for his reaction.[44] Meanwhile Beadle had visited with Weaver in New York while in the east to attend a meeting of the National Research Council's Committee on Growth in Washington. Weaver had not yet seen the draft proposal, but Beadle came away convinced that "It is clear that he [Weaver] is already completely and absolutely agreed with our claim that the Institute is the best place in the world for the kind of work we propose." But Beadle was not one to put all his eggs in one basket. He told Pauling what he had learned at the committee meeting. If Rockefeller didn't come up with enough money, they stood a good chance of getting more from the American Cancer Society. Also, the National Foundation for Infantile Paralysis (The March of Dimes) was another possible source of funds for fundamental research.[45] All this was good news. On the discouraging side was Weaver's indication that Rockefeller would not be able to come up with $6 million over 15 years. Not only was the sum very large but, with the end of the war, the Foundation was reconsidering its own programs.

Early in December 1945, the era that had made possible Beadle and Pauling's vision of a fusion of chemistry and biology came to a symbolic close with the death of Morgan. Later that month, Weaver fell seriously ill, dashing their hopes for prompt consideration by the Rockefeller Foundation. They continued working on the draft proposal and by early March it was in good shape.[46] Besides all this, that year Beadle was president of the Genetics Society of America. He "had never been so busy" in his life,[47] too busy to attend the annual Cold Spring Harbor Symposium that summer. David Bonner, then only 30 years old, had to defend the *Neurospora* group and the one gene–one enzyme hypothesis against Max Delbrück's vocal criticism.

In the spring of 1946 the Cal Tech trustees appointed Lee A. DuBridge, a physicist at the University of Rochester, as Cal Tech's president. He was just about Beadle's age and like Beadle was expected to bring fresh ideas to the Institute. DuBridge acknowledged Beadle's congratulatory letter and reassured the geneticist of his support: "...the knowledge that you were going to be at Cal Tech was an appreciable factor in my decision to go. I felt that with your being there, there would be at least one department on the campus that would not be giving a severe headache in the years to come."[48] Probably more important than Beadle to DuBridge's acceptance of the Cal Tech trustees' offer was the urging of Max Mason, who had taught mathematics to both Weaver and DuBridge at the University of Wisconsin.[49] Mason had since served a brief two years as president of the University of Chicago before moving to the Rockefeller Foundation as its chief executive.[50] When Mason left Rockefeller in 1936, he went to Cal Tech to work on the construction of the Palomar telescope and was still on the faculty when Beadle and DuBridge arrived. Under DuBridge's leadership, Cal Tech grew and became even more influential. The "desperate financial condition" of the university in 1946 was repaired and faculty compensation was significantly improved.[51]

The Sturtevants made their customary trip east in the summer of 1946, and in July the Beadles moved into their Pasadena house. In the fall, they settled into a small house three miles from the campus on Adelyn Drive in San Gabriel.[52] They got the house through a complex three-way swap involving houses in Palo Alto. Suitable houses in Pasadena cost $17,000 to $18,000, a price as daunting to an academic then as the 40-fold higher prices are now. Marion was not happy with the small place and although they planned to look for something better, they remained there for four years. She and David escaped the oppressive Pasadena heat during most of that first summer on the beach at Corona del Mar.

Before moving down to Pasadena, Beadle worried a good deal about how to integrate the existing biology faculty into his plans. E.G. Anderson, still an associate professor, was one challenge. Recently married, he, his wife, and young daughter lived rent-free at the Farm as he had done since 1929. The university maintained the property exclusively for his use. His monthly genetics seminars there attracted students from the main campus, but other than that and an occasional course he participated very little in departmental affairs.[53]Although Beadle and Anderson had known one another for more than 15 years, and were both scientific offspring of Cornell and Emerson, they had never gotten along well. As usual, Beadle tried to put the best face on things, but Anderson was not very happy with Beadle's return to Cal Tech. His discontent was exacerbated when Beadle tried to get him to yield some of the land at the Farm for construction of quarters for the young members of his laboratory group. Fortunately, this turned out to be unnecessary because everyone found satisfactory housing in town. The Farm itself seemed in good shape and Beadle was happy to facilitate Andy's expressed interest in being more involved with the department in town. Still, the Farm competed for departmental resources and its research did not easily fit with the new plans. Morgan's old hideaway, the marine laboratory at Corona del Mar, presented similar problems, especially since it was underutilized. Albert Tyler, who had followed the "Boss" into embryology, was the primary user of the marine laboratory. He was still an associate professor in 1946 and, stimulated by Pauling's ideas about antibody–antigen interactions, was studying the biochemistry of mutual recognition of eggs and sperm.

The two Dutch neurophysiologists, Cornelis Wiersma and Anthonie von Harriveld, were also still in the division as were the plant physiologists, Arie Haagen-Smit and Frits W. Went. Went would be "an albatross around Beadle's neck" for years.[54] The same age as Beadle, he had grown up in the privileged academic atmosphere of a mansion in the Utrecht Botanical Garden where his father was director and professor of botany. His discovery in 1926 of the plant growth hormone he called auxin established his reputation and secured a post at the Royal (Dutch) Botanical Garden in Java, the Dutch East Indies. Went was, according to some, a visionary with many inventive ideas, but he was unsympathetic to the new approaches to biology and genetics that Beadle represented. He had trouble, for example, accepting Kenneth V. Thimann's identification of auxin as 3-indoleacetic acid. This discovery was made shortly before Went came to Pasadena where Thimann was then a young instructor. [55]

Years before Beadle returned to Pasadena, Went had become interested in applied research and the effects of environmental factors on plant growth. Some even thought he was a Lysenkoist.[56] His strong ties with the California agricultural industry provided major funds for his facilities and research.

Beadle agreed to help Went raise money and cooperated in his arrangements with for-profit organizations. But Went was not easily satisfied. Finally, in 1958, after Beadle denied his request for expansion of the plant growth chamber facility, Went left to head the Missouri Botanical Garden in St. Louis. Like James Bonner, Sterling Emerson changed his mind about leaving when Beadle decided to come to Cal Tech.[57] Beadle was enthusiastic about Emerson's work on a mutant strain of *Neurospora* that required sulfanilamide for growth rather than being inhibited by it, as the normal mold was. Besides, Emerson could be counted on to support Beadle's plans and begin reorganizing the laboratories and equipment well before the move from Stanford occurred.

All these people were accounted for in Beadle's draft of the proposal to Rockefeller, but the success of the grand plan really depended on new recruits. Beadle was interested in reinforcing genetics and in adding research programs in virology, microbiology, enzymology, nucleic acids, and nucleoproteins as well as assuring expertise in the use of isotope tracers. Immediately after his decision to come to Pasadena, Sturtevant suggested appointing his former student Ed Lewis, then just completing service in the Army, as an instructor.[58] Besides being a "good guy," Lewis could tend to the *Drosophila* stock collection. Beadle agreed. On one of his frequent trips to Pasadena in the spring of 1946, Lewis told him that the collection required a full-time curator. It was the largest in the world and a very precious resource. Back at Stanford, Beadle identified a gifted undergraduate student, Pam Harrah, and asked her if she would like a summer job at Cal Tech after graduation. "I could never in my life have said no to Beadle about anything. If he'd said 'hey, we're going to join this cocaine-shooting club,' I'd have gone," she recalled. A few days after she accepted, Beadle called her back into his office and said "Hey, Pam, how tall are you?" She answered, "5'3." Beadle said, "Your new boss is 5'4" and he's 28. Maybe you'll like him so much you'll fall in love and decide to stay there at Cal Tech."[59] And so she did; they were married in September 1946. Lewis became a glorious legacy of the Beadle era at Cal Tech. His work formed a critical bridge between classical *Drosophila* genetics and the molecular approach to early development, the goal that Morgan had set so many decades before. In 1995, after the full significance of his work was realized, Lewis received the Nobel Prize in physiology and medicine.

Two other brilliant young scientists were appointed to the faculty in 1946: Ray Owen and Norman H. Horowitz. Owen was already a research fellow at Cal Tech, recruited by Pauling because of his interest in the genetics of the immune system. Horowitz, who at age 31 had proved his talent in Beadle's Stanford lab, moved to Cal Tech as part of the *Neurospora* group in 1946.

Other, more junior people mainly from Stanford, filled out the *Neurospora* group in the Kerckhoff building, including Herschel K. Mitchell and Mary Houlahan, who would soon be married, and graduate students Adrian Srb and August Doermann. Mitchell would in time become a professor in the division. Srb had just completed his Ph.D. at Stanford and had an appointment there that required some shuttling between Pasadena and Palo Alto. He left California in 1947 to join the Cornell faculty, where he had a long and very productive career. By then, Beadle and Srb, two Nebraskans and both students of Keim, had formed a close friendship that would be lifelong. The first of many cross-country mutual visits came in summer of 1949 when Srb was in Pasadena to work on a book with Owen.[60]

The young people needed little convincing to move to Pasadena. For scientists at the start of their research careers, the Beadle–Pauling vision was a dream come true. Enticing senior people away from good positions elsewhere was, however, a bigger challenge. C.B. Van Niel, the world leader in the study of microbial photosynthesis working at Stanford's Hopkins Marine Station in Monterey, was high on Beadle's list for the proposed microbiology program, but he chose to remain at Stanford.[61]

Max Delbrück was Beadle's first choice for a virologist.[62] Starting with his term as a Rockefeller Foundation fellow at Cal Tech in 1937–1939, Delbrück's work with bacterial viruses, called bacteriophage (usually abbreviated phage), had opened a door to understanding the replication of genetic information. Beadle's own work was on how genetic information was actually used. Beadle knew that Delbrück, who was very different in personality and scientific outlook from himself, would help sustain the vitality of the division and inspire students with his highly critical intelligence.

Delbrück was born in 1906 into a prominent Berlin family.[63] In 1930, he received a Ph.D. in theoretical physics, and for the next two years he wandered among the theoretical physics laboratories in Europe where he came in contact with most of the important physicists of the time.[64] Everywhere, everyone was impressed with his intelligence and erudition and most especially with his gregarious personality and sense of humor and fun. Of all the great physicists Delbrück met, it was Niels Bohr who inspired his interest in biology and turned him into one of the trailblazers for the migration of physicists into biology.[65] Bohr's thinking about biology derived directly from contemporary physics and Heisenberg's uncertainty principle.[66] In applying this idea to biology, he argued (according to Delbrück) that "you could look at a living organism either as a living organism or as a jumble of molecules; you could do either, you could make observations that tell you where the molecules are, or you could make observations that tell you how the animal behaves, but there might well exist a mutually exclusive feature, analogous to the one found in atomic physics..."[67]

Delbrück became Lise Meitner's assistant at the Kaiser Wilhelm Institute for Chemistry in Berlin in 1932.[68] It was an exciting place because Meitner and Otto Hahn were characterizing the products of the bombardment of uranium with neutrons. However, neither Meitner, Hahn, nor Delbrück realized in those years that they were observing nuclear fission. Meitner would figure that out only in 1938, after she had been forced to flee from Nazi Germany. Later Delbrück would say "...I should have guessed what was really going on, namely fission, but I, like everybody else, lacked imagination to see that."[69] "Thus at age 30, Max Delbrück had all the earmarks of the mediocre son of a famous father."[70] However, he was spending a lot of his time in N.W. Timoféeff-Ressovsky's genetics lab in the neighboring Kaiser Wilhelm Institute for Biology where he joined the effort to learn about the nature of *Drosophila* genes from the mutations produced by X rays. How, he wanted to know, could physics explain the fact that though genes were very stable through many generations, they could nevertheless change by spontaneous or X-ray-induced mutations that were themselves as stable as the original form? His theoretical contribution to a paper with Timoféeff-Ressovsky and K.G. Zimmer went straight to the atoms and the effect that the energy of the X rays had on them.[71] He implicitly converted the abstract gene of the geneticists into a physical entity composed of atoms.

Delbrück's fertile ideas underlay his uncanny skill at inspiring others to pursue his proposals. In Berlin he held weekly discussions at his mother's home because by this time the Nazi regime barred many people from attending official seminars. Of particular importance to Delbrück was his assumption that by studying biology, physicists would make profound new discoveries about physics, an idea that motivated many who came under his influence. It was a wrongheaded idea, and no new physics has come from biology. Plenty of new biology, however, emerged from the work of Delbrück and others trained in physics, because their different way of thinking about living systems was productive. In fact, it was chemistry, not physics, that proved essential to the advance of genetics, although Delbrück had "reservations about the powers of biochemistry."[72]

American geneticists first became aware of Delbrück through the paper with Timoféeff-Ressovsky and Zimmer, and the Rockefeller Foundation suggested that Delbrück apply for a fellowship to work in biology.[73] The idea of learning more about *Drosophila* genetics appealed to him, and Cal Tech was clearly the place to go.[74] Besides, it seemed wise to leave Germany because "probably as a result of too much frankness" he had performed unsatisfactorily at the Nazi indoctrination sessions required of candidates for university positions.[75]

Delbrück's enthusiasm for *Drosophila* genetics was short-lived. Even with the tutelage of Sturtevant and Bridges he could not cope with the complexity and dense terminology. Perhaps he would have been attracted to Beadle and Ephrussi if they had still been in Pasadena when he arrived in 1937. He finally found what he wanted in Emory L. Ellis's laboratory. Ellis was studying the infection of bacteria by bacteriophage, which had been discovered two decades earlier. It was already clear that phage were particulate and that, after infection, the bacterial cell was destroyed and many new phage particles released. Ellis took up the phage work hoping that it would help in understanding the role of viruses in cancer. He had isolated from filtered sewage a phage that infected the bacterium *E. coli*. Shortly before Delbrück paid a visit to his lab, he had confirmed an observation made by the codiscoverer of phage, Félix d'Herelle, namely, that new phage are produced in spurts during the course of infection. That is, new phage appear only some minutes after the bacteria and phage are mixed; then, if uninfected bacteria remain in the test tube, they are infected by the new phage and, after a similar amount of time, another burst of new phage occurs. Thus, the release of phage occurs in stepwise fashion. Delbrück was skeptical and said, "I don't believe it."[76] Eventually he would be infamous for loudly declaring the same thing to visiting seminar speakers.[77] But he was intrigued enough to join in Ellis's work. The physicist's mathematical skills in dealing with particles and populations of particles put the work on a quantitative basis and permitted important conclusions. They made accurate measurements of the time lag between adsorption of the phage onto the bacteria and the disruption of the bacterial cells, as well as the average number of phage liberated by a population of infected cells and the number produced by single bacterial cells. These methods, in turn, allowed them to study the effect of various conditions on phage production.[78]

By the time this work was published, Ellis had returned to the study of cancer, and Delbrück, who was hooked on phage, was unable to return to Europe because of the war. His Cal Tech and Rockefeller Foundation friends recognized the tremendous significance of his work and arranged for a position at Vanderbilt University. Not long after, he teamed up with Salvador Luria at Columbia and, together with Alfred Hershey then at Washington University, formed the nucleus of what would become the influential "phage group." This potent threesome firmly established phage as the ideal tool for the advancement of genetics and proselytized others to join them. Their most effective recruitment tool was the summer course in phage biology that Delbrück initiated at Cold Spring Harbor in the summer of 1945.[79] One of their most important findings was reported by Delbrück and Hershey at the 1946 annual Cold Spring Harbor Symposium that Beadle missed; phage could undergo mutations and perhaps even genetic recombination and must therefore have their own genes.

Beadle knew that Delbrück was a vocal skeptic about the one gene–one enzyme theory and unenthusiastic about the *Neurospora* work. Later, Delbrück said, "You could learn an enormous amount about actual biosynthetic chains and their interrelations, you did not learn at all how the enzymes came about...the question remained, how does the gene make the enzyme, and how does the gene make the gene, and this was in fact not answered at all by any of the biochemical approaches."[80] Although Delbrück would give him a hard time, he was unquestionably brilliant and would be a terrific if contentious stimulus to everyone in the division. Pauling, who knew Delbrück from his earlier years at Cal Tech when they had even published a paper together, supported the invitation; both he and Beadle were strong enough to live with a challenging colleague. Delbrück had fine memories of Cal Tech and loved the nearby desert. His wife's parents lived close to Pasadena, and besides, Tennessee and other potential locations were uninspiring for a sophisticated European.[81,82] When the Cal Tech offer arrived, Delbrück promptly visited with Pauling and Beadle. He drove a hard bargain regarding salary and research support. A generous offer for a full professorship was made in December of 1946, which relied on a five-year $300,000 grant from the National Infantile Paralysis Foundation.[83,84]

A month later, Delbrück moved to Pasadena. The phage group was beginning to take seriously Avery's claim that nucleic acid, not protein, carried genetic information. Nevertheless, Delbrück remained unconvinced that biochemistry had much to offer genetics, while Beadle, who was never part of the phage group, continued to emphasize the importance of building connections between the two fields. As late as 1949, Delbrück was still skeptical about the importance of chemistry to understanding genetics.[85] Only a few years later Watson and Crick proposed the double-helical structure of DNA and Arthur Kornberg demonstrated that genes could indeed be replicated in a cell-free system. By then, Delbrück's work on phage and genetics was over; he was moving on to something new. For the next 20 years or so he tried unsuccessfully to develop a model system for studying perception in fungi.[86]

Beadle had, nevertheless, correctly esteemed Delbrück's intellect and leadership qualities, cloaked though they were in what has been described as a "ruthless and demanding exterior."[87] Anecdotes about his "autocratic rule and sharp tongue" abound.[88] In spite of this, everyone seems to agree that he could be charming. The overnight camping trips he organized and led in the

California desert were stamped with his idea of a good time and they built a strong group spirit, although some recall that attendance was not voluntary. Those outside the circle sometimes described the group as a cult and the phage workers as disciples. Apparently, Delbrück exploited Germanic romanticism to build an intense group ethos of intellectualism and loyalty. Luria at least would have none of it. He "would not come to Cal Tech unless guaranteed immunity from camping."[89] But Delbrück's presence in the biology division helped assure excellence and vision.

The reorganized biology division faculty had three groups: genetics (including phage, *Drosophila*, and *Neurospora*), plants, and neurobiology. By 1947, four of the longtime associate professors had been promoted to full professors and Horowitz and Owen became associate professors. DuBridge helped the process along with significant raises in faculty salaries. All of this activity depended on Beadle's energetic and successful fund-raising both in conjunction with Pauling and on his own, and always with DuBridge's full backing.[90] The enormous Rockefeller grant never materialized. With Weaver's advice, Pauling and Beadle substituted the original request with a more modest one for $60,000 per year for five years, to be equally divided between the chemistry and biology divisions.[91] But even this request was set aside.[92] The proposal was once again downgraded, this time to $55,000 for one year. They were finally given $50,000 to divide.[93] The Foundation spent several years deliberating on how to reshape its programs in the postwar world. Finally in 1948 it provided $100,000 per year for seven years to the Cal Tech program, a magnificent sum but considerably less than the $6 million over fifteen years that Pauling and Beadle had originally requested.[94] The continuing struggle for funds led DuBridge to establish an Industrial Associates program; by the end of 1952 it generated more than $250,000 a year for the university.

Beadle gave all his energy and commitment to encouraging and supporting what quickly became once again a world center for biology. When DuBridge arrived in the fall of 1946, Beadle made sure he met the biology faculty.[95] When Barbara McClintock wrote that she was on to something interesting—transpositions—he promptly arranged for her to come to Cal Tech as a visiting professor.[96,97] Beadle's challenge was to build not only the biology division at Cal Tech, but a new field, variously called chemical biology, biochemical genetics, or, as Weaver preferred, molecular biology. Excellent research was essential to accomplish the goal, but so was the training of young people who would then take the new science to other institutions. Except for the program at the Cavendish Laboratory in Cambridge, England, Beadle's vision was unique. Within a year of his arrival, the number of research associates and fellows in the division doubled.[98] Beadle was also determined to integrate biology more fully into the Cal Tech scene, something Morgan had not succeeded in doing. The number of graduate students in biology had to be expanded and undergraduate interest stimulated. Beadle and Horowitz offered a course in advanced genetics for undergraduates and Delbrück offered biophysics.[99] Starting in the 1948/49 academic year, Cal Tech offered undergraduates the opportunity to concentrate their studies in biology in addition to the long-established concentrations in various physical sciences and engineering. That year, Beadle himself began teaching undergraduate elementary biology, frequently preparing elaborate illustrative demonstrations for the lectures. He continued his teaching until he left Cal Tech in 1961 and often taught the labs, leaving the lecturing to Horowitz and Owen. Through it all, Beadle maintained his characteristic "hands-on" style. "I've not seen anyone else who did the job like he did. He walked around to see what needed doing. I think that was the farm experience coming out... He asked people what they needed and he'd try and get it for them. And if there was a spill of water or something, he would be the first one in there with a mop and a mop-bucket."[100]

With all this activity, Beadle had little time left for his own research or for supervising graduate students. "In my own situation, I tried...what I thought of as an experiment in combining research in biochemical genetics with a substantial commitment to academic administration. I soon found that, unlike a number of my more versatile colleagues, I could not do justice to both."[101] When David Hogness wanted to start his doctoral thesis research on *Neurospora* in 1950, he decided that it was best to work with Mitchell, who was in the lab, rather than Beadle, who was rarely around.[102] Beadle considered the division his primary responsibility. Perhaps he would have persevered in the laboratory if he had seen a way to approach the critical question of the nature of the gene. Perhaps he was stymied, because he continued to emphasize the idea that proteins were important components of genes. "There is evidence that genes themselves contain proteins combined with nucleic acids to form giant nucleoprotein molecules...And it has been suggested that genes direct the building of non-genic proteins in essentially the same way in which they form copies of themselves."[103] As late as 1951 he wrote, "for genes are made up of the most complex compounds known to chemistry—proteins and nucleic acids."[104] Decades later he returned to experiments, but the last research paper he would publish for a while described work done at Stanford.[105]

Meanwhile, the move to Pasadena did not help the situation at home. Beadle was not only totally engaged by his new responsibilities, but he traveled even more than before to give talks, serve on important committees, and pursue funds. Marion no longer had students and postdocs or their wives to befriend, and, although she attended divisional social events and saw Sturtevant's and Emerson's wives, she was unhappy and troublesome.[106] David's difficulties at school continued, and he remembers that she was never satisfied with him and believes that it was she who put a distance between himself and his father. He warmly recalls two memorable summers during his high school years when Marion went off to Mexico for obscure reasons and father and son were together, on their own.

More often, it was Beadle who was away. Even in 1946, the year he moved back to Cal Tech, he traveled to colleges and universities all over the country as a Sigma Xi lecturer and attended three meetings of the National Research Council Medical Fellowship Board in New York between August and the middle of November. Although his own research had ceased, he was still a major figure whose views on developments in genetics were of great interest. For years to come he would be honored for what he had accomplished. Carl Lindgren described Beadle's position in a letter congratulating him on receiving the Lasker Award in 1950. "You are really the only person who represents the entire field of biochemical genetics and who has carried the message...to those far corners where it will do most good...your efforts have brought prosperity to all of us."[107]

CHAPTER 13

Postwar Science and Politics

Profound changes reshaped the nation once the war was over. Science and technology had played a crucial role in the victory. That experience encouraged expanded government research support in peacetime in the expectation that it would promote the national interest. Scientists responded enthusiastically. They had been accumulating new questions that begged for investigation during the long years when defeating the enemies was paramount. The rejuvenation of the Cal Tech biology division under Beadle's leadership was buoyed by national prospects such as the establishment of a National Science Foundation and the increasing significance of the National Institutes of Health.[1]

The scientific community that emerged after the end of World War II was different from the one that Beadle and his colleagues knew before 1940. Morgan was gone, and R.A. Emerson died in December of 1947. Leadership had passed to Beadle and his generation. The vestiges of that earlier, inner-directed community show in the letters Sturtevant and Sterling Emerson wrote to Beadle in the spring and summer of 1945. Allied victories over Nazi Germany in May and Japan in August went unmentioned. No one commented on the atomic bomb attacks on Hiroshima and Nagasaki although they marked extraordinary scientific achievement and presaged deeply troubling changes in the world. Yet before too long, geneticists would find that these events had profoundly affected their lives and institutions. Science became a public issue. Beadle glimpsed the future soon after he accepted Cal Tech's offer when Pauling asked him to support the Magnuson bill's version of the envisioned National Science Foundation.

Matters of direct interest to scientists soon demanded attention. They would eventually occupy much of Beadle's thought and time. One was a consequence of the atomic bomb explosions. Even before the atomic attacks on Japan, physicists were troubled over their roles in bomb development and the implications of the new weapons for world peace. But they could not respond knowledgeably to questions about the effects of radiation on human and other life. Physicians were needed to evaluate the immediate health hazards of radiation exposure and geneticists to estimate the more subtle, mutagenic hazards. Twenty years earlier, Muller and Stadler had proved that X rays caused mutations, and Beadle himself had used X rays to induce mutations in *Neurospora.* It was inevitable that geneticists would be called upon for advice as the nation struggled to understand the significance of the new weapons and their testing.

On July 1, 1946, the day that Beadle took up his position at Cal Tech, the first of the Operation Crossroads nuclear bomb tests took place on Bikini Atoll in the South Pacific. Three weeks later, a second explosion was detonated at Bikini. Cal Tech was already involved. E.G. Anderson had access to corn seeds irradiated during the explosions and initiated a detailed study comparing their mutations and chromosome aberrations to those of X-rayed material.[2] His extensive data indicated that the Bikini exposure produced effects equivalent to 15,000 roentgens.[3]

In November of 1946, President Harry Truman established a committee on atomic casualties at the National Research Council with a charge to carry out a long-term study of the effects of the atomic bombs on the Japanese. Beadle's nomination to this committee kicked off the first of several investigations of his loyalty by the Federal Bureau of Investigations (FBI). Investigators visited colleagues, neighbors, and teachers in Nebraska, Ithaca, Cambridge, Pasadena, and Palo Alto. Everyone spoke highly of Beadle's integrity, character, and loyalty. No one had a clear idea of his political views. Pauling described him as "mildly liberal" but apolitical and "unwilling to take time from his work to engage in political affairs."[4] Others suggested that perhaps he was mildly conservative. Plainly, Beadle did not often discuss his political views with colleagues and acquaintances. Beadle's application lists the organizations to which he belonged. All but one of them were scholarly scientific organizations, such as the National Academy of Sciences and the American Philosophical Society. The exception was his membership, in 1945–1946, in the Independent Citizens Committee of the Arts, Sciences, and Professions (ICCASP), a certain amount of whose activities had been criticized "as being 'pro-Communist'" according to one of those interviewed by the FBI at Stanford. This person added that he was certain that if this were the case, Beadle would doubtless sever his connections with the ICCASP, if he had not done so already, "in view of the fact that the applicant had always expressed anti-Communist sentiments."[5] A decade later Beadle reported to the FBI that he had indeed resigned after a year, but not because he was aware of any Communist sympathies. He disapproved of the fact that "the policies and programs of the ICCASP were being formulated and accepted without prior notification or approval of the members."[6] In July of 1947, Beadle was given a Q clearance, which authorized his access to all types of information related to nuclear weapons and energy, and he was promptly appointed a consultant to the Atomic Energy Commission (AEC). Apparently the FBI did not consider a few other activities, including attendance at a reception honoring three visiting Soviet scientists, serious enough to warrant denial of clearance.

There were some scientific advantages to being an AEC consultant. In 1953, Beadle was able to attend an atomic bomb test in Nevada and expose *Drosophila* to radiation for Ed Lewis, who wanted to measure the relation between the dose of neutron exposure and the frequency of chromosome translocations.[7] Lewis's application for Q clearance, which was required if he was to take the flies to the test site himself, was denied probably because he had signed a petition denouncing the loyalty oaths required of the University of California faculty. Although Beadle had been assured that the aluminum tubes (which blocked radiation by gamma rays) containing the flies were not contaminated, Lewis showed him otherwise when he returned to Pasadena. The neutron bombardment had caused formation of a radioactive aluminum isotope and Beadle received 10 roentgens of radiation exposure.

Substantial and growing federal grant and contract programs also required extensive interactions with the government in Washington. Wartime experience had demonstrated that supporting research in nonfederal institutions was an efficient and productive way to advance science and technology. If scientists were to influence the way the funds were appropriated and spent, they had to cooperate in the political process. Scientists' growing public activities coincided with the emergence of the Cold War and the resulting national concern, even paranoia, with security and loyalty. The ensuing national turmoil engulfed the scientific community. Geneticists studying the effects of radiation were subject to investigation and the threat of public charges of disloyalty, even if their work was unrelated to important national secrets. Previously, as Pauling had noted, Beadle had limited his political activities to typical scientific and academic affairs. Now he had to engage in matters in which national policies conflicted with the integrity of science and the interests of universities.

Beadle accepted the necessity of keeping the nation's nuclear secrets, but he was troubled by the government's procedures for routing out security risks and espionage agents. Many individuals of prominence in the arts and sciences were under suspicion, and the cases against them were often built on rumor and innuendo. Guilt by association was almost taken for granted in certain quarters, including influential parts of the U.S. Congress. Anyone with liberal or left-wing leanings was so suspect that the taint projected onto their colleagues, friends, and institutions. Most disturbing to Beadle were instances when close colleagues were embroiled in public accusations of disloyalty.

All of this offended Beadle's strong and rigorous sense of justice. His positions were reminiscent of his father's clear-minded approach to questions of right and wrong. His adopted son, Redmond Barnett (see later in this chapter), remembers discussing with "Pops" his response to Harvard's requirement that faculty sign an oath swearing loyalty to the U.S. and Massachusetts constitutions. Beadle said that it had not been a problem for him because the Massachusetts constitution preserved the right to revolution.[8] In 1948, when David Beadle campaigned for Norman Thomas in the presidential election and was suspended from high school for refusing to salute the flag, Marion worried about the effect on her husband's career. Beadle himself seemed unconcerned; "Pop stuck up for people."[9] He would continue to defend friends and colleagues because he found the loyalty problems "difficult, complex, and terribly important."[10] However, he preferred to act on his own rather than associate himself with like-minded groups such as the Scientists Committee on Loyalty Problems: "To do otherwise would imply my approval of Committee actions that must necessarily come in future and in which I will not take an active part.[11]

One of those he stuck up for was Edward U. Condon, the distinguished and popular physicist who had worked on the atomic bomb project and was now director of the National Bureau of Standards. Condon had been accused of being a serious security risk because of his "contacts with Russian scientists and pro-Communist sympathizers in this country."[12] In 1948 the National Academy of Sciences deliberated whether to send a public letter to the House Committee on Un-American Affairs chairman, J. Parnell Thomas, defending Condon in the face of the Committee's unsubstantiated public charges of disloyalty. The statement was sufficiently controversial among the Academy's members that the entire membership was polled. Of the 310 members (out of 401) who voted, 275, including Beadle, favored publication while the rest opposed. He wrote to the Academy president that the proposed letter was "a fine statement of the situation & my feeling in the matter."[13] Then, in 1949, he stuck up for Lewis Stadler at Missouri: "I do not know the charges against him but if they are of the nature I have heard...I am confident that any fair-minded loyalty board can not do other than find him completely innocent of disloyalty." Fortunately, Stadler was soon cleared of the charge.[14]

Beadle had become all too familiar with the issues surrounding allegations of disloyalty because Pauling was the center of a maelstrom that began about the time that Beadle arrived in Pasadena. By late 1946, the Cal Tech trustees were already critical of Pauling's political activities, especially his public support of Henry Wallace, who was forced to resign as Harry Truman's Secretary of Commerce when he spoke out "for the need to accept balance-of-power politics and Soviet hegemony over Eastern Europe as realities of the postwar world."[15] As Pauling's outspoken political activities increased, several trustees became convinced that he was a communist and called for his dismissal. The situation deteriorated when he took Sidney Weinbaum back into his lab in 1949. Twenty years earlier, Weinbaum had worked in Pauling's group and in the 1930s joined the Cal Tech Communist Club. After the war, he received security clearance for classified work at the Jet Propulsion Lab, but, in 1949, he lost his clearance and his job. Pauling intended to help Weinbaum while his appeal of the security decision was pending, but in the spring of 1950, Weinbaum was convicted of having lied to the security review board and imprisoned. As far as many of Cal Tech's trustees were concerned, Pauling's effort to help raise money for Weinbaum's legal expenses confirmed their worst suspicions.[16] The situation was further exacerbated in 1950 when he openly defended the Berkeley faculty's protests against dismissals for failure to sign the loyalty oath demanded by the regents of the University of California. Two committees, one of trustees and one of faculty members, were appointed to investigate Pauling. Neither they nor the FBI, which was investigating him at the same time, could find any evidence that he was a Communist. By then, however, Senator Joseph McCarthy had made Pauling's alleged communist connections front page news. He was denied a passport for travel abroad even to scientific meetings.

Little wonder that the trustees, who were both conservative and afraid for Cal Tech's reputation, wanted to avoid additional publicity about Communists at Cal Tech. In 1950, they declined to renew biochemist Jacob W. Dubnoff's appointment as a senior research associate in the biology division. Dubnoff, who was born and educated in California, had spent 2 years working as a biochemist in the Soviet Union before joining Borsook's lab at Cal Tech in 1936.[17] As a student, he had joined the Communist club in Pasadena.[18] By the early 1940s, Borsook and Dubnoff, then a graduate student, were carrying out important work investigating the source of the methyl groups on the creatine molecule.[19] After his Ph.D. was awarded in 1944, Dubnoff, who was "a first class biochemist," stayed on as a valued research assistant in Borsook's laboratory.[20] "We want you to go to the meeting in September and give'em the works" Beadle told Dubnoff in a memo saying that he would have funds to attend the meeting.[21] The senior members of the biology division all urged reconsideration of the denial of Dubnoff's reappointment, especially since he had ended the Communist association in 1937 when the Pasadena Communist Club disbanded. It was a tricky position, because the division, for scientific reasons, had no intention of appointing Dubnoff to a regular academic position. Beadle wrote "that as soon as he gets a suitable opportunity to move elsewhere he should do so." He pointed out that "the Institute will be subjected to criticism from some quarters if Dubnoff is kept" but "we will also suffer, particularly in academic circles, if his appointment is terminated."[22] DuBridge had Beadle's letter read to the trustees.[23] Based on Beadle's statements, Cal Tech's president opined that Dubnoff was a "loyal, hardworking, and valuable citizen."[24] Eventually, he was reappointed but he did not find a suitable position until 1962 when he became a professor at Loma Linda University.[25]

DuBridge realized that problems with loyalty issues would be ongoing. In 1950 he established a committee on academic freedom and tenure and appointed as chairman Professor Robert F. Bacher, head of Cal Tech's division of physics, mathematics, and astronomy. Bacher's earlier experience in Washington as a member of the first Atomic Energy Commission meant that he would be a savvy chairman. The Cal Tech faculty elected the other members of the committee and Beadle served throughout most of the 1950s. Three matters were paramount for the committee: keeping the trustees informed, the situation on the Berkeley campus, and continuing tensions over Pauling. The chemist was again accused of being a Communist party member, this time by the Army-Navy-Air Force Personnel Security Board and the California State version of the House Un-American Affairs Committee. In the summer of 1951, Pauling signed an affidavit stating that he was not and never had been a member of the Communist party, but the times were such that even that did not stop the campaign against him. He was simply too outspoken.

Security issues came even closer to Beadle when they threatened David Bonner, one of his outstanding students and James Bonner's brother. After leaving Stanford in 1947, Bonner had established an independent reputation for his work on *Neurospora* at Yale. An outspoken and informal person, his students thought the world of him.[26] When the biology division at the Oak Ridge National Laboratory convinced him to move to Tennessee, a security investigation for a Q clearance was initiated. Questions were raised about his and his wife's alleged communist friends, colleagues, and students. Supposedly self-incriminating statements made to an FBI agent investigating one of his former students, and his brother James's fund-raising for Weinbaum's defense, also came under scrutiny.[27] Alexander Hollaender, who recruited Bonner to Oak Ridge, alerted Beadle to the situation months before Bonner received formal charges.[28] Writing promptly to Hollaender and Shields Warren, then director of the Atomic Energy Commission's division of biology and medicine, the parent organization for the biologists at Oak Ridge, Beadle recounted his long association with David Bonner, his wife, brother, and other members of the family. Beadle believed that Bonner was "a thoroughly honest, conscientious and reliable person of complete loyalty to the ideals of our country...It is conceivable to me that in David's student days at Cal Tech he may have inadvertently associated with Communist party members...Certainly during the years I have known him I have never had the slightest basis for doubting his loyalty or his complete honesty even though I might not always have agreed with him on political or social issues."[29] Beadle also acknowledged in this letter that Bonner was seriously ill with Hodgkin's disease.

Bonner asked Beadle to testify at his hearing at Oak Ridge in midsummer, offering to pay his expenses.[30] Beadle, who had been advising him all along, agreed. Hoping to relieve the burden of the travel expenses, Hollaender invited Beadle to give a seminar at Oak Ridge when he came to testify. However, Warren realized that the plan could backfire if the prime reason for Beadle's trip was to testify for Bonner,[31] and Beadle paid his own way to the hearing on August 2, 1951. While there he did give a seminar and discussed science with Oak Ridge scientists as well.[32] Beadle sat through the entire 11-hour hearing thinking that it was "very fair and complete" and turned up "no evidence that David has ever done more than be somewhat unfortunate in his associations and make a few foolish statements from time-to-time—most of them not in a serious vein."[33]

Some years later in an article decrying the security clearance system because "it violates the basic principles of justice in a free society," Beadle described the string of questions he was asked during a hearing that most likely was Bonner's at Oak Ridge.

"Question: "Do you believe your friend A is a Communist?"
Beadle: "No."
Question: "Did you know B?"
Beadle: "Yes."
Question: "Did you know he was a Communist?"
Beadle: "No."
Question: "Then how do you know A is not?"
Conclusion: A is not a good security risk."[34]

Beadle's effort was to no avail. Security clearance was denied in March and Bonner not only lost out on the position at Oak Ridge, but after June of 1952 was denied his grants for unclassified work at Yale.[35] Although he had withdrawn his job application, he appealed the decision in order to clear his name.[36] Beadle supported the appeal and accompanied Bonner and his brother, James, when they consulted an attorney in Washington. He also tried to convince Shields Warren that an official recognition of no evidence of disloyalty but only a concern that Bonner was a poor security risk might preserve the grants at Yale.[37] Warren's response was disappointing, especially since it was identified as classified material. "It would be very hard for a tax-supported agency to justify the use of the people's dollars to support a project where not only the principal investigator has associated himself with Communists over a period of years at widely spread geographic sites, but has among his graduate students who are supported from the project persons with close Communists ties."[38] Many colleagues besides Beadle were aghast at Bonner's experience and the kinds of attitudes displayed in Warren's letter, even about unclassified research. H.H. Plough, a professor of biology at Amherst, who was temporarily assistant chief of the biology branch of the AEC in Washington, sent Beadle a private note saying that Bonner was denied clearance "in spite of the moral certainty on the part of all of us, that he has no connection whatever with any subversive organization."[39] Plough, who received his Ph.D. with Morgan in 1917, crossed the Atlantic with Beadle on the ship the *City of Flint* on the difficult journey back to the States in September of 1939. And it was Plough who arranged a job for Muller at Amherst when he returned from Edinburgh in 1940 with no prospect for employment. Eventually, all the efforts on Bonner's behalf yielded fruit and his good name was restored when, in 1953, the appeal process reversed the earlier decisions. No doubt, Beadle learned from this experience that in political situations, unlike scientific research, straightforward analysis of facts is compromised by attention to a variety of other considerations. His strong sense of fairness and justice was surely offended.

Regardless of all his other activities, the biology division still had first call on Beadle's attention. He was determined to make it a major force in the Cal Tech community as well as an international leader in research. Recruitment of new faculty was ongoing and time-consuming. Courses had to be organized and teaching schedules arranged. Graduate students and postdoctoral fellows had to be screened for admission. All Cal Tech students, undergraduate and graduate alike, were male, and there were many more of them than ever before because of returning veterans and the GI Bill. In May of 1947, the biology division faculty voted unanimously to "go on record as favoring the admission of women graduate students."[40] Beadle anticipated that very few women would want to undertake graduate study. Those who did, however, should be allowed at the discretion of the individual divisions, although he recommended against women having teaching duties for undergraduates. His caution did not help. It wasn't until six years later that the first woman was accepted as a graduate student.

Another change was the provision of stipends for graduate students from government grants and contracts to their professors.[41] While this policy was widely welcomed in U.S. universities, the biology division had trouble with it and developed its own rules.[42] "No work done by a graduate student employed by his major professor on a project for which the latter acts as responsible investigator will be accepted by the Division of Biology as material for the doctoral thesis."[43] The division held that work for a doctoral thesis must be the student's own. This, it concluded, cannot be the case when the thesis work is part of a project for which the major professor is responsible because it requires either that the professor closely direct the student's activities or that he transfer his own responsibility to the student. Either way was unacceptable. Moreover, "since in general the student depends on this source of income for his living, he loses much of his freedom of choice" in case he considers changing his research direction. The division was in the enviable position of being able to provide independent funds or help students find outside fellowships including those awarded to individuals by the National Science Foundation (NSF). Eventually the division sought and received from the NSF in 1957 the Foundation's first "training grant," which allowed the division itself to select and appoint fellowship recipients. Although this type of "block grant" was extensively and successfully used by the National Institutes of Health, the program was short-lived at the NSF because of both internal and external political considerations.[44]

Beadle and Delbrück were delighted when a friend of DuBridge offered $150,000 to initiate a program in animal virology. This wealthy man had suffered from shingles, which is caused by the latent herpes zoster virus often decades after a childhood infection and consequent chicken pox. When DuBridge told him that little was known about that or any other animal virus, he decided to support the research.[45] A search for a scientist who could initiate the program began in about 1950. Weaver, who knew everyone, everywhere, was usually a good source of advice but his suggestion to appoint Jonas Salk was rejected as unsuitable, probably because Salk was not concerned with fundamental science.[46] They found the right candidate in Renato Dulbecco, who had come to the United States from Italy in 1947 at the age of 33 to work on phage recombination in Salvador Luria's lab at Indiana University. The two had been in medical school together in Turin before the war. When Dulbecco went to Cold Spring Harbor for the summer meetings of the phage group, Delbrück suggested that he come to Cal Tech as a senior research fellow.[47] He was already in Pasadena when Delbrück offered both Dulbecco and Seymour Benzer, another research fellow, the opportunity to work on animal viruses. Benzer was uninterested, but Dulbecco was enthusiastic. He started to work out a quantitative assay for animal viruses analogous to that used so successfully with phage. Beadle worried that the viruses might infect people working in the building and hid the lab in the subbasement, away from the main laboratory traffic. Dulbecco succeeded magnificently with western equine encephalomyelitis virus and chicken embryo cells growing as a single sheet of cells (a monolayer). Any cell in the monolayer infected by the virus produced additional virus that then infected and killed the surrounding cells. The live cells were stained with a dye, leaving colorless circular patches of cell debris from the killed cells on the otherwise colored, intact monolayer. The number of such patches (called plaques) was a measure of the amount of virus added. This finding allowed an important conclusion: one virus particle was sufficient to make an infection.[48] Dulbecco's method put the investigation of animal viruses on a quantitative and fruitful path. Because he planned to extend the methods to the study of poliovirus, the National Foundation for Infantile Paralysis promptly began generous support for the work. Beadle was impressed and appointed Dulbecco to a regular faculty position. In 1975 Dulbecco received the Nobel Prize in physiology or medicine along with two of his former students, David Baltimore and Howard Temin.

In the midst of all these consuming activities, the situation at home had become untenable. Beadle was away much of the time and distracted when he was in Pasadena. The small house never satisfied Marion, and after David went off to Reed College in Portland, Oregon, in 1949, they moved into an apartment on South Marengo Street. Familial relationships could not have been helped when, in 1951, David flunked out of Reed, married, and joined the Air Force while Pauling's daughter was succeeding at the same college.[49] In April of 1951, Beadle moved out.[50] By July, he had settled into the Athenaeum, Cal Tech's faculty club, where he lived for two years. Marion sued for divorce in August of 1952.[51] Beadle confided in Sturtevant and worried whether, in view of the divorce, he should resign as chairman. Sturtevant, perhaps more attuned to changing views about divorce, "assured him that of course he shouldn't."[52] Within the Cal Tech community, the situation was neither secret nor widely known. There were no outward changes in Beadle's activities except that he seemed to work even harder and spent many nights at the lab laboring over speeches and talking with students.[53] Delbrück wrote from Cold Spring Harbor "wanting to say something friendly to you about your divorce. I will at least say that we very often think of you and wish you well."[54] Beadle responded with his customary unrevealing reserve: "thanks for the kind personal words. In a situation that's not easy, they are appreciated."[55]

Living on his own, Beadle was free to explore new people and activities.

He took up rock climbing, which he enjoyed in the company of a Cal Tech physical chemist, Gunnar Bergman, and the visiting Swiss biochemist, Alfred Tissières. In late June of 1953 they took off for several weeks to climb 8,800-foot Mount Doonerak in Alaska. It was the kind of challenge Beadle liked, because an earlier visitor had claimed the mountain was almost impossible to scale and they would be the first to do so. In the end, the 65-mile hike from the Nolan Creek gold mine to the mountain in constant rain and deep mud was a tougher challenge than the one-day climb up the mountain itself. The first day's hike turned out to be typical. "The result was fatigue and discouragement such as I have never before experienced."[56]

The Doonerak expedition marked the start of a special summer. During the time that Beadle lived in the Athenaeum, friends had introduced him to interesting women. One of these, Muriel McClure Barnett, was a lively young widow with a small son named Redmond. A successful journalist, she was editor of the *Los Angeles Times* Women's Section and, like Beadle, a lover of cats. The matchmakers, according to one story, were neighbors Vernon and Virginia Newton.[57] Another story claims that Barnett and Beadle were introduced by Jess Williams, a highly educated widow who prepared food for the flies in the lab and supervised the dishwashing. [58,59]

Though born in California in 1915, Muriel Barnett was raised in Chicago, her mother's hometown. When it came time for college, she wanted to establish her independence and returned to California and Pomona College. As soon as she left Chicago, she plucked her eyebrows, cut off her long hair, and started smoking. After graduation in 1936, she started her career by writing advertising copy in Chicago. She married Joseph Y. Barnett during World War II and when the war was over, they moved to Los Angeles with their infant son. There, Muriel became a feature editor on the *Los Angeles Times*. Fifty years later, at her college reunion, she discussed the then unusual path she took: "When the guns fell silent, 'back to normal' for me meant a role as wife, mother *and* [sic] career woman. I discovered not only that I could do it, but that I enjoyed it."[60] She was fortunate because when Joseph Barnett died in 1951 from problems arising from a WW II wound, her job at the *L.A. Times* helped support her 7-year-old son, Redmond. By then, she was editor of the Women's Section and a well-known, popular figure in a city that thrived on celebrities.

The developing relationship between Beadle and Muriel Barnett was not secret, and his colleagues gave her high marks.[61] His divorce from Marion was final after the legally required wait of one year on August 10, 1953.[62] Two days later, Muriel and Beadle were married and went off to Jackson Hole, Wyoming, for their honeymoon. The new family of three settled into the old Morgan house on San Pasquale Street and Muriel continued working.[63] A year later, Beadle adopted Redmond "Red" Barnett. David visited only occasionally. He had divorced his first wife and remarried, and his new wife gave birth to a son, John Vincent, a month after the Barnett–Beadle wedding. Redmond, like David, remembers "Pops" as not very affectionate, but colleagues sensed a more relaxed Beadle who now had two sons, a grandson, and a friendly, successful, and witty wife.[64] He even took up scuba diving, although previously he never ventured into the water.[65]

Meanwhile, Beadle continued to build the biology division. In 1953 he set his sights on appointing Roger Sperry. Sperry was 40 years old and well known for his outstanding research concerning the relation of brain structure and neuronal specificity to behavior. He was planning to move from the University of Chicago to the National Institutes of Health (NIH) but was delayed by slow building construction at the Institutes' Bethesda, Maryland campus. Beadle proposed that the division of biology take advantage of the situation and recruit him. Most of his colleagues were enthusiastic about the idea and DuBridge, who had the final say, approved.[66] Sperry moved to Pasadena as professor of psychobiology early in 1954 and remained until his death 40 years later. He added luster and substance to the division and received a Nobel Prize in physiology or medicine in 1981.[67]

Under Beadle's guidance, the biology division was beginning to reflect his vision of a biochemical genetics. Meanwhile, extraordinary developments in genetics occurred in other places and unanticipated ways. Late in 1951, at the Cavendish Laboratory in Cambridge, England, Frances Crick and his younger American colleague James D. Watson started trying to model the structure of DNA. Watson and others in the phage group had begun to take Avery's 1944 experiments seriously and, if genes were made of DNA, knowing its structure would be the key to understanding how genes actually worked.[68] Solving the structure of DNA became even more urgent in 1952 when Alfred Hershey and Martha Chase at Cold Spring Harbor confirmed Avery's earlier conclusion that DNA, not protein, was the genetic material in phage. Hershey and Chase made two preparations of phage containing either a radioactive isotope of sulfur in the proteins or a distinguishable radioactive isotope of phosphorus in the DNA. After allowing the phage to initiate infection of a suspension of bacteria, they subjected the mixture to a whirl in a Waring Blendor that sheared the phage off the bacterial surface, and then collected the bacteria by centrifugation. The phosphorus isotope and thus the DNA were mainly associated with the bacteria, while the sulfur isotope and thus the phage protein were largely in the solution freed of bacteria. Within the bacteria, the infection and formation of new phage proceeded normally. Thus, it seemed that the DNA alone was needed to supply the genes required to make new phage.[69]

Watson and Crick were not the only ones intent on determining the structure of DNA. Maurice Wilkins and Rosalind Franklin at King's College, London were working on the X-ray analysis of DNA and, by the end of 1952, Pauling, who had recently announced the helical structure of certain polypeptides, was also trying to solve the DNA structure.[70] The story of this race, brilliant and sordid, was famously told by Watson and in scientific detail by others.[71] Both discouragement and progress were rapidly communicated between Pasadena and Cambridge. Pauling's son, Peter, then studying for a Ph.D. at the Cavendish Lab, shared the information in his father's letters with Watson and Crick. Meanwhile, Watson was writing frequently to Delbrück, which meant that everyone at Cal Tech had frequent updates on the race. When Pauling rushed to publish a paper describing a triple-stranded structure that was embarrassingly wrong, Watson and Crick knew of his error months before the paper appeared in print.[72] Delbrück, after hearing Pauling present his structure at a Cal Tech seminar, bluntly told Pauling that he was wrong. Watson and Crick had the advantage of knowing the critical crystallographic data obtained by Franklin in London, although they utilized it in a less than honorable way and failed to credit it properly. They were also clever enough to pay attention to Erwin Chargaff's analytical data showing that in all DNA preparations, the amount of adenine equals that of thymine and the amount of guanine equals that of cytosine.[73] With this information, Watson and Crick got the DNA structure brilliantly correct.[74] Their elegant double helix with its two deoxyribonucleotide chains wound around one another is now a familiar popular icon.

When the triumphant Watson returned to the United States in 1953, his close relationship with Delbrück clinched his acceptance of Beadle's invitation to join the biology division as a senior research fellow. The biological community had already elevated Watson to its supreme ranks, but the draft board in his hometown of Chicago had other ideas. The Korean War was on and Watson, age 25, was called for induction into the army. After a series of appeals for a deferment were denied, Beadle called on DuBridge to pull out all stops to have the case reconsidered. "Because there are literally no other James Watsons in the world, he is absolutely essential to our research work... He is simply irreplaceable."[75] Finally, a deferment was granted by the highest possible authority, the Presidential Review Board. Years later Beadle reminisced that "Knowing Jim, I'm sure that the military service was saved plenty of headaches."[76] Instead, Beadle had the headaches.

By the spring of 1954, Delbrück wanted "to wash my hands off [*sic*] the responsibility for Jim" even though he was "still one of the apostles of the creed which consider Jim the Einstein of Biology." The "fright" Watson experienced at the prospect of being drafted explained why, in Delbrück's view, Watson "completely repressed and forgot any commitments he had made" for that first year at Cal Tech. The failure to remember responsibilities was only one of Delbrück's problems with Watson. He also thought his young colleague to be "ruthlessly egocentric in scientific matters," faulting him for "publishing prematurely, putting his name as senior author, accepting too many public lecture obligations, etc."[77] Actually, the new work Watson did that year at Cal Tech was published either on his own or with a sole coauthor, Alexander Rich, who was then a postdoctoral fellow at Cal Tech.[78]

Delbrück was even "furious with Beadle for having played up so much to Watson."[79] He was no longer willing to have Watson receive payment from his National Foundation for Infantile Paralysis grant and dumped the problem in Beadle's lap. Beadle, whose goal was to have "all the good people and all the good work be at Cal Tech,"[80] and the rest of the division, wanted Watson to remain. Beadle wrote to Watson who was summering at Woods Hole: "Max...feels like you and I thought he would, i.e., he's a strong booster for Jim Watson in principle but still expresses his disappointment about the details of the last year."[81] He arranged a three-year appointment in which Watson would continue as a senior research fellow with support from either Merck or Rockefeller funds. The remaining National Foundation funds would support Niels Jerne, a promising European biologist.[82] Undoubtedly it was a great disappointment to Beadle when, after only one additional year, Watson accepted an appointment at Harvard.[83]

With or without Watson, the biology division thrived as everyone had hoped it would when Beadle was recruited. It received large grants from the Atomic Energy Commission and from 1952 to 1954 ranked sixth, after much larger institutions, in support from the National Science Foundation.[84] Robert Sinsheimer, who spent six months in Delbrück's lab in 1953, recalled that "Cal Tech was one of the world centers, and everybody came through Cal Tech. In a year there, or six months, by going to seminars you could learn what was going on everywhere. It hadn't been true at MIT either."[85]

National concerns with loyalty and security continued to plague the scientific community and Cal Tech. Pauling's NIH grants were suspended late in 1953 and he was denied new ones. Others at Cal Tech were having similar problems. The Cal Tech trustees as well as the university's wealthy benefactors grew increasingly skittish about having Pauling on the faculty. Each time his passport was denied or his grants rescinded, some trustees called for his removal. It fell to Beadle to tell Pauling that his U.S. Public Health Service research grants were suspended because the agency feared it would be accused of supporting a Communist. DuBridge had a difficult time, but with Beadle's support he stood by Pauling, although relations between the president and the chemist grew so touchy that they barely communicated. It all came to a head in early November of 1953 when Pauling was awarded the Nobel Prize for chemistry. Although the FBI had unearthed no evidence that Pauling was or had been a member of the Communist party, Ruth Shipley, the chief of the State Department's passport division, saw no good reason to allow him a passport for travel to Stockholm to accept the prize. Wiser and more powerful people than Shipley realized just how terrible the international publicity would be if Pauling was barred from going to Stockholm and she was overruled. Herbert Hoover, Jr., who had been hoping for years to see Pauling dismissed from Cal Tech, was sufficiently angry to resign from the board of trustees.[86]

Beadle's judicious views and leadership at Cal Tech earned him a reputation as a wise and prudent defender of science and scientists. The American Association for the Advancement of Science (AAAS) voted him president-elect in 1954. His own confidence grew and he began to speak out more broadly as the national frenzy over security matters increasingly affected the scientific community. He objected publicly and forcefully to the AEC's treatment of Robert Oppenheimer. In the magazine distributed throughout the Cal Tech community, including alumni,[87] he put his own name to ideas developed in an AAAS report. Part of his criticism concerned the procedures being used to establish qualifications for access to classified information, including guilt by association and the lack of access by an accused person to allegedly derogatory information. He understood the need for some national secrets and argued for a sense of balance in achieving security. "Allowing reasonable freedom of communication involves a risk of leaks and this risk must be balanced against the gain from more freedom." He criticized the "extension of security clearance to unclassified areas" and especially noted this practice by the U.S. Public Health Service (USPHS) and thus the National Institutes of Health. He insisted that "the responsibility for determining that the investigator is a person of good character should rest with the institution" not with a government agency. Although Pauling's name is not mentioned, Beadle must have had him and other colleagues in mind. The article ended with a call for university faculties to study this problem carefully and take responsibility to make the public understand their conclusions and concerns.

By the time Beadle's article was published, everyone knew about Pauling's Nobel Prize. Cal Tech could celebrate unabashedly because its president and trustees had taken no official actions against Pauling. As Beadle knew, however, they had privately pressured Pauling into withdrawing from public political activities. Pauling was quiet, but frustrated, and the celebrations around the Nobel Prize renewed his confidence. After the ceremonies in Stockholm, he made a trip around the world, including Japan. He became convinced that radioactive fallout from tests of a new type of hydrogen bomb at Bikini posed an unacceptable threat to human health.[88]

Beadle decided that it was time to defend Pauling in public. He explained his views in a talk on March 25, 1955 to the National Nephrosis Foundation dinner in Beverly Hills and repeated them in an article about the chemist directed at the entire Cal Tech community.[89] He set the stage by summarizing his earlier description of the troubling aspects of the security systems then in place. He then analyzed the "ingredients of greatness" in a scientist, concluding "But whatever are the ingredients of greatness in science, there can be no doubt that Linus Pauling has them." He didn't restrict himself to Pauling's science. "Pauling possesses an array of additional traits that make him an all-around 'great guy.' I admire the courage with which he stands by his convictions even at times when his views may not meet with popular favor. I am proud to belong to the faculty of an institution with the foresight to see his greatness and the wisdom to give its development full freedom." Neither DuBridge nor the trustees could miss the point of Beadle's conclusion: "I am proud that Cal Tech has a president who knows the true meaning of academic freedom and who has the courage to speak and act accordingly...I am grateful for a Board of Trustees that has not succumbed to the disease of mistrust and suspicion that could so easily undermine their faith in the wisdom of academic freedom and rightness of liberal decency." As president of the AAAS, Beadle's remarks were newsworthy, and the *Washington Post* and *Times Herald* reported on this talk to the nephrologists. This brought national attention to the USPHS's policy of denying research grants to investigators who were the subject of loyalty investigations.

In the same month, Beadle publicly agreed with Erwin Griswold's defense of the Fifth Amendment to the Constitution.[90] At the time, the integrity of the amendment was challenged by the cold warriors in Congress and elsewhere who interpreted its use as an admission of guilt. The AAAS had commented to the Senate Committee on the Judiciary on the "unjustified inferences drawn from the use of the Fifth Amendment."[91] Beadle wrote, "Let us hope that all members of Congress will read Griswold's little book."

Although Pauling had been singled out in 1953, the USPHS determined in the spring of 1954 to rescind any NIH grants to individuals alleged and "established to the satisfaction of the Department [of Health, Education, and Welfare]" to be disloyal to the United States.[92] The policy was applicable to unclassified work and no formal proceeding or opportunity was provided for the accused to counter the allegations. This contrasted with the policy of the

National Science Foundation, which only rescinded or denied grants to those demonstrated to be Communists by admission or formal judicial proceeding.[93] Late in November of 1954, Cal Tech learned that Borsook's two NIH grants had been prematurely terminated with no explanation.[94,95] Beadle explained to Leonard Scheele, the Surgeon General, that the institute was "forced to assume that the cause of the terminations was the existence of derogatory information about" Borsook.[96] Borsook himself believed that the derogatory information was related to his raising money for the Spanish Loyalists.[97] With Beadle's support, the Cal Tech faculty courageously resolved on December 6, 1954 to recommend that "no new Public Health Service research grants be accepted until such time as the present policy is appropriately modified."[98] Beadle warned the Surgeon General by wire of the proposed public protest.[99] The statement by the biology division in support of Borsook was drafted by Sturtevant, von Harriveld, and Sperry and the biology division voted in writing; Borsook abstained and only G. Alles disapproved.[100] DuBridge discussed the issue with the trustees. At their direction, he informed the faculty that one or more of the trustees were willing to guarantee $1 million for at least one year if all their current grant money were returned to the NIH. The faculty agreed. However, when DuBridge called the NIH director to arrange return of the funds, he was told that NIH had no mechanism for receiving returns.[101] Within weeks, the Surgeon General wrote stating that Borsook's grants would be reinstated.[102]

Beadle's scientific reputation, his skill as a university administrator, success as a fund-raiser, and judicious approach to difficult issues invited frequent requests for participation in the development of science policy. Perhaps his admiration of Pauling convinced him of the value of taking public stands on important issues, although always in his own, less flamboyant way. Increasingly, Cal Tech and national affairs kept him away from the campus. As an advisor to the biology division of the National Science Foundation, he spent occasional weeks in Washington, often stopping at other places on the way to help out other institutions. He even seriously considered spending the 1952–1953 academic year on the NSF staff in Washington but declined, citing personal as well as other reasons.[103] The growing convenience of air travel made California seem within easy reach of the east coast. At the end of the 20th century, the constant travel of scientists between the country's two coasts would become an all-too-familiar phenomenon. Beadle's travels were prototypical.

CHAPTER 14

Genetics and the Nuclear Age

People all over the world were, by the early 1950s, increasingly troubled about nuclear weapons. The implications of their destructive powers for world security in the growing Cold War between the United States and the U.S.S.R. were almost incomprehensible. Dangers inherent in the building, storing, and testing of the weapons also began to be widely appreciated. Radiation was especially frightening because it could not be sensed without special instruments. Most emphasis was on the immediate threat to human health. Neither the public nor its governmental representatives thought in terms of mutational effects on genes. Yet, as Muller and Stadler had learned decades earlier, radiation causes mutations. Genetics was about to become a public issue of major importance in the U.S. Geneticists would no longer be obscure, ivory tower scientists.

In the first decade after the explosions at Hiroshima and Nagasaki, the Atomic Energy Commission (AEC) had the primary responsibility for policies and public information about radiation hazards in the U.S. and kept secret much of the data on which its pronouncements were based. Beadle, in his first foray into issues of national scope, joined the commission's Ad Hoc Committee to Evaluate Effects of Atomic Energy on the Genetics of Human Populations in 1947 and authored a report on radiation genetics.[1] After the testing of the "superbomb" at Bikini Atoll on March 1, 1954, however, people everywhere realized, in spite of denials by the AEC, that radioactive particles were being spread worldwide in the aftermath of the explosions. A new word, "fallout," was on everyone's lips. Pauling quickly made his concerns about the unknown and known radiation hazards public, although it was only a few months since he had once again resolved to lay low and tend to research in the hope of placating the Cal Tech trustees.

Pauling was not the only member of the Cal Tech faculty to speak out publicly about radioactive fallout. Sturtevant took advantage of being president of the Pacific division of the American Association for the Advancement of Science (AAAS) to present his views to its June, 1954 meeting. He stressed the differences between the immediate, sometimes reversible, effects on individual health of high doses of radiation for short intervals, and the undetectable and irreversible genetic changes, or mutations. He believed these occurred without threshold at all doses and cumulatively.[2] According to Sturtevant, there was "no clearly safe dosage." He was disturbed that AEC chairman Lewis L. Strauss stated in a White House press release that the increase in natural background radiation resulting from fallout after U.S. and Soviet tests was "far below the levels which could be harmful in any way to human beings."[3] Sturtevant thought that the benefits from the atomic explosions might be worth the risk, but knowing the risk to future humanity was now impossible to estimate, especially since most of what was known about the genetic effects of radiation came from experimental data on animals, not humans. Moreover, the overwhelmingly deleterious effects of the mutations were unlikely to emerge for many generations and might result in only slight decreases in the efficiency of reproduction. It was improbable that anyone could ever know with certainty how much the testing added to the load of undesirable genes already accumulated in the world's population. Sturtevant went even further, stating that there was "reason to suppose that gene mutations, induced in an exposed individual, also constitute a hazard to that individual—especially in an increase in the probability of the development of malignant growths, perhaps years after the exposure." As early as 1927, Muller had suggested that the induction of cancers by X rays might be associated with mutagenesis.[4]

Dr. John C. Bugher, who had no background in genetics but was then the director of its division of biology and medicine, responded for the AEC. Bugher agreed with Sturtevant that radiation can cause permanent genetic changes, usually negative ones, and that all doses can have an effect. Illogically, however, he then concluded that the radiation from fallout had been genetically "insignificant."[5] Beadle was in substantial agreement with Sturtevant regarding these issues.[6]

Late in 1954, when Beadle was president-elect of the AAAS, he helped draft the organization's statement on "Strengthening the Basis of National Security."[7] The crux of the statement was contained in one sentence: "Continued scientific progress provides a better guarantee of military strength and security than does excessive safeguarding of the information we already possess." The AAAS board strongly endorsed the concept that "a security-screening program is made necessary by the peril of the times." But it emphasized that "The basic fact—the frightening fact if you will—is that there are simply no such things as permanent scientific secrets." The report asked for a "positive program of preserving national security by" substituting "the question *How can we best aid national progress?* [*sic*] for the negative question *How can we avoid the danger of leaks?* [*sic*] " It urged that more weight be given, in evaluating security risks, to the potential contributions of a scientist and to the nature of the work to be done.

The AAAS statement attracted press attention. An editorial in the *New York Herald Tribune* misinterpreted the statement as suggesting that scientists, regardless of the kind of work they did, should not be subject to security procedures.[8] Warren Weaver, who was president of the AAAS that year, felt obliged to send a strongly worded letter to the *Herald Tribune*, complaining about the newspaper's faulty characterization of the statement.[9] Despite strong support for the statement within the AAAS, some affiliates expressed "disapproval and indignation" during the AAAS annual meeting and were even angrier when the meeting adopted a resolution recommending "that government funds for research not involving national security be granted to all except proved or avowed Communists, regardless of whether they are denied security clearance on other grounds."[10] A year later, a committee of the National Academy of Sciences published a similar statement.[11]

As 1955 began, Beadle replaced Weaver as president of the AAAS and moved to the front line in the Association's affairs. The presidency was actually a three-year stint, the first year as president-elect, the second as president, and the third as retiring president and chairman of the board. Perhaps Beadle counted on his presidential duties being largely ceremonial especially because of the highly competent professional staff under the wise Dael Wolfle at the headquarters in Washington, DC. Still, the president had to appear at various events and Beadle even traveled to Alaska for an AAAS meeting in June and went back and forth to the east coast a great deal.

Besides the security and loyalty issues, he was faced with continuing problems concerning the construction of the new AAAS headquarters building in Washington, DC and the major public dispute engendered by the Association's plan to hold its 1955 annual December meeting in Atlanta, Georgia, a racially segregated city. Objections were raised as early as 1953 when the plan for the 1955 meeting was first made. The clamor grew as the date of the meeting approached. Opinions in the African-American community as well as the AAAS membership varied. Some agreed with the view of the board that "the advantages—in terms of breaking down segregation barriers by example as well as other ways—outweighed the disadvantages."[12] But at least one AAAS section decided to boycott the meeting. "It is a crude and gross insult to the whole membership of the AAAS as well as to its Negro members for an announcement to appear in *Science* such as is found on page 752 of the May 27, 1955 issue, listing the gala array of hotels and hotel headquarters and facilities available for the convention when it must be understood that these are 'For White Only.'"[13] Trying to change the meeting site only six months before its December date was problematic and the board decided to proceed with plans for Atlanta.[14] Wolfle attempted to ameliorate the situation by arranging for the program, exhibits, and social events to be held on the campuses of Negro universities and other desegregated facilities.

Through all this national activity, Beadle was devoting a great deal of time to his new family and the biology division. In midsummer of 1955, the Beadles drove to Nebraska so that Muriel could meet his father, now age 88 and living at a nursing home in Ashland. From there they drove west to an institute for college biology teachers in Wyoming and a visit with their son Redmond in a northern California summer camp before returning to Pasadena at the end of July. On the Cal Tech campus, Beadle and the others were busy with planning for the new Church Laboratory of Chemical Biology. This project had been proposed years before by Morgan and again by Pauling and Beadle when Beadle first returned to Pasadena. Finally, in 1954, the Rockefeller Foundation granted $1.5 million, an equal match to the funds bequeathed by Norman Church. Church himself had become a fan of Cal Tech years before, when he was accused of "doping" one of his race horses and Arnold Beckman, then assistant professor of chemistry, proved that it wasn't so.[15] By the time the Church Lab was finished in 1955 it was already crowded and Beadle began to lobby for a new wing connecting Church to the old Kerckhoff biology laboratory, as envisioned in 1931 by Morgan.[16]

In 1956, as retiring president of the AAAS, Beadle gave the major address at the annual meeting in New York.[17] His scope was on a grand scale. He swept through thoughts about cosmology, the origin of life, biological and cultural evolution, food and population problems, the threat of nuclear warfare, and inequality of opportunities and resources among nations. Significant in view of his earlier reserve concerning the chemical nature of genes, he said, " These properties, self-duplication and mutation, are characteristic of all living systems and they may therefore be said to provide an objective basis for defining the living state. Evidence is accumulating that the nucleic acids of present-day organisms possess these two properties." He ends with a sweeping vision for the promise of genetics. "He [man] is in a position to transcend the limitations of the natural selection that have for so long set his course. But knowledge alone is not sufficient. To carry the human species on to a future of biological and cultural freedom, knowledge must be accompanied by collective wisdom and courage of an order not yet demonstrated by any society of men."

When he delivered this statesmanlike speech, Beadle had been occupied for more than a year with a panel of distinguished geneticists considering the genetic consequences of atomic radiation. Detlev Bronk, the president of the

National Academy of Sciences (NAS), appointed the panel in October of 1955, and the work was financially supported by the Rockefeller Foundation. The Foundation's officers appreciated the need for advice from independent, nongovernmental sources concerning the consequences of radioactive fallout. They also had a sense of responsibility because the Foundation had supported research in nuclear physics in the 1930s. The Foundation's trustees pressed for "a broad appraisal of the effects of atomic radiation on living organisms" including the identification of "questions upon which further intensive research is urgently needed."[18] After consultation with the AEC, the Foundation asked the National Research Council (NRC), an arm of the NAS, to undertake the study. The panel on genetic effects was the first of several panels established by Detlev Bronk, president of the NAS, under the general heading of Committees on the Biological Effects of Ionizing Radiation (BEIR). Later, others on, for example, pathological effects, the safety of the food supply, and the disposal and dispersal of radioactive wastes were established. The emphasis on genetics was a departure from the approach of the AEC, which had been chiefly concerned with assuring the public that the low level of radiation associated with fallout from bomb tests was different from the large exposures that caused radiation sickness, leukemia, and other cancers. The genetics panel faced the added challenge of trying to overcome the troubling reputation genetics suffered as a consequence of Nazi experiments during World War II.

The first panel's roster was a who's who of genetics and included besides Beadle, James Crow, Demerec, Dobzhansky, Bentley Glass, Muller, James V. Neel, Tracy Sonneborn, Sturtevant, and Sewall Wright. William L. Russell and Alexander Hollaender from the Oak Ridge National Laboratory, both experts on mouse genetics and radiation, also became members. Shields Warren, who was not a geneticist, provided experience in statesmanship.[19] Bronk's appointment of Warren Weaver, a mathematician, to chair the genetics panel was bold and, as it turned out, brilliant. Bronk correctly believed that Weaver would deal judiciously with the expected disagreements among the panel members. "The two most prominent geneticists on the Committee on Genetic Effects refused to speak to one another."[20]

Muller's personal history was problematic, but he was already a Nobel laureate and had thought as much about the genetic effects of radiation as anyone on Earth. His longstanding views were well known. In his Kimber lecture in April of 1955 he argued against both of the extreme views then being expressed; one claimed that there were no risks from continued testing of nuclear weapons and the other that all testing should be stopped immediately because of the risks. Muller believed that weapons development was necessary. No doubt his earlier firsthand experience with the Soviet dictatorship afforded a realistic view of the threats posed by the Soviet Union. However, he objected to the AEC statements that erroneously reassured the public about the hazards of fallout. Work with *Drosophila* had convinced him that genetic damage occurs in proportion to the total accumulated dose. From the available data on gamma ray exposures, he calculated that each person in the U.S. had been exposed to an average of 0.1 roentgens of radiation from testing; with 160 million people, this added up to 16 million roentgens total. According to the available figures, the 160,000 people at Hiroshima had each received an average of 100 roentgens for a total of the same 16 million roentgens. Thus, an equivalent number of mutations were expected in the two populations. Muller was also very concerned about unnecessary radiation exposures from medical and dental X rays and the disregard for appropriate shielding.[21] This speech landed Muller in a very public dispute with the AEC.

Muller had left Indiana for Europe in June 1955, under the impression that he had been invited to give a talk at the International Conference on the Peaceful Uses of Atomic Energy in Geneva in mid-August. After his departure, the AEC sent him a letter saying that the United Nations (UN) had not requested his paper and he was not to be part of the U.S. delegation. A few days later, they sent him another letter apologizing for the late notice and blaming the UN. Nevertheless, Muller went to the conference. When he appeared at the session on Genetic Effects of Radiation: Human Implications, the embarrassed panel gave him a standing ovation. According to the UN, and contrary to what the AEC told Muller, the AEC had informed the UN by letter in June that Muller would not be a delegate and that the Commission did not want Muller's paper on the program.[22] Later, the AEC not only confirmed the UN version of the story but gave as its reason that it was "definitely inadmissable" to have a paper that mentioned the Hiroshima bombing at a conference devoted to peaceful uses of nuclear energy. AEC chairman Strauss took responsibility for what he called a "mistake."[23]

As president of the AAAS, Beadle voiced the scientific community's dismay over Muller's treatment in a *Science* magazine editorial in October.[24] The facts of the case were presented in an accompanying news article.[25] Beadle wrote that "many persons will regret the affair because devious methods appear to have been used to keep Muller off the program and because his viewpoint, which happens to differ significantly from that of the commission (AEC), was apparently not as fully represented at the conference as most geneticists would have wished." Beadle was right about devious methods. To the distinguished chemist Willard F. Libby, who had worked on the Manhattan Project and was an AEC commissioner, Communists and anyone even suspected of Communist leanings were anathema. He did not trust Muller's loyalty and anti-Communism and had convinced the other AEC commissioners to override the recommendation of its own division of biology and medicine to invite Muller to the UN conference.[26]

The first meeting of the BEIR panel was held at Princeton just before Thanksgiving in 1955. Bronk charged the panel in person and the geneticists agreed to remain quiet about their deliberations until the final report. At Muller's urging they agreed not to use any classified material and refrain from contact with the AEC weapons group and division of biology and medicine.[27] Weaver recognized that the panel had a special opportunity to "seize this dramatic occasion" for a beneficial effect on the relation of science to the public.[28]

Together, the panel represented the world's expertise concerning the mutagenic effects of radiation on experimental animals under laboratory conditions and after exposure at the Eniwetok test site. Data about humans came largely from the unclassified surveys of 80,000 people carried out in Japan by Neel, Crow, and others under the auspices of the Atomic Bomb Casualty Commission established by the NRC. These data, however, largely cataloged immediate injuries, many of which were reversible in time unlike human mutations, which would be irreversible. The distinction was poorly understood by both the public and, apparently, the AEC. Most problematic of all, data that would permit the calculation of the dosages received in Japan were classified secrets. Once that information was available, there would be the task of interpretation, an issue whose difficulty the panel could only dimly suspect. In the absence of data, the panel could only define the central matters for public concern, assuming it could reach agreement. Contentious matters began to emerge at this very first meeting.

The AEC had introduced the concept of a "permissible dose" of radiation, but the significance and reliability of its estimate were at best unclear and at worst much too high according to panel members. Realistically, there had to be some such number in order to advise the public not only about radiation from fallout, but also that from X rays and the then popular fluoroscopy. For example, should there be special permissible doses for those whose jobs required radiation exposure? Panel members agreed that some balance was required between the dangers of radiation, the nation's need to build bombs for its own security, and the possible use of nuclear energy for civilian purposes. They disagreed on just where to strike this balance. Weaver already knew that Neel would not sign a report that recommended any permissible dose at all while Muller would not sign without such a recommendation. Everyone agreed that any radiation to gonads would yield an increase in mutations, there was no threshold for damage to genes as there might be for tumor induction, and genetic effects would be cumulative. They also agreed that more fundamental knowledge was needed about the structure of genes and mutations if the public was to receive sound advice. For now, they had to write a report and make recommendations in the face of ignorance and uncertainty. Weaver proposed a concentrated effort, with a report to be published as soon as possible. He would meet with the chairmen of the other panels who would, together, be responsible for coordinating all the work. The panel members were on notice that they would be very busy with this effort in the coming months. Beadle, for example, explored ways to engage Japanese geneticists in studying the hazards of human radiation when he attended the International Genetics Conference in that country.[29]

The panel met again in February 1956 in Chicago. Weaver understood that in spite of the disagreements within the committee, they had to construct a report that all panel members could approve. He came to Chicago prepared to succeed. He proposed that the initial reports of all the panels, including genetics, be brief and incorporated into a single, plain-spoken document; a draft of the genetics section's report was in his briefcase. He also held out a bribe; the Rockefeller Foundation would provide substantial support for fundamental genetic research so that in future, more knowledgeable recommendations could be made.[30] Weaver read from his draft at length, with minor interruptions from the geneticists and a few on his part to apologize for any errors in the science. Beadle, who was pretty quiet during most of the presentation, finally gave him the go-ahead: "You just keep going. We don't need any professional popularizers, all we need is you. We will put the commas in and make the corrections and put the qualifications in and we will have a fine job here." Muller agreed. Then Weaver knew that he would have a finished, brief report within months.

Certain issues were central to the broad-ranging discussions in Chicago. As expected, Neel argued that radiation at all levels causes mutations and opposed endorsing a value for a "permissible dose."[31] The others didn't disagree, but were attentive to political ramifications. If the panel set no value, the AEC would, and set it too high. If the panel set a very low value, it would be ignored by the AEC. If it set a useful value they would ensure a future role for geneticists in the policy debate. Beadle insisted on the importance of setting an exposure limit. However, he objected to the term "permissible dose" itself because it was scientifically inaccurate and confusing. He sharply criticized AEC commissioner Libby, who had been using the term as if it meant something. Instead, he favored the term "maximum permissible dose." Most people would rarely be exposed to the maximum but it would be a useful benchmark for people working with fissionable material. Perhaps, he foolishly commented, such jobs might be limited to those past the reproductive age. His colleagues quickly reminded him that men are reproductively competent for their entire lives. Muller worried that research physicists would not take the limit seriously either for themselves or their students because they had no understanding of genetics. A compromise was reached. They would publish a range of values reflecting the best current knowledge accompanied by a strong statement about the uncertainties and extent of ignorance.

During the two days in Chicago, the panel members frequently returned to the problem of the relative significance of radiation doses to individuals compared to that received, on the average, by the entire world population. Assuming that most mutations are recessive, they would be unlikely to affect the immediate descendants of a particular individual. Mutations that caused severe abnormalities in the first generation were expected to disappear quickly. But more subtle, recessive mutations could have a profound effect on future human populations. Wright maintained that "from the genetic standpoint I don't think that the dose received by an individual really means anything."[32] Muller in contrast worried about the effects of radiation-induced mutations on individuals and their progeny. Muller also maintained that all mutations were deleterious, while Wright had concluded that many mutations led instead to neutral alleles. Wright was correct on this point, although there was little relevant evidence at the time. Looking back on the discussions 40 years later, James Crow considered Sturtevant's position the best because it recognized that most radiation-induced mutations were not small chromosomal changes such as most spontaneous mutations, but deletions and broken chromosomes.[33]

The reports of all the panels were published in a joint, brief summary document and released at a press conference in June of 1956.[34] Neither of the terms "permissible dose" or "maximum permissible dose" appears in the report. The recommendation that had the greatest public effect was to reduce the medical use of X rays as much as consistent with medical necessity. The panel had estimated that the average exposure to the gonads of individuals in the U.S. up to age 30 was about 3 roentgens resulting from medical and dental uses and about 4.3 from inescapable background radiation. In comparison, the panel estimated that even if weapons tests continued at the rate of the past five years, the additional comparable exposure would be in the range of 0.02 to 0.5 roentgens. The report recommended that the total average radiation to an individual's gonads from conception to age 30 should be limited to 10 roentgens and no individual should experience more than 50 roentgens to the reproductive cells up to that age. Although these limits were not expected to yield pathological effects in people, they could yield on the order of 5 million mutant genes in the total U.S. population. To assure the collection of important data, individual records of accumulated lifetime exposures should be kept.

Public interest in the report was high. Newspapers covered the findings extensively and a Congressional hearing was held at which Bentley Glass represented the geneticists. The effect on medical practice in the United States was immediate. By July, the National Tuberculosis Association had sent advice to all its affiliates to give up the use of X rays and fluoroscopy for young people and to be conservative using them with all others. A *Science* magazine editorial noted that the report of the Medical Research Council of Great Britain, published almost simultaneously, had come to some similar conclusions.[35] On the question of how much of an increase from current levels of radiation exposure would double the human mutation rate, the best estimates were from 30 to 80 roentgens in both reports. But on the politically charged question of the effect of fallout, the Americans emphasized that all radiation is damaging while the British concluded that increased hazards from fallout are "negligible." Publication of these reports coincided with the increasing effort by many scientists and citizens, including Pauling, to obtain a halt to testing.[36]

The day after the report was released, Weaver submitted his resignation to Bronk.[37] The work was taking too much time and besides, he argued, the future agenda was more technical and required an expert chairman. Weaver also believed that he had worn out his welcome with the geneticists. He sent Bronk his candid thoughts on a replacement.[38] Ten of the panel members were "definitely ineligible" and four others "not ideal" to take over. There was only one person left and that was Beadle, who agreed to become chair, to Weaver's delight.[39] Whatever relief Beadle and the panel may have felt with the completion of their "short report" was quickly dissipated by the serious work still to be done.

Beadle's commitments would have overwhelmed most people and surely protected him from his tendency to be "easily bored."[40] Often he was a fleeting presence at home and in the biology division, although his Cal Tech business always got done. His exhausting schedule and frequent flights to the east coast benefited the airlines as well as the scientific community. The Genetics Panel met in May of 1956 in New York. Later, after the spring 1956 meeting of the AAAS board, he spent more than two weeks crisscrossing the country. The first stop was the Cold Spring Harbor Symposium for four days. Then, after flying home for commencement and a day of gardening, he flew east to receive an honorary degree at Wesleyan University and rejoin the symposium for two more days. Returning to Pasadena, he worked for four days on his paper for another symposium and then attended that four-day meeting on genetics and biochemistry in Baltimore before finally settling down.[41] On one occasion, Muriel Beadle scolded him for returning one morning from Washington on an overnight flight and leaving again for the east at the end of the same day.[42]

At the same time, Beadle was teaching the general biology course at Cal Tech and would deliver the Condon lectures at the University of Oregon in March and May. He also had continuing concerns about his father who wanted to leave the nursing home in Ashland, Nebraska and return to Wahoo to live on his own.[43] Ruth, Muriel, and George worried about how the 89-year-old man would take care of himself and were concerned about the costs. They thought that CE would reject suggestions that he come to live with Ruth or George and Muriel in California, or move to the Lutheran House for the Aged in Fremont, Nebraska. But their father surprised them. He decided that he didn't want to "batch any more at my house" and by May was pleased with his new quarters in Fremont.[44] Beadle was apologetic about being unable to stop off in Nebraska on one of his many trips across the country or to take care of his father during a proposed visit to Pasadena during Christmas week because he would be away at the AAAS meeting.[45] Beadle was dutiful, but apparently had few compelling connections with his father.

At its very first meeting in November of 1955, the Genetics Committee commissioned a full report documenting the nature and magnitude of atomic radiation and the associated hazards. When Beadle took over as chair in midsummer of 1956, the report was still unfinished. He also had to worry about organizing a longer follow-up to the panel's own report, which would include a description of the research agenda that was required if the genetic effects of radiation were to be better understood by scientists and the public.

A draft of the Laughlin-Pullman report was ready for review in the spring of 1957.[46] The intention was to keep it confidential during the review process, but with 150 copies in circulation, the press had little trouble getting access. Unauthorized publication of some of the data on gonadal exposure caused a big stir.[47] The new estimate of radiation exposure from medical and dental procedures was 50% higher than in June 1956. Professional and health organizations weighed in as did private physicians, all concerned that the public would now take a negative view of the use of X rays. Bronk heard from irate physicians claiming that the report would undermine confidence in the medical profession and even in the integrity of the Academy. The Picker X-Ray Company, worried what the report might do to its business, sent objecting telegrams to the committee members and to Bronk.[48] Bronk took over the task of dealing with most of these problems while Beadle contended with the controversy that was heating up among the panel members.

A subcommittee chaired by Muller and including Beadle had been appointed at the panel's first meeting and was charged to prepare a blueprint for a future research agenda. The panel struggled with this off and on, trying to come to grips with its internal disagreements. One contentious issue was whether to recommend research with mammals. Radiation effects on humans would be of key interest to the general population, but experimental approaches were impossible and epidemiological data were scarce and difficult to interpret even when reliable. The mouse was well developed as a model mammalian organism, but no one knew at this point whether the mouse data were relevant to humans. Demerec in particular argued that the research should focus instead on experimental microorganisms, plants, and animals such as *Drosophila*.

The limited scope of the human epidemiological data became apparent when Beadle circulated a draft paper on the quantitative relation between leukemia and exposure to ionizing radiation by his Cal Tech colleague Ed Lewis.[49] Lewis had gathered the available data including those that related radiation doses to the induction of leukemia in radiologists and, with Beadle's help, those collected by the Atomic Bomb Commission in Japan from the survivors at Hiroshima and Nagasaki. Regardless of the population considered, his calculations showed that "the probability of developing radiation-induced leukemia is estimated to be two in a million per rad (a unit of adsorbed dose of radiation) per year." He suggested that the explanation for this "linear relationship between the incidence of leukemia and the dose of radiation" may be radiation-induced mutations in somatic cells, a hypothesis first discussed by Muller in 1927[50] and alluded to by Sturtevant in his 1954 paper. The result was consistent with the linear relation between radiation dose and mutations that had already been established in *Drosophila* and was therefore taken to support those on the panel who argued that model organisms should be emphasized in a proposed research agenda.

Lewis, a *Drosophila* geneticist, was skeptical about several commonly voiced opinions about radiation effects in humans. His leukemia data contradicted, for example, the view of Cal Tech physicists who were convinced that there had to be a dosage threshold to the effects of radioactive fallout rather than the linear relation Lewis found. Similarly, he worried about Libby's public pronouncements on behalf of the AEC regarding the effects of strontium-90 fallout: "the worldwide health hazards from the present rate of testing are insignificant."[51] Lewis challenged this conclusion head-on with his own analysis of the strontium data.[52] The response from the AEC and the Congressional Joint Committee on Atomic Energy was angry denouncement in the pages of *Science* magazine, and DuBridge and Robert Bacher took the

AEC's part.[53] Two years later, Lewis evaluated the radiation dose to the thyroid glands of infants and children from the excess iodine-131 ingested by drinking milk from cows grazing on grass contaminated by fallout. He concluded that the youngsters experienced annual radiation doses as much as two times the annual dose from natural background radiation.[54] After seeing Lewis's manuscript, Bacher warned him not to publish it. DuBridge also tried to influence Lewis and arranged for him to talk with a radiologist who, in addition to his closed mind on radiation hazards, also expressed racist thoughts. When Beadle heard about the interview, he expressed disgust and wondered "how can he (DuBridge) have such friends?"[55] Beadle, in his customary behind-the-scenes way, supported Lewis.[56] Pauling used the strontium-90 analysis in his public campaign to halt testing of nuclear weapons.[57]

Beadle, Sturtevant, and Crow worked on a draft report in the spring of 1957.[58] Muller revised the draft and Demerec sent his own views to Beadle in July: "at the present stage of our knowledge, research in human genetics and research with mammals could not make a significant contribution toward solution of the problem of genetic effects of radiation, because of the tremendous complexity of the mechanisms involved." He described a visionary research agenda and correctly predicted that it would require large sums of money and several decades to accomplish. It took another 25 years to answer even partially the fundamental questions Demerec raised. They included the chemical and physical structures of genes, the structure of chromosomes, gene action, the interaction of genes within a genome, the chemistry and mechanism of both spontaneous and radiation-induced mutations, and experimental tests of theoretical population models.[59]

When the Genetics Panel met in August of 1957 they had new worries and disputes. If their report described what research ought to be done, would it be assumed that anything omitted ought not be done? The only workable solution to all their disagreements was to describe the scope of the proposed research in a general way.[60] On the way to the meeting, the Beadles stopped in Ithaca so that Muriel could meet Jo and Adrian Srb.[61] It was to be the first of several visits between the two families on both coasts.

Another problem confronting the committee was the estimation of the number of mutations expected from a given radiation dose. Data existed for *Drosophila*, but they concerned mutations in selected genes with easily identified phenotypes, and there was no reason to believe that mutation rates would be the same for all genes or even that all mutations in a given gene would have the same phenotype. Additionally, no one knew whether mutation rates for humans and for experimental animals such as *Drosophila* or even mice were comparable.

In 1958, while at Oxford on sabbatical (see Chapter 15), Beadle was still trying to complete the report. Then, in December, Sewall Wright proposed including an appendix concerning "some possibilities of appraising genetic effects of radiation on man, especially with respect to their contributions to social burden."[62] He developed a "classification of human phenotypes with respect to social value" and treated the question "in terms of the balance between contribution to society and social cost." Such controversial issues as the heritability of mental disorders, intelligence, and the cost to society of physical disability were at the core of his analysis.[63] The proposed appendix caused a storm in the committee. A particular bone of contention was Wright's insistence on stating that an increased rate of mutation may be beneficial in some instances because it leads to diversity in the human population.

Muller was adamant against including the appendix, considering it "irrelevant to the matters of radiation damage actually at issue" and tending "to mislead the public by inducing them to think that the genetic radiation-damage is less extensive and less well demonstrated than has been claimed by most of those working in the subject."[64] Beadle concluded that the only way to finish the report was to eliminate all contentious matters. The draft was reduced to only ten lines reaffirming the 1956 conclusions. When the committee of the chairmen of the various panels learned of this at its meeting in October it refused to accept the short report and Beadle had to go back to work.[65] Finally, the Genetics Committee forwarded its report and the appendix with the statement that "No satisfactory and practicable way of resolving this difference of opinion has been found."[66] The executive secretary of all the BEIR Committees decided that the appendix should not be published with the report.[67] Beadle warned him that Wright and other panel members might dissent if the appendix was omitted, although he himself would not. He recommended that the entire committee be polled by wire. He also cautioned that "the main body of the report was beat out word by word and that feelings may run high if there are editorial modifications that could change meaning or even slight shades of emphasis. Geneticists are touchy it seems."[68]

By the time the report was forwarded, Beadle had already told Bronk he was ready to resign from the chairmanship and the panel.[69] He recommended that Bronk rotate the panel membership provided that Wright and Muller were rotated off at the same time to avoid the impression that Bronk was favoring one or the other.[70] But Beadle was still chairman in May of 1960 when the updated summary reports were published.[71] The serious lack of knowledge about the effects of low doses of radiation was a theme common to the reports of all the panels. All the members of the genetics panel signed the report, which itself reflects some of Wright's ideas. The contentious appendix, signed by Wright alone, was included. Later that summer Beadle was finally relieved of his responsibility when, at his suggestion, Bronk appointed James Crow as the new chairman.

Depite all the uncertainties, the report stated "that from a genetic point of view there appears to be no threshold level of exposure below which genetic damage does not occur." The word "appears" was an important hedge because the absence of a threshold was controversial and is disputed even today.[72] Coming down on the cautious side, the panel confirmed its earlier conclusion that "the average gonadal dose accumulated during the first thirty years of life should *not* [*sic*] exceed 10 roentgens of man-made radiation." The strongest and longest section of the report is that detailing a future research agenda that includes both practical aspects such as dosages experienced by the U.S. population and basic research. Research toward a broad understanding of mutagenesis by radiation and chemical mutagens is singled out for special attention. Altogether, the report reflects a sophisticated and up-to-date understanding of genetic research, including careful attention to the then still somewhat controversial concept "that some malignant neoplasms owe their origin to somatic mutations."

Beadle took a leading role in convincing the American Cancer Society to adopt a broad view of the kinds of research that were relevant to fighting cancer. Then, in 1956, in addition to his other tasks, he was named chair of the Scientific Advisory Council that was to administer the society's new policy including the review of grant and fellowship proposals. This program marked the beginning of the appointment of American Cancer Society professors, which allowed research universities to enlarge their permanent faculty with external funds.[73] In April of 1958 Beadle went from England to a UN meeting on radiation effects and he and Muriel managed some time off in the Alps. A week later, he attended a National Academy of Sciences Symposium in Washington on genetics and radiation hazards that he had organized.

Most significantly, chemical genetics, the concept that Pauling and Beadle had envisioned more than a decade earlier, was flourishing at Cal Tech. In 1957, Robert Sinsheimer was appointed professor of biophysics in the biology division. After Sinsheimer's six-month-long stint learning about phage in Delbrück's lab in 1953, he had returned to the department of physics at Iowa State and began to characterize a very small phage, ϕX174, that was discovered in the 1940s. Beadle admired this elegant research and invited Sinsheimer to give a seminar at Cal Tech late in 1956. Later published, the talk included a rigorously argued summary of the eight different experimental findings that linked DNA and genes.[74] Beadle promptly recruited the 37-year-old Sinsheimer. In Pasadena he discovered that the ϕX174 genome was a single, circular strand of DNA rather than a double helix.[75] One of the clues was finding that the amounts of adenine and guanine did not equal the amounts of thymine and cytosine, respectively, as required for a double helix.[76]

Perhaps the most cogent affirmation of Beadle and Pauling's original vision was the elegant experiments by Matthew Meselson, a graduate student in chemistry, and Franklin Stahl, a postdoctoral fellow in biology. They proved that the replication of a DNA double helix is, at least in bacteria, semiconservative. That is, each of two new daughter double helices contains one strand from the parent DNA double helix and one newly synthesized strand.[77] This demonstration of an important prediction from the double-helical structure of DNA is described in virtually every textbook on molecular biology and is one of the most famous experiments in the field.[78]

By the fall of 1958, the biology division was all that Beadle had hoped for. Cal Tech was widely acknowledged to be one of the world's foremost centers of molecular, cellular, and developmental biology. Funds for research and fellowships were generous by any standard, teaching loads were not onerous, and by and large students and faculty formed a collegial group.[79] Beadle's leadership had fostered a collegial atmosphere among a group of outstanding scientists. He established consensus regarding divisional policies by speaking individually to people before holding a meeting for decision. Consequently, division meetings were without rancor or major disagreements.[80] The laboratory buildings were spacious and modern. Faculty members were invited to describe their work to potential donors at frequent evening meetings at the Beadle home.[81] Muriel participated in all this with enthusiasm and pride, and her own professional activities must have lessened the impact of Beadle's frequent absences. When his major outside activities began to wind down, he had established a reputation nationwide as a thoughtful, nonideological, and pragmatic voice on challenging questions of science policy. Even in policy debates, he was rigorous in respecting scientific knowledge. His dedication to duty cannot explain his intense, even frenetic, engagement in national affairs. It seems likely that he garnered a great deal of satisfaction from the contributions he made, just as he had done with experimental science.

Beadle accomplished a great deal without ever losing his identification as a Nebraska farm boy. This is apparent even in the tape of a lecture about genetics he gave to the Los Angeles general public. His voice is high-pitched with a nasal twang. The words maternal and paternal are pronounced "may-ternal" and "pay-ternal," for is pronounced "fer," and there are four syllables in the word interesting.[82] To explain mutations, he described them as analogous to the infrequent errors a typist makes in copying a text, a device he and

Muriel later used in their popular book.[83] Everyone in the audience received a paper strip containing phenylthiocarbamide to chew on and then he counted how many people found the taste bitter and how many tasteless. As expected from the known genetic frequency of the dominant "taster" trait, 70% of the audience experienced the bitter reaction. Beadle's style during the talk and the question period following was typically simple and good-humored. When asked about the perceived conflict between science and religion, he responded that there was no conflict because the two deal with different questions. R. Metzenberg remembers that "You were kind of aware all the time that he was from Wahoo, Nebraska...In the academic world we always try to seem as smart as we can. I think that caused a lot of people to underestimate him. They bit for it and he was really a very shrewd man."[84]

CHAPTER 15

Oxford and the Nobel Prize

By the summer of 1957, the task of restoring the biology division to its Morgan-era eminence was near complete. Although Beadle's own laboratory work had ended several years before he arrived in Pasadena, he had become an active and articulate expositor of the discoveries and events that were transforming biology. Public lectures and meetings with community groups and even with his own nonscientific Cal Tech colleagues helped raise the awareness of the "biological revolution." But most of his energy and wisdom during the 1950s had gone into assembling a collegial group and that had prospered scientifically, with many becoming leaders in their respective fields. Although teaching undergraduates was valued and acknowledged, the primary focus in the division was research and training of graduate students and postdoctoral fellows. Cal Tech was widely acknowledged as a major center of molecular, cellular, and developmental biology, and students and visiting scientists competed for the opportunity to work with one or another of the faculty. Beadle's day-to-day oversight of the biology division's activities had given way to commitments to commissions, committees, and societies that sought his guidance. But these, too, were ending, winding down, or could be postponed. He could seriously, perhaps enthusiastically, consider an invitation from Oxford University's Balliol College to accept the George Eastman Professorship for the 1958–1959 academic year. He might well have seen this as an opportunity to take stock of what was next for his career: continuation in science, education, or public service.

DuBridge approved Beadle's request for a sabbatical and designated James Bonner as acting chairman of the division in his absence. The Eastman Visiting Professorship, endowed in 1929 by George Eastman, founder of the Eastman Kodak Company, was intended to permit senior American scholars of the "highest distinction" to spend a year at Oxford. Nominees are selected by the University's vice-chancellor, representatives of the Eastman Trust, and, because Balliol College is the designated college affiliation for the visitor, by the Master of Balliol College. The list of previous Eastman Professors was a veritable honor roll of American scholarship. Ten years before, Linus Pauling held that fellowship, and the year before Beadle arrived that honor was accorded to George F. Kennan, the distinguished statesman.

The Beadles traveled to England by ship, arriving in Liverpool and continuing by train to London.[1] Beadle picked up a new car, a German Borgward, he had ordered before leaving, and drove to Oxford while Muriel and Redmond went on by train. Muriel's first impression as their train pulled into Oxford was that "it looks like the outskirts of Peoria," a sentiment that was not improved by the seemingly chaotic setting when they disembarked. Their taxi ride through the commercial traffic of Oxford to what would be their home in the suburb of Headington failed to dispel the gloom. The cottage, variously estimated to be 150–300 years old, had been rented for the Beadles by the college. Muriel immediately dubbed it "a no-nonsense stone box devoid of frills," but a tour of the rooms revealed that, although somewhat charming, it was weak on the amenities Americans had come to expect. The "Rayburn," their landlady Lady Headington's name for the plain old coal stove, was intended for both cooking and producing hot water for the entire house. Learning to feed and care for the Rayburn was only the first of many chores the family acquired. Staying warm required tending three paraffin stoves, three electric fires, a fireplace, and the Rayburn.[2] In time, they adjusted to the English ways, especially after Beadle put his own distinctive signature to the house's garden by planting ten dozen newly purchased spring bulbs. Muriel was prompted to remark "No soil is alien once you've planted bulbs." Redmond was enrolled in one of Oxford's private schools where he was expected to make up for subjects he had not yet taken in his Pasadena school.

The Borgward automobile proved to be nearly useless in Oxford because of both the awkwardness of getting in and out of their garage and the horrendous traffic congestion. The local people were amused when they learned that Beadle preferred the two-mile walk along Cuckoo Lane, Woodland Road, and Mesopotamia Walk to his lab in the botany department. However, the car provided the Beadles with "wheels" to explore the English countryside and the historical sites they had only read about before. Brief trips to Stonehenge and Bath introduced them to the charms of regional pubs, two-star inns, and the perils of English drivers.

The first visit to Balliol also jarred the Beadles' image of Oxford. "It was just a chain of dirty, gray stone buildings with a Scottish baronial tower on one end, at the base of which was a door that looked as little as a mousehole."[3] A tour of the college soon taught Muriel some of Balliol's quirks: non-Fellows are excluded from parts of the college grounds, and it's an unbreakable rule "that women may not dine in Hall during term." It was a rule that was to keep Muriel and Redmond eating alone during many nights while Beadle went off to enjoy the offerings of a good chef and a fine cellar. Muriel's solution, as it was for other college wives, was to have an independent social life, most often participating in women's clubs activities. "Sherry party after sherry party" begins the college social season and continues with little relief until the term ends a few weeks before Christmas. The social hiatus between terms is temporary as the partying is resumed as each term begins. During the year, the University Vice-Chancellor and Master of Balliol hosted formal dinner parties to which Muriel was invited and the Beadles, as college protocol dictated, took their turn at entertaining the friends and acquaintances who had entertained them.

Eastman Professors are expected to give a regular schedule of lectures during the Michaelmas (fall) and Hilary (winter) terms and a set of informal seminars throughout the academic year. Beadle initially proposed to give sixteen lectures each term, a number he was told was required by the statutes. However, his host professor thought that number excessive and encouraged Beadle to keep the number to eight for each term, which meant one lecture per week. The general theme of the lecture series was "topics in genetics" and judging from the titles, he described the emerging concepts of genetics and their impact on society. When the time came for him to give his public lectures, he pondered whether they should be general or "meaty," something about DNA. He began his first lecture; "I propose, ladies and gentleman, to lecture American style." At the end, one of his colleagues complimented him for having made it solid and meaty with enough showmanship to keep the audience awake. The professor asked what he had meant by the American style. "Not a thing," Beadle replied. "I was just protecting myself in case anything went wrong."[4]

Beadle, of course, had several encounters with Oxford's idiosyncrasies and Balliol's traditions. He never mastered the protocol of when and where to wear the gown emblematic of an Oxford don. Nor could he contain his amusement at being required to pledge to obey a set of what he thought were silly regulations about taking an oath before being granted eligibility for reading privileges. His eagerness to keep up with the world of science was also frustrated by the arcane arrangement for locating a book or periodical at the 50 Oxford libraries.[5] Finally, he arranged with his Pasadena office to airmail monthly editions of the *Proceedings of the National Academy of Sciences*. Although he often lunched and dined at the college, it was more a club than a headquarters for him. His attendance at college meetings was infrequent because he found it difficult to understand the Oxonian speaking style during the fellows' political debates.[6] Often such debates invoke obscure references to long-forgotten college history and people, and occasional attempts at erudition are laid on to embarrass or confuse the opponent.

As the year progressed, Beadle was less and less impressed with the quality of the science students. At one of the college gatherings, he remarked that one of the students he was tutoring who had "read" botany would be unlikely to be admitted to Cal Tech for Ph.D. study because of his hopelessly inadequate background in math and physics. He was disappointed that students rarely asked any questions during his genetics course lectures, even though it was clear later that they had not understood what he was saying. But several Balliol fellows who had taught at American universities confessed that they were annoyed at having been interrupted by questions during their lectures. Beadle admired one chap, admittedly drunk, who grabbed him by the lapels after one of his lectures and said "Professor Beadle, that was a most enlightening lecture. Far better, if I may say so, than the series you gave at University College, London. Much less turgid."[7]

Not long after the Beadles' arrival at Oxford, rumors began to circulate about a Nobel Prize for Beadle. The physics and chemistry prizes are selected by the Royal Academy of Sciences, the medicine or physiology prize by the Royal Karolinska Institute, and the prize in literature by the Royal Swedish Academy. Nobel gave the responsibility for awarding the peace prize to the Norwegian Parliament. The Nobel committees, which actually evaluate the nominations, pride themselves on maintaining utmost secrecy. Consequently, in most years, the announcement of the prize recipients in Stockholm and Oslo during mid-October is a surprise. Occasionally, however, gossip as well as leaks about the probable winners circulate, as they did in 1958. Sometimes the rumors were unfounded, so as the usual date for announcing the recipients approached, Beadle was wary and determined not to celebrate prematurely. He was not, however, immune to anxiety as he discovered that he'd given his last lecture backward.

In spite of the Nobel committee's care against premature disclosure, news was leaked to newspaper reporters in Stockholm, who alerted the international wire services. For seven long days a nonstop press conference went on in the homes and labs of that year's Nobel laureates. Reporters and photographers hounded Tatum in New York, Lederberg in Wisconsin,[8] and Beadle in England. Reporters asked the same inane questions such as, How do you feel? and, How will you spend the money? Beadle's answer was fine and none of your business. Finally, on October 30, two days after the announcement in Stockholm, Beadle received the official telegram from the Nobel Foundation that he would share the Nobel Prize in physiology or medicine with Edward L. Tatum and Joshua Lederberg. Commenting on the drawn-out publicity, Muriel felt that getting the Nobel Prize "would be an unalloyed joy if it came like a bolt from the blue."[9] Beadle's still sharp-minded father, now 91 years old and living in a rest home in Fremont, Nebraska, quipped that "if heredity is responsible for any of George's fame, it must be on his mother's side. Hattie would have been so proud of him today." Recalling that Beadle kept pet coyotes and opossums and loved experimenting in the garden, he commented that the farm was "the best place you can raise a boy."[10]

Beadle's prize was greeted with pleasure and considerable creative humor by his friends and colleagues at Cal Tech. DuBridge and the chairman of Cal Tech's board of trustees sent congratulatory telegrams expressing their pride.[11] Beadle's former student, Adrian Srb, eschewed the usual congratulatory message for one that signaled the close personal relationship they had formed over the years: "I'm boxed, Boss. The only thing it seems for me to say is that I'm enormously happy about it all, mostly because you must be also." Srb recalled that when J.B. Sumner received the Nobel Prize for the crystallization of urease, he didn't care what people thought. The prize had come to him and he was glad, and didn't care who knew it. Srb added, "If a funny old guy like Sumner was glad, why shouldn't you be?"[12] In replying, Beadle expressed "real doubts about deserving such an honor" and observed "it's too bad that when so many people contribute in essential ways to a given development in science, one or a few get picked out to be wined and dined, and given medals and fat checks."[13]

One congratulatory cablegram was signed N. Crassa, the fungus Beadle made famous. A group of Cal Tech biology graduate students wired, "Tremendously pleased and proud of you. Come home. Bring money." Beadle's response was reputed to be "Thanks fellows You too can win Nobel Prizes. Study diligently, Respect DNA, Don't smoke, Don't drink. Avoid women and politics. That's my formula."[14] Perhaps the most whimsical message was expressed in a nonoverlapping code based on then current speculations about the genetic code. After a good deal of effort, Beadle deciphered the message as "Break this code or give back Nobel Prize." Beadle responded: "Thanks for your telegram. I'm sure it's a fine sentiment." Soon after, Beadle admitted in an air letter to Max Delbrück that "I'm such a poor cryptographer and so many friends and relatives have already helped spend the money that I can't give it back. What shall I do?"[15] At the Nobel ceremony, the coding game continued when Beadle received a special double helix made of toothpicks in four colors, the colors representing four elements of still another coded message: "I am the riddle of life, know me and you will know yourself." Beadle's reply after returning to Oxford was "Know my self but;" the rest was undecipherable.[16]

This American jocularity contrasted with the response of his English colleagues. To spare him the embarrassment of public attention, none of them mentioned the award in company. Those who happened to run into him when he was alone cleared their throats diffidently, offered a mild handshake, and murmured "'splendid news about your-er-honor; or simply, Good show!" or well done. There might even have been a call for a round of sherry at Balliol College's High Table to mark the occasion. Two former English Nobel laureates, Sir Hans Krebs (1953) and Willis Lamb (1955), briefed Beadle and Muriel on the ceremony and festivities they would encounter in Stockholm.[17]

Besides Beadle, Tatum, and Lederberg as recipients of the physiology or medicine prize, Fred Sanger of Cambridge University won the prize in chemistry for having determined the amino acid sequence of insulin, the first protein sequence to be solved. Sanger had developed methods for smashing the insulin molecule into small fragments, separating the fragments, determining the amino acid sequence of each fragment, and then determining the correct order of fragments in the full-length molecule. That achievement established one of the fundamental axioms of modern biology: A protein possesses a unique sequence of amino acids rather than being a mixture of related sequences.

The physics prize was awarded to three Soviet scientists, P.A. Cerenkov, I.M. Frank, and I.E. Tamm, for their discovery and interpretation of the bluish light emitted by charged particles moving at speeds approaching the speed of light. Cerenkov made the initial observation, which is why the effect is named for him, but it was Tamm and Frank who provided the theoretical explanations of the phenomenon. The literature prize was awarded to Boris Pasternak, a noted Russian poet and author for his remarkable contributions to contemporary poetry as well as to the grand tradition of the Russian novel.[18] No particular mention was made of his book, *Dr. Zhivago,* which detailed the betrayal of the socialist vision and the continuing brutality of Stalin's stifling rule. The book had been banned in the Soviet Union but appeared in the west after the manuscript was secreted to publishers in Italy and France. It was an embarrassment to the Soviet regime, and the recognition by the Swedish Academy triggered a backlash that embroiled the Soviet literary world in an internecine war with Pasternak in the crossfire.

Elated by the prize, Pasternak telegraphed the Swedish Academy: "Immensely thankful, touched, proud, astonished, abashed."[19] Within days, however, the board of the Soviet Writers Union met to condemn him, and the Union's Literary Gazette mounted a violent attack on him and the Swedish Academy, charging them with "a well conceived ideological act of sabotage ...permeated with lies and hypocrisy." The Union, which until then regarded Pasternak as one of their most distinguished colleagues, expelled him, thus depriving him of the perks to which he was entitled, and all chances for literary earnings. Even his close friends turned on him, calling on the Union to "deprive the traitor of Soviet citizenship and expel him from the country." Pasternak had to send a second message to the Swedish Academy: "In view of the meaning given to this honor by the community to which I belong, I should abstain from the undeserved prize that has been awarded to me. Do not meet my voluntary refusal with ill will." In a humiliating letter to Krushchev, Pasternak was forced to admit "mistakes and errors" and to plead that he not be exiled because "leaving his native land would be the equivalent of death." This was the second time in the history of the Nobel Prize that a recipient declined to accept the prize; the first occasion was in 1939 when Hitler forbade A.F.J. Butenandt (chemistry) and G. Domagk (medicine or physiology) from accepting their prizes.

Krushchev had reason to regret the way the affair had been managed. "Now that I am approaching the end of my life as a pensioner, I feel sorry that I did not support Pasternak. I regret that I had a hand in banning his book and that I supported Suslov. We should have given the readers the opportunity to reach their own verdict. Judgements should depend on the readers. I'm truly sorry for the way I behaved toward Pasternak. My only excuse is that I didn't read the book."[20] Pasternak died in 1960, and later under Gorbachev's leadership, the Writers Union repealed Pasternak's expulsion and permitted publication of *Dr. Zhivago* in the Soviet Union.

Muriel and Redmond accompanied Beadle to Stockholm, but neither his sister Ruth nor his son David, who was living in Arizona then, made the trip. Tatum's wife Viola and his two daughters, as well as Lederberg's wife, Esther, rounded out the biology contingent. The six or so weeks following the announcement of the prize were taken up with preparations for the trip. Beadle, Tatum, and Lederberg had to prepare the required lectures and everyone had to arrange for the formal evening wear ("white tie and tails" for the men and long gowns for the women) needed for three separate occasions. Muriel was accustomed to dressing up, but Esther Lederberg found it a challenge to find something suitable to her taste. Because Tatum's first wife's concession allowing their daughters to accompany him to Stockholm came at the last moment, they were unable to be "properly attired" and stood out at the formal ceremonies in their short dresses. [21]

When they arrived in Stockholm, the Beadles were introduced to a young Swede from the Foreign Service, whose responsibility it was to brief them on the time, place, proper dress, and etiquette for the various social events. The central heating at the Grand Hotel, where they all stayed, was for the Beadles a luxury they had foregone since coming to Oxford. The days were filled by scheduled obligations interspersed with luncheons and informal gatherings.

There were several press conferences at which the Swedish press wanted them to explain the meaning of their accomplishments. One of the highlights for the Swedish populace was the television roundtable, during which all of that year's laureates engaged in a broad give-and-take discussion of such matters as foreign affairs and futurology.

The formal award ceremony took place in Stockholm's elaborately decorated Royal Concert Hall where, besides the family and friends of the laureates, Sweden's scientific elite and the royal family were gathered. Following a trumpet's blare, accompanied by a triumphal march, the laureates filed onto the stage and took their red plush seats facing the King and Queen seated on their thrones flanked with numerous Swedish and foreign diplomats and leading Swedish scientists representing the various Nobel committees. By tradition, the physicists lead the entry followed by the chemist(s), then by the winner(s) of the physiology or medicine prize. Normally, the literature winner, absent in this case, would have been last in the file. Each laureate, in turn, was summoned to center stage, for a recitation of their accomplishment and the bestowing of the gold medal and certificate from King Gustav VI. Later they received the monetary award. In this instance, Beadle and Tatum shared half of the $41,250 prize and Lederberg received the other half. Ironically, the Tatum's divorce decree had stipulated that his former wife, June, would receive half of his prize money if he were ever to win the Nobel Prize. Marion, to her last days, resented that she had not made a similar pact.[22]

The ceremony was followed by a dinner at Stockholm's grand City Hall. Its Blue Hall had a golden glow enhanced by the candlelight lighting, and flowers provided by the citizenry of San Remo, Italy, where Nobel died, decorated the hall and dining tables. Orchestral music signaled the time for the King to escort Mrs. Cerenkov down the grand stairway where nearly 1500 notables were already seated. Then the laureates entered, each with a lady of the royal family on his arm, followed by their wives escorted by men of the King's family. Orchestral music and student-led choral groups serenaded the assembled diners throughout the meal. Customarily, the laureates address the gathering near the end of dinner. Tamm spoke for the physicists and startled the audience by beginning in Swedish before continuing in English. Sanger spoke briefly, then Lederberg added some comments pertaining to his half of the prize, and Tatum, speaking on behalf of Beadle and himself, followed.

The next night, the laureates and their wives dined in the Royal Palace with the King and Queen; Muriel called it "a real party" probably because it permitted more relaxed conversations with the guests. The remaining days were filled with shopping, being tourists, and partying. Early one morning, George and Muriel were awakened by a parade of beautiful young women, wearing crowns adorned with burning candles, bearing coffee and cakes to their room, and singing the Santa Lucia hymn that celebrates the beginning of winter's Festival of Light. Crowning the social events of their week's stay was the traditional formal dinner dance and hijinks with Sweden's university students. With the close of the festivities and fatigue setting in, the Beadles headed back to Oxford. Unfortunately, fog-bound London required their flight's diversion to Manchester, from which they traveled by slow train and, after a considerable delay, arrived in Oxford, at 2:45 a.m. Even a night's rest could not dissipate the let-down feelings from the previous week's celebrations. Beadle tried to lessen Muriel's gloom by commenting, "Wow! another week like that one, and I'd never have gotten you into the kitchen again."[23]

Over the course of 100 years of the Nobel Prize, there have been frequent accusations of omissions and misjudgements by the selection committees. Because Nobel's will insisted that no more than three recipients be cited for any one prize, the list of candidates has to be pared down to the one, two, or three individuals who are considered most closely associated with a particular scientific achievement. The records of the Nobel committees' deliberations culminating in the selection of Beadle, Tatum, and Lederberg will be, according to their 50-year rule, unavailable until 2008. But almost surely, the committee realized that there were other key contributors to the one gene–one function idea. One such claimant was Boris Ephrussi, Beadle's colleague in the studies of the genetic control of eye color in *Drosophila* during the 1930s. That partnership led them to propose that each step in the pathway for producing the mature eye pigments was controlled by a single gene. In their writings, they hinted that the gene might be acting through enzymes, but they recognized that *Drosophila* was entirely unsuitable to pursue the gene–enzyme connection at the chemical level. The inability to go further with the analysis was the inspiration for Beadle and Tatum to change strategy and adopt a new biological system.

The Nobel citation emphasized that the prize was intended to recognize the development of the approach that Beadle and Tatum had employed to connect genes with their metabolic functions. Speaking for the Nobel committee, Professor Torbjörn O. Caspersson put their contribution in perspective. "The characters, which are transmitted by the genes from generation to generation, present a picture of bewildering multiplicity. This very multiplicity of the genes' effects made it difficult to attack experimentally the problem of their structure and manner of functioning. The situation was radically changed by Beadle and Tatum, who, through a daring and astute selection of experimental material, created a possibility for a chemical attack upon the field. The discovery provides our best means of penetrating into the manner in which the genes work and has now become one of the foundations of modern genetics. Its importance extends over other fields as well."[24] Thus, it was not only the formulation of the one gene–one enzyme paradigm that warranted the prize, but also the creation of the new strategy and experimental approach that made it possible to explore gene function in greater detail than had been previously possible.

Ephrussi was initially deeply disturbed at having been excluded from the prize. After all, he and Beadle had provided the first insight that genes acted through the production of proteins. Those who knew him well were aware of how disappointed he was with the Swedish Academy.[25] Some French biologists, and particularly the press, regarded the Academy's decision as being "partial or insufficiently informed." *L'Express*'s comments were typical of French press reaction: Ephrussi and Beadle, they claimed, had "opened a new path demonstrating that genes...act by modifying chemical reactions within an organism." The article continued, "If the Nobel Prize is to reward finished work, the award was fair. If, however, it is to recognize those pioneers who have opened a new path towards the understanding of life, the name of Ephrussi should have been associated to that of Beadle."[26]

Soon after Ephrussi learned of the news, he voiced his distress in a letter to his friend, Tracy Sonneborn: "I must admit I was disturbed these days by the Nobel Prize. I hate to admit it: I suddenly felt my life wasted."[27] Concerned for his friend's reaction, Sonneborn encouraged Ephrussi to "see the whole thing in its proper perspective" and that his life and work were "not to be evaluated as success or failure on the basis of winning or not winning a Nobel Prize." He concluded, "I hope you can reconcile yourself to yourself and still preserve the zest and drive for discovery which has in the past resulted in a degree of fame and glory for you which should also be a great satisfaction."[28] Ephrussi was moved by Sonneborn's soothing encouragement and responded at length about his second thoughts. "First of all, I want you to know that my momentary feeling of distress about the Nobel Prize is all over and left in me only a burning desire to do more work... I can, in all conscience [*sic*], assure you that external recognition in any form has never been my motivation. (This does not mean, of course, that when it does come, I do not enjoy it, or am not stimulated by it as much as most of my professional colleagues). In particular I never dreamed of the Nobel Prize. Why was I then disturbed by the announcement of the award to Beadle and Tatum (who I agree, did deserve it)? The upsetting thing to me was not so much the fact that I did not get it, as the suspicion that I have been kidding myself about the importance of the *Drosophila* transplantation work. I do not mean to say that the *Neurospora* methodology does not appear to me as the decisive step in the development of biochemical genetics, deserving the Nobel Prize, but simply that it was the direct and logical outgrowth of the *Drosophila* work. These are my last words about the Nobel Prize."[29]

Two weeks before posting that letter to Sonneborn, Ephrussi was interviewed for an article in *L'Express*. "I am delighted," he said, "to see Drs. Beadle and Tatum being awarded the Nobel Prize, thus sanctioning the importance of this field of research to which I have been dedicating so many years of my life. The Swedish Academy is free to choose its Laureates and any claim would be out of place. On the other hand, the development of a branch of science is made through successive steps, each new progress relying on the previous one, each new discovery being the starting point to the one yet to come. Each person is entitled to his own judgement which will be inevitably somewhat arbitrary. What is important is that the judgement be made as objectively as possible, and I am sure that the Swedish Academy has tried to do that."[30]

Years later, Beadle acknowledged to Herschel Roman, a noted geneticist, that had it not been for the work that he did with Ephrussi he would never have gone on to do the other work.[31] In the waning years of their lives, Beadle told Ephrussi that "I've many times thought how unjustly is credit assigned in science—and I suppose elsewhere too. You had more than your share of ideas. You and I had the basic idea of one gene–one reaction. *Neurospora* only confirmed it—actually only confirmed Garrod who had it all straight 40 years earlier. Ed and Josh were incredibly lucky in being at the right place and the right time to get more than a fair share of the credit. I've many times tried to tell the story as it was but one can go only so far."[32] In crediting Ephrussi's critical role in the origins of the gene–enzyme relationship, Beadle revealed not only a basic sense of fairness and honesty but also more. He understood and sympathized with his friend and former collaborator's lingering feelings that Beadle himself had not fully valued the importance of his contributions to the discoveries. Yet there are others who dispute Ephrussi's claim to any parenthood of the one gene–one protein concept. During his collaboration with Beadle, and in the years following, Ephrussi focused very little attention on the way in which genes influenced metabolic reactions; more specifically, there is little evidence that he thought much about a possible one-to-one relation between genes and enzymes.[33]

Having been "passed over" by the Swedish Academy did little damage to Ephrussi's reputation, for after the war he had discontinued the *Drosophila* work and embarked on a very different set of investigations. Recognizing that Beadle had been successful in exploiting *Neurospora* for his genetic work, Ephrussi chose yeast as his experimental organism. Together with a student, Piotr Slonimski, he obtained a novel class of yeast mutants that grow poorly, so-called "petite mutants," because they are unable to use oxygen as an efficient energy source. Ephrussi and Slonimski proved that these mutants are unable to produce a normal mitochondrion, the cytoplasmic organelle that had been identified as the cell's energy-producing center.

Leaders of France's biology establishment were delighted with the discovery of nonnuclear inheritance, for they had continuously rejected the widely accepted view that nuclear genes were the sole determinants of a cell's heredity.[34] Now the nucleus was no longer the sole repository of an organism's genome; indeed, Ephrussi and Slonimski established that the mitochondrion contained DNA and that the respiratory deficiency in yeast resulted from mutations in the mitochondrial DNA.

Equally surprising at the time was their finding that unlike the yeast's nuclear genes, which are inherited by Mendelian rules, mitochondrial genes are inherited unchanged during matings.[35] In mammals, this asymmetric mode of inheritance is referred to as "maternal inheritance," a fact that is now used to trace inheritance patterns from conservation of the mitochondrial DNA sequence between mother and offspring. About the same time, Ephrussi moved out of Paris to direct a department of genetics in a government-supported research center in the suburb of Gif-sur-Yvette. In 1966, he was enticed to become chairman of the developmental biology department at Case Western Reserve University in Cleveland, Ohio. There he initiated the field of somatic cell genetics, an entirely new line of mammalian genetic research. This relied on the creation of so-called somatic cell hybrid cells by fusing either mouse or hamster cells with human cells in tissue culture. This enabled him and others to assign particular genes to specific chromosomes. He returned to Gif-sur-Yvette during 1971, but illness plagued him thereafter and he was unable to continue with his research. When Ephrussi died in 1979, he was viewed as one of the giants of French genetics and developmental biology. Indeed, Salvador Luria, who had shared the Nobel Prize with Max Delbrück in 1969 for their work with bacteriophage, nominated Jean Brachet, T.O. Caspersson, and Ephrussi for the 1976 Nobel Prize, the latter for creating "the new field of genetics of cultured (animal) cells, which promises further advances in the analysis of the human genome."[36]

As Eastman Professor, Beadle was obliged to be in residence during the university's three terms, but he managed several brief trips to California during the spring. There was time during the break between Oxford's winter and spring terms for the Beadles to take advantage of invitations for him to lecture in Rome, Naples, and Strasbourg. Muriel welcomed these opportunities to plan an extensive tourist itinerary. Traveling by bus, the family journeyed through northern Italy stopping in Venice, continuing to Rome, then

Florence and the Tuscan hill towns, and ending in Naples. Visiting the well-known tourist attractions thrilled Muriel and Redmond but Beadle had had his fill of cathedrals and museums soon after the travels began. Their tour of Pompei's treasures including the pornographic wall paintings in the House of the Vetii failed to amuse Muriel although she insisted that Redmond be permitted to accompany them. As Muriel fumed, Beadle chortled: "There aren't many boys," he said wiping his eyes, "whose first visit to a whorehouse is engineered by their *mothers*!"[37] Near exhaustion from their travel schedule, they had still to endure French hospitality and food in Alsace before returning to England. Other opportunities for travel took them to Edinburgh and Glasgow, where they met and enjoyed the company of Guido Pontecorvo, a transplanted Italian fungal geneticist.

Soon after their return from Rome, the Beadles received news that Chauncey had died on April 5, 1959. At 92, he had lived to a "ripe old age" and seen his son succeed beyond his imagination. While planning for his trip to Wahoo for the funeral, Beadle was assured by one of the women who had been looking after his father in the retirement home that "he sure had a good mind, remembered everybody and knew everybody" right to the end. On their return from Wahoo, Beadle and Ruth corresponded extensively about the disposition of Chauncey's properties in and around Wahoo. After considering various options, they accepted an offer to sell part of the property in Wahoo to a local benefactor who would in turn join them in gifting the entire parcel to the city for use as a public park or, if ever needed, for schools. Beadle's astute business sense produced some financial gain along with a charitable tax deduction. The houses in the surrounding towns were given to the people who had lived in and cared for them over the years.

Near the end of their stay at Oxford, Beadle was informed that he was to receive an Oxford honorary degree, a distinction that had previously been awarded to only a select few of distinguished scholars, scientists, and politicians. While not as festive as the Nobel Prize, receiving an Oxford honorary degree comes close in its pomp to the Stockholm occasion. The traditional strawberries and champagne started the day, ostensibly to make the processional march in full academic dress through Oxford bearable. Held in the Sheldonian Theater, the Oxford Encaenia, the occasion for awarding honorary degrees is the most ceremonial event of the year. Besides Beadle, the people honored that day were the ballerina Dame Margot Fonteyn, the prime minister of Australia Gordon Menzies, chief of the United Kingdom atomic weapons research Sir William G. Penney, British jurist Lord Summerville of Harrow, historian Peter Geyl, and German scholar Eliza Butler. Preceded by the University officers in their varied color gowns, the honorees followed in pairs; Beadle accompanied Dame Margot and claimed later that it was not luck that the two were paired but "fast footwork." One by one, each of the honorees' achievements were extolled in Latin by the public orator. The only words familiar to Beadle were "*fungum illium qui appelatur Neurospora crassa,*" then the final pronouncement that "*Georgium Wells Beadle....Collegio de Balliolo ascriptum, Praemio Nobeliano donatum, ut admittatur honoris causa ad gradum Doctoris in Sientia.*"[38] It was a fitting end to their year at Oxford.

While the year abroad had been valuable, the entire family was looking forward to their return. They planned to leave England on August 10th and arrive in Pasadena on the 22nd, with a brief stop in New York to visit with David and his family before their move to England. The Beadles had had their fill of travel and were eager to take up a normal lifestyle. But they were amused at how the time away had blurred their memories of the familiar. Each of them "swore" that they remembered seeing, even dusting, andirons in their California home's fireplace, but Beadle's earlier photographs of the house interiors proved that the fireplace had never had andirons.

CHAPTER 16

Becoming a University President

While Beadle was abroad, the biology division functioned well and there were few major problems that needed his attention. He was kept abreast of what was happening in Pasadena through regular correspondence with Delbrück. In one letter, Beadle was shaken by Delbrück's request for a two-year leave of absence to participate in the creation of an Institute of Genetics at the University in Cologne, Germany.[1] Several months later, admitting that he would miss him when he was gone, Beadle approved the request. Concerned that Delbrück might remain in Germany and that several of the division's senior faculty would soon be retiring or leaving, Beadle urged Delbrück to think about new faculty appointments, and emphasized that the division needed a better balance between molecular- and physical-chemical biologists.[2] They exchanged several letters about possible candidates but agreed that decisions would have to wait until Beadle returned to Pasadena.[3]

Beadle thrived on new challenges, and after his year at Oxford he was on the verge of being bored by the predictable issues in the division. Since Lee DuBridge, who had become Cal Tech's president the same year Beadle arrived, was about his age, it seemed most unlikely that he could aspire to become president. Sensing his restlessness and his desire for more participation in the institute's academic affairs, DuBridge invited Beadle to be the acting dean of the faculty when he returned from Oxford. Presumably, DuBridge planned to appoint him dean when E.C. Watson retired at the end of 1959. Beadle was extremely pleased by DuBridge's confidence in his ability "to fill a bigger pair of shoes," adding that "it's a pretty damned fine feeling... despite my doubts."[4] Two weeks later, while visiting Rome, Beadle urged DuBridge to allow Watson to retain the title of dean until he retired, out of consideration of his many years of service in that position; he was content to serve with the title acting dean.[5]

When Dean Watson retired at the end of the year, the word acting was dropped from Beadle's title and he was officially dean of the faculty. He thanked DuBridge for his confidence but wondered "if I'm up to the job." His substantially increased salary both pleased and embarrassed him. His acceptance ends "As far as I'm concerned there's no better place in the world to be" than Cal Tech. Besides claiming to be unworthy of the responsibility, Beadle was wary that his statement faulting the trustees for their treatment of Pauling's political advocacy might be a problem. He was particularly concerned that a *Los Angeles Mirror* article quoting his defense of Pauling's activities in a Cal Tech publication five years earlier might embarrass DuBridge with the trustees and he offered to resign the deanship.[6] DuBridge, however, would have none of it, and the matter was dropped.[7]

As a Nobel laureate, Beadle was in even more demand as a lecturer than before. Now, his topics were increasingly tangential to his "hard core" science. Frequently, he spoke out on how advances in biology could be made more meaningful to the "person in the street" and how they might address societal problems. At a forum on Resources for the Future held in Washington, D.C. soon after he returned from Europe, he observed that "man has learned how to direct the evolution of plants and animals to his own purposes and could apply this knowledge to direct his own evolutionary future." But, he added, "unless we were to do this with a great deal more wisdom than we have so far demonstrated, we would surely fail miserably."[8] He believed that the "science-society gap" needed to be bridged and the place to begin was in academia. Cal Tech, however, seemed an unlikely place to initiate that effort because its emphasis was on science and engineering while the humanities and social sciences were only marginally represented.

In October 1960, the opportunity to begin bridging the science–society divide arose from an unexpected quarter.[9] One afternoon, Beadle received a call from DuBridge's office. "Could Dr. Beadle see Mr. Glen Lloyd, who has dropped in from Chicago, for a few minutes?" Beadle agreed, but it would have to wait until the end of the seminar he was planning to attend. The seminar ran late, and it wasn't until 5:30 that he got to his office. Patiently waiting for him was "a dapper little man obviously an 'Easterner' because he carried a hat and was wearing a suit with a vest." Embarrassed, Beadle invited Lloyd home for a drink and dinner, hoping that Muriel could scramble together enough food for an extra person. Fortunately, she was able to add a few more pieces of chicken to the pot to accommodate the visitor. Because Beadle didn't know whether Lloyd was a potential Cal Tech donor, a salesman, or a poll-taker, the conversation remained polite but vague. Soon Lloyd identified himself as the chairman of the University of Chicago board of trustees and said that the university was searching for a new chancellor and he wanted Beadle's advice and suggestions about possible candidates. Beadle offered several well-known names, notably university administrators and public figures, and Lloyd dutifully made note of his suggestions and left.

The search committee consisted of five trustees and five senior faculty elected by the council of the faculty senate to serve with the trustees. From the beginning there was a strong consensus that the university should search for a "top scholar in his own right; a bright light that would lure other scholars to Chicago."[10] Customarily in such searches, trustee–faculty committees cast a wide net for the names of suitable candidates. By the time Lloyd contacted Beadle, the committee had reviewed 375 individuals; included on its list were national political figures as well as prominent university presidents and administrators, most of whom were deemed to be unavailable or inappropriate. Beadle was nominated by the distinguished biologist, Lowell T. Coggeshall, chairman of the division of biological sciences, and several distinguished scholars, scientists, and deans at leading institutions. Edward Levi, apparently the leader of the search committee's faculty group, notified Lloyd that there was considerable enthusiasm for Beadle although he wondered whether Beadle was sufficiently "broad gauged" for Chicago's presidency. Accordingly, Beadle's candidacy languished until shortly after Lloyd's visit to Cal Tech.[11]

Evidently Lloyd was not put off by Levi's reservations or by the informality of his meeting with the Beadles, because a month later they were invited to Chicago so that Beadle could meet with the entire search committee. At their meeting, Beadle neither encouraged nor discouraged the committee regarding his interest in the position. He and Muriel were impressed by the trustees they met, for they saw in them a cross section of Chicago's most vigorous, sophisticated, and public-spirited business and professional men with an intimate and committed familiarity of the university. On the other hand, the prospect of moving to Chicago seemed unappealing.[12] During several additional visits, Beadle met with selected faculty members from the different academic divisions and spoke plainly of his ambitions should he accept. Apparently, Levi's initial reservations were muted, for the proceedings progressed rather quickly to the point that Muriel was invited to visit the campus to assess the prospects of living in the chancellor's house. To avoid rumors about the university's interest in Beadle and to protect the university's image should the negotiations break down, their visits were conducted on the sly. Indeed, only a few selected members of the faculty were aware of his candidacy.

During her initial visit, Muriel was definitely "under-whelmed" by the campus and the surrounding areas. She did not warm to the "gray Gothic pile" of campus buildings, the neighborhood with its "soot-begrimed brick and stone buildings bounded by streets littered with rubbish," or the Victorian house in which they would live if they were to come to Chicago. Despite their continuing reservations about leaving Pasadena, Beadle could not refuse an invitation for one more visit. The search committee felt he should meet with Levi, who had been unavailable on the previous visits. Instead of playing up the university's virtues, Levi stressed its many problems, literally daring Beadle to take them on. The University of Chicago had been a place of continuous ferment for 70 years and Beadle was being challenged to participate in that ongoing adventure. Evidently, Levi had figured out one of Beadle's weaknesses: his inability to turn away from a challenge. It was not the last time that Levi would have an influential impact on Beadle's life.

The search committee agreed to offer the position to Beadle late in December. They were "impressed by his personal qualities of modesty and integrity and with his analytical ability, decisive and direct response, and action." They were taken by the "breadth of his scholarly understanding" and "were especially attracted by his convictions on the interrelationships of knowledge and the necessity of relating and focusing all areas of scholarship for the effective resolution of the problems of the modern world."[13] When he received the offer to become chancellor of the University of Chicago, the seventh chief executive, he promptly accepted. Levi wrote immediately saying "how delighted everyone is that you have accepted the leadership of what is now your university."[14] Beadle replied, "Had it not been for your forthrightness in stating the problems to be faced there, and your enthusiasm and encouragement about solving them, we would not be leaving this gorgeous sunshine! And, believe me, I mean that."[15] The board of trustees made the appointment official on January 12, 1961. With the exception of the biologists, few members of the faculty knew who Beadle was. Nevertheless, learning that he was a noted geneticist and a Nobel laureate, they believed that he would bring distinction to the university and help attract other scholars and scientists to the campus.

DuBridge knew of Beadle's discussions in Chicago and his initially mixed feelings about making such a move. Admitting that he probably had "holes in his head," Beadle acknowledged the challenge: $250 million worth of endowment and $350 million worth of problems."[16] Although forewarned, and aware that Beadles' departure would be a significant loss to the Institute, DuBridge did not try to persuade him to remain in Pasadena; perhaps he knew better. Just after the first of the year, Beadle informed DuBridge officially of his decision to accept the chancellorship at Chicago,[17] and soon thereafter Lloyd and E.L. Ryerson confirmed that decision to DuBridge.[18] DuBridge's public statement accompanying the news release admitted that "the loss of Dr. Beadle is a most serious blow to the California Institute of Technology...he has built here one of the greatest research centers in biological science in the country... In his new capacity as dean of the faculty, he was about to launch a vigorous new program of educational advancement."

Wishing not to be a "lame-duck" dean or chairman, Beadle immediately resigned from the Cal Tech faculty.[19] Less than a week later, he addressed the University of Chicago alumni of greater Los Angeles at the California Club. He identified the high-priority issues that confronted him: attracting top faculty and students, continuing the university's neighborhood renewal efforts, and increasing the endowment to support new buildings and increased faculty salaries. Characteristically, he was on the job even before he had been officially appointed.

Unaware that he was considering the position at Chicago, Beadle's Cal Tech colleagues and students were stunned by the news. Most were puzzled, wondering "Why would he want to leave the Institute?" There was no hint of discontent or pressure from the Institute, and nothing seemed to warrant so drastic an action! Indeed, six months earlier Beadle had told DuBridge that there was no place in the world he would prefer to be than Cal Tech. Answers to the question were the subject of considerable speculation because Beadle was spending much of his time in Chicago. His inimitable style over the 15 years of his chairmanship had come close to creating a "cult of personality" and his impending departure introduced a degree of uncertainty about what would follow.

Beadle understood that a leadership vacuum had to be avoided, and a few days after the announcement he recommended that DuBridge appoint Ray Owen to succeed him as chairman of the division and increase Owen's salary.[20] DuBridge also received a signed petition from the division faculty recommending that Owen be appointed as chairman; Beadle wrote again concurring in the division's selection and urging him to offer Owen the Morgan house as his residence.[21] Owen accepted the chairmanship on condition that he be permanent, not acting, chairman, and be able to keep his laboratory in operation for at least the first year.[22] Less than a fortnight after Beadle's departure was announced, the Cal Tech biology division had a new chairman.[23]

The faculty, staff, and students' affection for the Beadles was expressed in traditional Cal Tech style: a "comically tragic operetta" entitled "What Makes Beadle Run" performed by the Cal Tech Stock Company, a group studded with the institute's important scientists.[24] The humorous sketches and lyrics conveyed unmistakable evidence of the warmth and affection for both Beadle and Muriel. The opening scene's action takes place on the "blackest day of Cal Tech history, Jan. 5, 1961." "That day," the prologue intones, "we all remember all too well, George Wells Beadle decided to become Chancellor of the University of Chicago. The tragedy that has befallen Cal Tech is so stark, brutal, and basic that it makes Medea sound like Auntie Mame and King Lear like

One Man's Family. Who can forget the shock? It registered nine on the Richter scale. Full Professors wept. Secretaries fainted. Graduate students who hadn't seen daylight in three years rushed out of their laboratories screaming. And from every corner of Cal Tech arose the same agonizing question. Why? Why? ... Ruthlessly probing beneath the layers of fear, guilt and hysteria, our drama will show why George Beadle is leaving us; it will tell what makes Beadle run."

A chorus belting out songs in the style of a Gilbert and Sullivan opera interspersed with clever asides, acknowledged culpability in driving Beadle and Muriel to leave: "Let George Do It," "What Did We Do," and "Blue Genes." An Englishwoman sent from Oxford to "study the Beadles in their natural habitat" claims that she is the Empire's answer to Muriel Beadle. Her mission is to write a sequel to Muriel's bestseller, "These Ruins Are Inhabited," entitled "These Inhabitants are Ruined." After a good deal of hijinks, Muriel herself appeared on stage calling for a halt to the foolishness and promising to reveal the real reason Beadle is leaving. She explains that "It's Biological, it's Metamorphosis"; he's entering "a late [*sic*] executive phase." Relieved that they were not at fault, one of the characters responds with "You're saying that George actually likes us and that he isn't being driven away because we're a collection of odd balls, eggheads and kooks." Muriel's refrain of "George thinks you're the finest people in the world and he loves you" introduced the final song that has Beadle's colleagues coming to terms with his departure. The chorus then sings "We're glad we had him and we cheer him on his way; No use to whine or wheedle, we must say goodbye to Beadle, and his DNA. For what you have done for Cal Tech, we thank you. For what you are about to do for Chicago—we salute you."

Beadle's responsibilities were not to begin until March 16, 1961. Nevertheless, not being one to put off challenging opportunities, he took off for Chicago in February to immerse himself in the intricacies of his new institution and educate himself about the many issues that recruiting visits rarely disclose. Beadle settled in at the Quadrangle Club, the university's faculty club, where he could also have his lunch and dinner meals. An inveterate early riser, he was up and about long before the club served breakfast, so he walked the few blocks to the hospital cafeteria that was open 24 hours for his morning meal and to mix with other early risers and digest the morning newspapers. Invariably, he was the first one in the office and his customary casual dress led some to mistake him for one of the custodial crew. One morning, returning from breakfast, Beadle encountered a young man sitting on the steps of the administration building waiting for the building to open. He invited him in, offered him coffee, and directed him to the office where he was to have his scheduled interview. Imagine the young man's astonishment when he learned that the man who had invited him in was not one of the custodial staff but the university chancellor.[25] In those first few months on the scene, Beadle began to learn what made the university tick and a good deal more about the problems that Levi had merely hinted at earlier.

In his first interview with Chicago newsmen, Beadle was asked whether one of his priorities was to narrow the gap between the sciences and the humanities. The two, he said, are complementary and intermingled; all are part of human culture. In a later lecture, he voiced the same theme in responding to student fears that as a scientist he might place too much emphasis on science instead of the humanities: "The separation between the sciences and the humanities is a fallacy....Science is not opposed to culture any more than culture is opposed to science...intelligent people seek balance."[26] The nonscience faculty was delighted at his evenhandedness, while many biological scientists, concerned that their programs had remained stagnant at a time when others prospered, were counting on him to elevate Chicago's standing in biology to the level of Cal Tech's.

During this period of "getting acquainted" with the intricacies of the university, Beadle soon realized that it was considerably more complex than Cal Tech. It comprised an undergraduate college with its own dedicated faculty and four graduate divisions comprising the physical sciences, humanities, social sciences, and biological sciences which included the medical school and its 11 associated hospitals. There were also seven professional schools including graduate schools of business, education, law, and divinity, and the School of Social Service Administration. He quickly learned that the unique feature of having the medical school as an integral part of the division of biological sciences had several drawbacks. According to some, the association of the basic and clinical sciences slowed the flowering of molecular biology on the campus. There were also the Harper Memorial, the University Art Gallery, 17 departmental libraries supplementing the university's main library, and a bookstore. The Oriental Museum, the University Press, and a center for the study of disturbed children were autonomous but their activities served the academic and research needs of the faculty. An additional university responsibility was the management of the Argonne National Laboratory, a direct descendant of the Manhattan Project's Metallurgical Laboratory that had helped develop the atomic bomb, but was now committed to developing nuclear reactors for peaceful purposes. Clearly, managing this entire enterprise was going to be decidedly more challenging and far broader in scope than anything Beadle had previously experienced.

Beadle was well aware of the University of Chicago's lofty academic reputation, for even with its difficulties in the postwar era, it was still a formidable bastion of intellectualism. But he was probably less familiar with the troubles the university had experienced following the end of war. Serious financial shortfalls and the deterioration of the university's unique neighborhood had contributed to a dramatic falloff in undergraduate and graduate student enrollment. The loss of distinguished faculty had seriously eroded the university's prestigious academic reputation. During the 1950s, however, the local decay was stabilized[27] and even reversed. Now with new leadership the university could once again pay full attention to its primary role of education and research. As Beadle readied himself to accept the chancellorship, he understood that although stable, the neighborhood problem would need continuing vigilance and sustained effort. His own priorities would be to restore the university's financial security and its deteriorating physical facilities. Ahead but still unanticipated were problems arising from the politicization of faculty and students before and during the Vietnam War. Levi had not exaggerated the magnitude of the challenge that confronted him.

Because there was no reason to prolong the interval before he was officially invested as chancellor, he persuaded the trustees to schedule the inauguration for May 4, several months earlier than what had been traditional. He was raring to engage the challenges confronting his new administration. In the two months since his arrival, Beadle had set a fast pace and held conferences day and evening with the University's academic and administrative officers. Spending a night riding in a campus police squad car helped him get a first-hand view of the community's crime problems. By the time of the inauguration, he had formulated his priorities for getting the university's image and academic activities invigorated.

Academic inaugurations, like coronations, are invariably grand occasions, full of pomp, plans, expectations, and hope. The formalities of the occasion preserve a tradition and encourage the new president to lay out a vision and plan for the future. Beadle had been selected by a relatively small and elite constituency, and his inauguration provided an opportunity for the broader community to celebrate and welcome his arrival. Besides his future colleagues, delegates from 300 American universities and colleges, 50 foreign universities, and, at Beadle's request, 300 students turned up to celebrate the occasion. There were also many of his and Muriel's close friends and family, some having come from great distances, although notably his son, David, was absent.

The inaugural festivities began the night before the actual event with a black tie dinner at McCormick Place, a civic center situated along the magnificent Chicago lakeshore. It was an occasion for public officials in local, city, and state government and civic and financial leaders to meet the university's new leader. Mayor Richard Daley welcomed the Beadles and pledged his support for the university's activities. Beadle was especially pleased by the presence of his old friend, Warren Weaver, formerly associated with the Rockefeller Foundation but then with the Alfred P. Sloan Foundation. In his dinner remarks, Weaver recalled meeting Beadle nearly 30 years before at Cal Tech and afterward at Stanford, where, he noted, Beadle "constructed the elegant and revealing edifice of modern biochemical genetics." He had recognized that Beadle's discoveries ushered in a new era of biological discovery and had helped bring Beadle and Pauling together at Cal Tech, where what he envisioned as "molecular biology" became a reality. "Once," he recalled, during a discussion at the Rockefeller Foundation lunch table, "I remarked that it was easy to make grants to those who have already won a Nobel Prize. You only need the list in the World Almanac and a checkbook. But it is real fun, real satisfaction to aid young scientists who are going to get a Nobel Prize. I offered three to one odds that I could name two young United States scientists who would receive a Nobel Prize within ten years. It was a steal: for my two names were Vincent du Vigneaud [1955] and George Wells Beadle [1958]." In a more constructive vein, Weaver urged the new chancellor and the university to consider a "unity of learning" rather than promoting increasingly narrow divisions through the creation of institutes of X and programs of Y.[28]

Beadle's brief remarks recalled the great Chicago architect Daniel Burnham's admonition: "Make no small plans, for they have no power to move the hearts of men." He promised "great plans" for the university, worthy of the City of Chicago.[29]

The "modified Gothic" Memorial Chapel provided a traditional and splendid setting for the inauguration. It was built during 1926–1928 with part of John D. Rockefeller's final $10 million gift in 1910 and is located on the southeast end of the quadrangle facing the Midway Plaisance.[30] The chapel's interior is dominated by a soaring vaulted ceiling decorated with inlays of colored tiles depicting religious motifs and coats of arms. Filtered light enters through tall, narrow leaded windows, tinted in subtle shades of gray, green, and blue. Flanked on each side by the intricately carved choir with organ pipes overhead, the raised altar is crowned by a window with a brilliantly colored five-petal floral design reaching the ceiling. The chapel's organ, considered one of four great university organs in America, provided the musical accompaniment to the inauguration proceedings.[31]

The threat of rain, which never materialized, did not diminish the turnout for the convocation procession. An estimated 2500 people crowded the Midway to view the professors of the various divisions and guests in their brightly colored academic attire march to the chapel, with Beadle following. Unable to find his own academic hood, Beadle borrowed the one L.A. Kimpton wore at his inauguration ten years before, both having received Ph.D. degrees from Cornell University. Beadle took the "lecturer's chair" from which, according to tradition, he moved to the "ceremonial chair" symbolic of the office, after being designated as chancellor of the university by the chairman of the trustees. Following a musical interlude and his inaugural speech, he performed his first official act by conferring honorary degrees to a stellar group that included the chemist Robert B. Woodward and the biologist James D. Watson.

Beadle's inaugural speech relied on a time-honored rhetorical device of comparing the University of Chicago to an ideal university X.[32] "A fundamental feature of university X," he believed was "to preserve, evaluate, understand, and transmit to future generations the best of man's total accumulated culture—its history, religion, art, music, literature, science and technology." Research and scholarship, he added, "continuously modify and update this accumulated knowledge." Individuals, he felt, "had to experience the incomparable thrill of original discovery to participate in that process," and he added how difficult it is to make that point to "one who has never experienced the...satisfaction of original discovery." Believing that a liberal general undergraduate education could only occur in a setting where specialization and graduate instruction coexisted, Beadle was critical of top scholars who forsake undergraduate teaching to further their research reputations, individuals that Warren Weaver labeled "intellectual eunuchs." The gap in understanding between humanists and scientists—what C.P. Snow characterized as the "two cultures"—would not exist at Beadle's university X. There, all scientists would begin their training by acquiring a truly liberal education and its humanists and social scientists would have absorbed a deep understanding and appreciation of science.

Beadle reserved much of his passion for his concerns over threats to academic freedom. At university X, students receiving government financing would not be required to sign a disclaimer saying "I... do not believe in or support any organization that believes in or teaches the overthrow of the United States government by force." Beadle praised the trustees' decision to terminate the federal student loan program because of the government's insistence that students receiving such loans sign the disclaimer affidavit. "An academic institution, dedicated to the uninhibited search for truth," he proclaimed, "may not accept conditions that denies [*sic*] the right to question, to doubt or restricts what students may think about." He saw no wrong, however, in requiring students to take "an oath to uphold and obey the law of the land" and be prepared to accept the consequences of its violation. Colleagues and many in the community applauded his unswerving commitment to academic freedom. Some alumni and local military, however, were outraged by his unwillingness to defend the government's insistence that students supported by federal funds had to renounce any actions promoting the overthrow of the United States government by force.

In contrast to university X, Beadle believed that Chicago's financial and physical resources were grossly inadequate. Both deficiencies made it difficult for the university to retain and attract the kind of outstanding scholars it had always claimed. More money and improved facilities were to be his administration's highest priorities, for if he failed in that goal the university's intellectual capital would be dissipated. The agony endured by the university during the previous decade did not escape Beadle's comments. Believing initially that university X should be located in the country, distant from the exigencies of a bustling society, he conceded that "the University of Chicago would be a far less interesting place and a far less significant institution if it were so isolated." Until we "learned to rebuild and prevent slums, restore beauty to our cities, and provide education and social opportunities to people who have not had them—largely because of the color of their skins," Americans, he felt, would have failed those who laid the foundation of our nation. "We cannot," he said, "do it by running away or burying our heads in the sand. If a great university will not stay and use its knowledge, wisdom and power to help solve a critical problem, who will do it?" The reactions to his speech from the local and national news media and his university colleagues were largely complimentary. He seemed ready to take on the challenges that confronted his presidency.

Perhaps the major challenge he faced was the one he and the search committee had identified, namely, restoring the university's academic eminence. That meant stanching the loss of existing faculty while recruiting top scholars to Chicago. At Beadle's first meeting with the trustees, only a few months after his arrival on campus, he presented his initial assessment of the strengths and weaknesses of the university and its faculties. This, he made clear, was the first step in establishing priorities for action. The changes and improvements he had in mind necessitated raising substantial amounts of new funds.

In his talk "Thoughts of a New President," at the annual dinner hosted by the trustees for the faculty,[33] he focused on how to increase the university's strengths at the expense of its weaknesses. The university's only legitimate measure of academic standing, Beadle offered, was "selective eminence," a metric his colleague the economist George Stigler defined as characterizing the top five departments in any field. Using that criterion, and applying his admittedly subjective assessment, Beadle estimated that of the 42 academic units in the four divisions and seven schools, ten were very good and five were tops. "That's pretty good, but not good enough." His and their goal had to be increased eminence across the board and for that to be achieved more money, much more, was needed. Beadle estimated that bringing the university's intellectual and physical assets to the level of eminence he sought would require generating 200 million dollars annually and, therefore, a staggering 4 billion dollars as an endowment! Clearly, that was an unattainable, even inconceivable, amount and, therefore, making prudent choices and insisting on the highest standards in appointments and promotions was imperative. His goal was to have 20 departments at the top in national competition. In closing, Beadle announced the creation by the trustees of a special fund, which he referred to as the "Selective Eminence Fund," for bringing additional outstanding scholars to the university and keeping those already there.

Raising those very substantial amounts of money to implement the academic renaissance was more easily said than done. Owing to the nearly single-minded focus on stemming the neighborhood decay during the preceding ten years, the university's fund-raising infrastructure had deteriorated. To make the efforts more coherent and consistent, his office was the logical place to begin, and its first task was to establish a "master plan" for the administration and the faculty. Creating an effective infrastructure for raising money and faculty recruitment proved to be more ambitious than had been supposed. Because it did not fit his style, Beadle requested that the "lofty" title of chancellor be abolished. The trustees agreed that the executive head of the university would be called president as it had been earlier.

Beadle soon acknowledged that the University of Chicago's organization and the unique social dynamics of its faculty were far more complex than those with which he was familiar at Cal Tech. His experience there had been one of managing a division of about 19 faculty members in an institution whose entire faculty numbered about 275. His concerns about students were limited to undergraduates majoring in biology or graduate students doing Ph.D. research in the division. Obtaining funding for the division's activities was important but not an overwhelming concern. At Cal Tech, academic proposals originated with the president and a single dean and were passed "down" to the faculty and students by established procedures. As chairman of the biology division, Beadle had instituted an intimate style with his colleagues, one in which things he wanted to see done were achieved by consensus, not fiat.[34] In contrast, Chicago's institutional and divisional initiatives emanated from the faculty.

Apparently, the pace at which Beadle was able to master and coordinate the respective activities of the disparate academic enterprise did not match the urgency the trustees believed was needed. The strain on Beadle was plainly evident to Muriel: "What a month! The inescapable pressures and obligations of the job here have mounted steadily...Poor George is so gray and tired and dispirited that I can't expect him to write...the minute he gets home at night—after a non-stop day—he goes into a sleep of utter fatigue."[35]

By the end of Beadle's first year, a decision was made to create a new office of provost to replace the existing position of dean of the faculty.[36] The provost was to assume responsibility under the president for academic administration and to have authority over academic budgets. Edward Levi was chosen and, after some hesitation, agreed to accept the newly created position. It is unclear whether the initiative for creating the position of provost and the selection of Levi originated with the trustees or with Beadle. Perhaps the trustees, who knew and respected Levi, were the prime movers. After all, he was exceedingly successful as an administrator and dean of the law school and had recently chaired a committee that reorganized the undergraduate college; he had also been on the selection committee that recruited Beadle to Chicago. However, the suggestion may well have originated with Beadle, for he is reputed to have told his Cal Tech colleagues that he had found in Levi the ideal person for his provost.[37] From the time Levi had persuaded him to accept the presidency, Beadle admired his deep and extensive knowledge of the university's academic, administrative, and financial activities and he was well aware of his high standing with the faculty.[38] Indeed, long a local favorite, Levi had been mentioned as a candidate for the presidency, but believing that the University of Chicago was not ready to accept a Jewish president, he made it clear early in the search that he did not want to be considered for that position.[39] Responsible for the academic side of the university, Levi would have a major role in rejuvenating the faculty through his control of the university's budget process and his influence with the academic leadership. Beadle would manage the university's "outside" affairs and fund-raising; he would in effect be the voice of the university, articulating a vision of the institution's values and mission to the outside world.

The announcement of the new academic partnership was hailed as an important advance in the university's renaissance. Almost immediately, Beadle and Levi began an aggressive recruiting campaign taking advantage of the trustees' commitment to their $3 million fund for attracting new distinguished faculty. Most members of the faculty admired Beadle's willingness to accept Levi's leadership of the university's academic dealings. Ego, it was clear, was not one of his weaknesses.[40] Moreover, there was never any evidence of competition or one-upmanship between them.[41] Occupying adjacent offices, they were very likely in frequent oral communication, since very few written memos or papers exist in Beadle's presidential papers.[42] Occasionally, Beadle would scribble a comment for Levi's attention on a memo he had received or a suggestion that they meet over some issue that concerned him. The partnership worked amazingly well, in part because "Edward found him easy to work with. He was very admiring of Ed and vice versa. It was a very good team. I think he needed Edward."[43] Their dual stewardship lasted throughout Beadle's term as president and culminated in Levi being selected as his successor, the eighth president of the university.

Growing up in the Hyde Park–Kenwood neighborhood, Levi entered the University Laboratory K–12 School, then he graduated from the university college and law school, both with honors.[44] He returned to the Midway as an instructor in the law faculty immediately after receiving a doctorate of jurisprudence as a Sterling Fellow at Yale in 1938. Twelve years later, his brilliant teaching and scholarship made him the logical candidate to be dean of the law school. As dean, Levi was remarkably successful in luring top legal scholars to the campus and diversifying the school's faculty to include economists and sociologists, thereby changing the nature of legal curricula nationwide. Beadle could attest to Levi's formidable powers of persuasion, talents that would stand them in good stead as they began their quest to attract top scholars to Chicago. Edward Levi's wife, the former Kate Sulzberger Hecht, a formidable personality in her own right, had grown up in Hyde Park in a family long associated with the university. Kate's family background had provided her with well-honed social instincts, and she was at ease with Chicago business leaders in civic and philanthropic activities. From the time that Levi teamed up with Beadle in the administration, Kate often collaborated with Muriel in persuading spouses of prospective faculty to the varied benefits of coming to the university.

To stay in touch with the day-by-day thoughts and happenings on campus, Beadle took his lunch regularly at the Quadrangle Club. There, he would generally participate in whatever discussions were ongoing, whether it was physics, Greek history, philosophy, or academic politics. Faculty felt free to stop by his table to chat or introduce a visiting colleague. He always viewed himself as one of the faculty rather than of the administration. In that light, the club frequently exhibited on its walls Beadle's photographs of the campus, botanical subjects, and places he visited on summer jaunts. Colleagues remarked that he took considerable satisfaction in having his photos displayed there.

Either before or after Beadle came to the university, a parking area near the president's house was set aside to create a field for him to grow corn. His lifelong love of farming had created an unquenchable need to work the soil, whether it was growing flowers or tending corn. Friends and colleagues happening by often stopped to talk and invariably learned something about corn. Beadle also had an experimental garden where he and a neighbor grew and hybridized daylilies, many of which he gave away to friends and colleagues. Strangers to the campus were often surprised when they learned that the man in coveralls tending the president's cornfield was the president! On one of his purchasing forays to the plant nursery for manure, his claim to be the university president was met with disbelief, probably, because of the way he was dressed.[45] At a presidential reception for foreign students, Muriel noticed that an Indian student dropped out of the long receiving line with a stricken look that worried her. Later she learned that he was a medical student who, when he passed by the president's house on his way to the hospital at 7 a.m. every morning, stopped to greet the gardener at work. He was totally unnerved as he proceeded along the line and realized that the gardener was the president.[46] Years later, one of Beadle's colleagues, paraphrasing an old Broadway tune, quipped "You can take the boy out of the country but you can't take the country out of the boy. Beadle never assumed to be anything but a farm boy, the unassuming boy from Wahoo, Nebraska." Not at all put out by this characterization, he conceded "that's what I am."[47]

Beadle's attention to the campus's physical plant, particularly the shabbiness of the campus, was often a subject of his trustee–faculty dinner speeches, and he chose to make his peevishness known. Students were notorious for taking shortcuts across the grassy areas, often trampling the flowers as well. Beadle referred to them as "horticulturally deprived." Believing that the campus center deserved better, he had a sprinkler system installed and the main quad freshly sown with grass seed. He contrived to issue a "Proclamation" from the "Committee on Grass," which was announced with a trumpet fanfare and then posted on Cobb Hall's front door. Following ten "whereas's," each attributed to a notable authority, e.g., the Bible, Chaucer, Shakespeare, Montgomery Ward, and the botany department, it was resolved "that young grass, like ideas, be allowed to grow freely on the campus of the University of Chicago." The resolution was identified as coming from the university president on behalf of the "Committee on Grass Interdisciplinary and International." Numerous signs appeared each day quoting obscure aphorisms such as Seneca's Latin admonition "shame may restrain what the law does not prohibit." Aside from becoming a topic of conversation and speculation about who was posting the signs, this humorous stunt revealed the light touch Beadle often used to make his point.[48] Good humor had long stood him in good stead when he encountered resistance to what he thought needed to be done, but that light touch was to be tested repeatedly in the years ahead.

CHAPTER 17

Restoring the University's Eminence

Soon after Beadle and Levi became a team, they began detailed planning toward their goal of restoring the university's former intellectual and scholarly eminence. Although Beadle had in mind certain academic areas that needed strengthening, no specific fields were given priority over others: The aim was to recruit top scholars irrespective of their specialty or expertise. The trustee's $3 million Selective Eminence Fund was the "carrot" offered to the deans and departmental chairmen for soliciting nominations for these distinguished positions. High on the compendium of financial needs to be presented to foundations and other granting agencies was the need for new research buildings, libraries, and equipment, as well as refurbishment and expansion of the existing physical plant. These, they reckoned, would figure prominently in the university's push to recruit top quality faculty.

All the while that Beadle was involved with fund-raising and recruiting, other trouble intruded. During the early 1960s, civil disobedience emerged as a powerful weapon of protest over the denial of civil rights for American blacks. College students, protesting discriminatory customs and laws, were often at the forefront of confrontations with local and state police. Emboldened by their success in contesting racial injustices, students brought back to their campuses an antipathy to authority and a belligerence at what they perceived as injustice on the part of their professors and administrations. The University of Chicago was among the first to experience such a confrontation, perhaps because of its efforts to deal with the Hyde Park neighborhood's decay. Triggering the student action was the university-led program of buying up properties for restoration and renting the renovated apartments to tenants other than faculty, students, and staff, but specifically, the university's decision to rent these apartments to whites in preference to blacks. The justification for that discriminatory policy and Beadle's defense of it was based on the premise that whites had to be encouraged to remain in the area at a time when the black population seemed to be overwhelming. The intent was a stopgap action to create and stabilize an integrated environment. But discrimination in university housing for whatever reason was sufficient to arouse the ire of student participants and supporters of the Congress of Racial Equality (CORE).

After interminable discussions pitting Beadle's pragmatism against students' idealism, some 30 students occupied the small reception area outside his office and refused to leave until the university stated publicly that it would not discriminate in renting, leasing, administering, or selling any property it owns or controls. By the standards of sit-in behavior a few years later, the students were relatively restrained. They ate, sang, and talked. Not surprisingly, the local press played up the students' action at the expense of the university's image. What to do? The most important of Beadle's decisions was not to use force to expel the students from the administration building and risk publicizing what would be construed as university-sponsored suppression of dissent. He favored doing nothing while making it clear to the students that the rental policy they were protesting would not be changed. In effect, he bet that the students would tire of protest and that the sympathy their stand had garnered initially would erode over time. After two weeks, and Beadle's warning that those who remained would face disciplinary action, the students left. Although intrusive and disruptive, the 1962 sit-in left no scars.

At a subsequent trustee meeting,[1] when Beadle reported the events leading to the sit-in and its resolution, chairman Lloyd "grilled" him about statements in the *New York Times* reporting that "Chicago Students Claim Sit-In Gain." "Was it true," he asked, "that he had agreed to negotiate university housing matters with the students?" Beadle denied making such an agreement, nor had he agreed, as the *Times* reported, to set up a negotiating committee that would include representatives of CORE, the Urban League, and religious and political officeholders to expedite integration of the university off-campus apartment buildings. Beadle told the trustees that he had appointed a committee of several leading faculty members to advise him on housing matters. The sharpness of Chairman Lloyd's questions made it clear that Beadle's liberal and conciliatory manner of dealing with student unrest was at odds with the views of at least certain trustees. The 1962 sit-in was the forerunner of things to come, and the precedents established during this baptism of fire served the university well in the coming years. The explosive "free speech" controversy at the University of California, Berkeley, followed soon after by the increasingly violent student-led antiwar protests against the government's and American universities' involvement in Vietnam, were the harbingers of what was to come.

As a Nobel Prize winner and president of a prestigious university, Beadle was sought for comments on a variety of worldly issues. Most often his talks concerned the revolution in genetics and its implications or the importance of learning and education in our increasingly complex society. Because he believed that he had an obligation to speak out on these issues, he rarely declined invitations for public appearances. Soon after he arrived in Chicago he spoke of his personal creed in the context of science and religion before the Chicago Sunday Evening Club. The topic was a challenge, considering that he had rejected a belief in God at an early age, possibly reflecting his father's strong bias against what he felt was the hypocrisy of formal religion. A reporter who attended his "this I believe" lecture commented that it elicited far greater interest on the part of the audience than if he had lectured on genetics.[2]

Beadle disagreed with the widely held view that "one can not accept the findings of science and at the same time retain faith in the existence of God." He led the audience through the scientist's view of evolution and argued that the emergence of life in all its manifestations was inevitable given the properties of the hydrogen atom and the elements formed from it. In accepting this "exciting and awesome" view of life's origins, Beadle admitted that his view rejected the faith-based view that "everything in the universe was created by the guiding higher intelligence we call God." "Is it," he asked, "any less awe-inspiring to conceive of a universe created of hydrogen with the capacity to evolve into man than it is to accept the creation of man as man?" While conceding that "the scientist does not invoke faith at the same level as the priest," he emphasized that his faith was equally insistent. "The problem of ultimate creation remains fundamentally the same for both." Beadle's views on this occasion were somewhat more tempered than David's characterization of him as a "vehement atheist," and from his earliest days "intolerant of religion and other forms of superstition." Although Beadle was not religious in the literal sense, David saw his father as a very moral person, basing his precepts in reason rather than a prescribed faith.[3]

One of Beadle's and Levi's early initiatives was to increase faculty salaries to reduce faculty resignations and improve the ability to recruit first-rank candidates. By the middle of 1963, applying his scorecard of the previous year, Beadle judged that several of the academic units had risen to Stigler's benchmark of eminence and that several more had achieved top rank. In the faculty search, Beadle sought advice from his former colleagues at Cal Tech to vet the credentials and standings of some of those being considered. Being a realist, he acknowledged that "25 years will say whether we were right in our choices."[4]

At his 1963 State of the University presentation, concerning the university's pressing need for expanded space and improved facilities, Beadle could only describe projects that were under way or planned.[5] The most notable of these was the agreement to build the $15 million "post-modern-of-all general libraries" on the site where Enrico Fermi and his colleagues initiated the first sustained nuclear reaction. Beadle also paid special tribute to Glen Lloyd, who was retiring as chairman of the trustees, for his wisdom in establishing the fund that facilitated the recruiting efforts. The new chairman, Fairfax Cone, a longtime trustee, assured Beadle and Levi of his commitment to their rebuilding efforts.

While Beadle was engaged with faculty recruitment, fund-raising, and managing student uprisings, grounds, and buildings, Muriel was busy with her own range of activities. There were the usual activities of a president's wife: weekly faculty-wives teas, arranging dinners for trustees, visiting academic and political dignitaries, as well as touring royalty (the Shah of Iran, the Norwegian King, and the Japanese Crown Prince). She reckoned that in the first three years of their tenure in Chicago, she had served food and drink to about 7,100 guests, and in one hectic month she entertained more than 850 visitors. Young faculty recall the pleasure of attending dinners at the president's house, especially those in which the Beadles entertained old friends such as Sturtevant and Ephrussi. Many of the faculty who attended these social occasions found Beadle generally reserved while Muriel's charm and exuberance dominated the evening.[6] Despite occasional irritations such as keeping elaborately detailed records of expenses, Muriel truly loved the job of being the president's wife. She enjoyed mixing with like-minded civic activists, especially those who took a keen interest in promoting and participating in the neighborhood renewal efforts.[7] Years later she became the unofficial chronicler of the Hyde Park–Kenwood renewal "miracle."[8] She also took major parts and often wrote the skits and "songs" in the musical comedy productions organized by the faculty wives. Her exuberant role in "Hips, Hoops, Hurray" granted Muriel a certain notoriety in Hyde Park circles.

Amidst all the demands on their time, George and Muriel were continuously troubled by David's seeming inability to settle down. It was a problem that had plagued David from his adolescence onward. David's early marriage to Ruth Axtell had failed, and soon after, he married Joyce Smith, the mother of Beadle's first grandson, John Vincent. The marriage to Smith failed sometime in 1957 when David took up with Jane McCandless and shortly thereafter moved to Europe. For years, David neglected to provide child support for his son and eventually, following a court order, Beadle took on the responsibility of sending regular checks until the boy could manage on his own.[9]

During their itinerant travels in Europe, David and Jane had two sons, George Press (1959) and Ian Christopher (1961). The Beadles were stunned when word arrived from the American consulate in Florence that David and Jane had abandoned their two children in the care of a maid who had run out of funds for the children's support, leaving a $4000 debt. Without hesitation, Beadle covered David's debts, as he must have done repeatedly; Muriel estimated that that month they had sent David in excess of Beadle's total take-home pay.[10]

As if David's financial irresponsible behavior was not enough, Beadle was utterly devastated when he learned that the two children from David's supposed third marriage were illegitimate. Uncharacteristically, he broke down and cried; his mid-West-honed sense of propriety could not deal with hearing the children referred to as bastards.[11] Muriel, being considerably more worldly, was furious at having been lied to and castigated David for not having been content with just having an affair. Nevertheless, because David's finances were a shambles, Beadle sent financial support for these children as well.[12] Eventually, David settled down in England and married again, this time to Jacqueline Isitt, a woman he'd met at the *Economist Magazine* intelligence unit, where they both worked. Her personality and the relative stability of their marriage, along with the news of a new grandchild, left the Beadles with renewed hope that David's troubles were over. Happier news on the domestic front was the completion of George and Muriel's coauthored book, which made molecular genetics accessible to a nonprofessional audience.[13] They were also exceedingly pleased and proud when Redmond graduated summa cum laude from Harvard and was awarded a five-year graduate fellowship for a doctorate in history.[14]

Summers at universities are a time for relaxation and recovery from the usual year-end academic activities. For the Beadles, summers most often meant travel together to conferences with other presidents and their spouses. But on one occasion Beadle went off alone on an expedition to the Air Force's "Cool School" in Goose Bay in the northernmost part of Labrador where a fleet of bombers and fighter planes were maintained on five-minute alert. On that trip, he was hoping they might make a diversion and circle over Mt. Doonerak to see if the Cal Tech flag he had planted on the peak years before was still there.[15]

Beadle and the trustees believed that the division of biological sciences' hospital and medical school were in the top rank nationally.[16] However, with the exception of biochemistry, the division's basic biology departments did not measure up to the eminence of the clinical enterprise. That had not always been the case. When the university was founded, and for three decades afterward, the zoology department was especially notable for its important contributions to embryology.[17] But its influence and reputation suffered when some faculty insisted that cytoplasmic and environmental influences rather than genetic contributions were paramount in guiding embryonic development.[18] Although rejecting "Morganism," the department's appointment of Sewall Wright to its faculty introduced more enlightened views on the physiological action of genes in Mendelian terms. Nevertheless, the department failed to integrate genetics into their views, and its standing in embryology waned. A proposal to the Rockefeller Foundation's general education board to create an Institute of Genetic Biology aimed at studying the social problems of the human race rather than problems of the human individual got little support. Warren Weaver, director of Rockefeller's natural sciences division, was by then committed to the emerging field he called molecular biology and was convinced that Chicago had limited expertise in that area. Shut out of any Rockefeller funding, ecology, once a mainstay of the zoology department's research, declined and never recovered its former prominence.[19] Perhaps the university's preoccupation with the neighborhood problem throughout the 1950s caused it to underestimate the transformation of biology and therefore fail to recruit and retain their few rising stars in molecular and genetic biology.

A principal concern of the biologists ever since the division of biological sciences had been created following the university's acquisition of the medical school and hospital stemmed from the division's administrative structure.[20] Shortly before Beadle became chancellor, the university appointed H. Stanley Bennett, a physician and a highly regarded specialist in electron microscopy, as dean. His principal responsibilities were for the hospitals, clinical faculty, and the training of medical students. Consequently, botany, zoology, and other nonclinical departments had to compete for resources and attention with departments of medicine and surgery. Many of the biologists had hoped that there would be a turnaround in their status and national standing after Beadle's arrival, for they were well aware of his accomplishments at Cal Tech. But Bennett's first appointment to the faculty was Humberto Fernandez-Moran, also an electron microscopist, a choice that did little to instill confidence in his leadership or contribute to or complement the needs and interests of most biologists at the university.[21] Many of the biologists believed that aging facilities and a lack of adequate research space had contributed to the departure of several of their distinguished colleagues. Moreover, they agreed that recruiting new faculty and expanding graduate training in the newly emerging areas of biology was nigh impossible without a new research building. At Bennett's request, a committee of senior faculty proposed an ambitious and rather specific plan to build a new biological science complex that would also include a library and space for the chemistry department. Although funds were available for undertaking such a project, little was done to implement the plan until well after Bennett's five-year term as dean ended.

Many of the biologists who were at the university in the 1960s thought that Bennett's five-year reign as dean of the division was ineffective, even counterproductive. Failing to appreciate the forces that were transforming biology into a molecular science, he was refractory to recommendations and advice from those who were clamoring for change. With few exceptions, appointments to the science faculties failed to raise that part of the university to the level of eminence for which Beadle was striving, and none were as distinguished as those who had left. Curiously, the division of biological sciences was not among those that had taken advantage of the trustees' fund for restoring eminence to their faculty.[22]

Why did Beadle and Levi permit the biological sciences to languish? Quite possibly, they had decided in the first few years of their joint tenure that solving the plight of the basic biology departments was secondary to strengthening the university's faculty across all the divisions, professional schools, and the college. More likely, however, their efforts were thwarted by the almost unique prerogatives of the faculty in academic matters, particularly where new appointments were concerned. These were strongly guarded, and intrusions on that responsibility by the president's and provost's offices were likely to be ignored if not resented.[23] Beadle understood and appreciated the depth of that feeling of independence on the part of the faculty. On one occasion he commented, "We expect deans, departmental chairman, directors of various academic and administrative units, and faculties to assume responsibilities commensurate with the authority they enjoy by statute or assume in practice." The outcome, he believed, resulted in "the kind of community of free and able scholars that is ideal." There were few rules and regulations at Chicago, and except for Cambridge and Oxford Universities, he knew of no place where "so little attention is paid to those it does have."[24] With such a constituency and tradition, Beadle was virtually powerless to impose his judgements for new faculty.

Beadle insisted repeatedly that his and Levi's primary mission was to advance "Stiglerian eminence" in all the university departments and schools. While they stressed new recruitments as the route to that goal, others advocated nurturing the budding talents already at the university. The failure to do that aggressively was brought home by a memo from Earl A. Evans, the long-time chairman of biochemistry.[25] He pointed out that Konrad Bloch, who had left the biochemistry department for Harvard University, had just received the Nobel Prize in medicine or physiology. While absolving Beadle's administration of responsibility for allowing Bloch and several other likely Nobel Prize winners to "get away," Evans charged that the "university administration has not and still does not recognize the strength and potential capacity of its biological scientists. A first-class university should not serve as a training ground for the 'big leagues.'" After bewailing the fact that his department's space was grossly inadequate, he added that "If my language is intemperate, it is because we are losing a golden opportunity to make basic biology outstanding at Chicago." Beadle's response a week later was, "I've been worried about non-clinical biol [*sic*] since I came. Edward and I have discussed the situation many times and what to do about it. We have funds for facilities but no plans on what to do from Biol [*sic*]."[26] A year later Evans asked Beadle to facilitate the return of Konrad Bloch, but the possibility faded in spite of Beadle's encouragement and offer of support.

Evans's concern was not the first warning Beadle received about the sad state of the basic sciences in the division of biological sciences. An "off the record" memo from Leon A. Jacobson, professor of biological and medical sciences and director of both the medical school's department of medicine and the Argonne Cancer Hospital, sounded a clear alarm.[27] Although he had been a strong advocate of a unified division, he expressed dismay at the plight of the basic science departments. In Jacobson's view, Bennett's preoccupation with the administrative and oversight demands of the clinical enterprise left him inattentive to the needs of the traditional and emerging basic sciences. He concluded that the division had grown to a point where "it is beyond one man's capacity to manage it well." Furthermore, he noted, "the clinical and basic science activities have unique requirements and each requires full time thought and action." Jacobson was quite blunt in saying that Bennett was "OK but overwhelmed and inept." Although fund-raising pronouncements had continuously touted how well the division's unified biological sciences served medicine, the actual interactions were feeble and rare. Maintaining the basic and clinical sciences in a single division no longer made scientific, programmatic, or administrative sense. Intending to soften his criticism of Bennett's leadership, Jacobson believed that Chicago's eminence in biology had begun to decline when the clinical departments were added in 1927, the year the medical school became part of the university. Many, he said, agreed "that biology needs leadership unencumbered by worry of the clinical areas and vice versa." Jacobson's solution was to abandon the nonclinical designation in favor of basic science departments and to separate their administration from the traditional medical departments. Fully aware that such a solution could be disruptive, he suggested that the issue might be discussed when Bennett's reappointment as dean was reviewed.

Beadle relayed Jacobson's memo to Levi with an appended note that said, "Let's discuss. I agree there is a real problem but I'm not sure what the solution is."[28] Further concerns about the state of the biological sciences on campus came in the form of a report of graduate student admissions. While there had been about a 30% increase in graduate enrollment in the humanities, physical sciences, and social sciences divisions between 1963–1964 and 1964–1965, graduate enrollment in the biological sciences division dropped about 10%.[29] In response, Beadle fired off a note to Bennett asking "Shouldn't we be doing something about recruiting in Biology areas or making them more attractive to grad students?"[30]

Many in the division shared Jacobson's concerns about Bennett's stewardship. Almost certainly, Bennett was aware of his faculty's discontent and he must have anticipated or been informed that he would not be reappointed for another five-year term. Given those circumstances, he asked to be relieved of the major administrative burdens of deanship in order to resume his research activities. In 1965, acknowledging the university's "indebtedness" for his service as dean, the trustees appointed Bennett to a newly created division professorship and director of the newly established laboratories for cell biology.[31] Jacobson was appointed dean of the division in Bennett's place.

Concluding that a complete separation of the basic science departments from the division might be too radical a solution, Jacobson adopted a more practical approach. He appointed Richard C. Lewontin, an innovative geneticist who had come to the university from Rochester a few years earlier, associate dean for basic sciences. Lewontin was given the authority for reviewing faculty appointments and promotions and, more importantly, access to the divisional "treasury" for administering the "subdivision." One of his early actions was to merge the departments of botany and zoology, his own department, to form a new department of biology.[32] Nominally, the division remained intact but now the basic sciences were no longer the "poor cousins"; they had the full support of Jacobson and the rigorous oversight and management of a first-class basic scientist.

Did the new organizational structure lead to a renaissance of biology at Chicago? Not according to the biologists on campus or to Lewontin. By 1966, Lewontin confessed that the basic science departments remained "rather undistinguished" and that "it would be a very serious matter if the president and provost were to assume from the start that the division will settle for mediocrity." Lewontin's efforts at recruiting the division's top choices had been only partially successful, ostensibly because of insufficient support from the administration. Finally, he projected a bleak future: with 85 full-time faculty, 58 tenured members, and few retirements over the next five years, the chances of achieving any distinction "without making about three new appointments in each department" were nil.[33] But Lewontin's concerns and pleading went unanswered and he departed for Harvard soon afterward.

Beadle and Levi believed from the beginning that they would have to mount a major funding campaign to finance the growth they anticipated. Their ten-year projection of the university's needs was $360 million, but the decision was to reach that goal in two stages: The first stage set a three-year target of $160 million, the largest amount of money ever sought in a university campaign; approximately $1 million per week for the next three years.[34] Beadle and Levi spent three years persuading the Ford Foundation that the university was in sufficient need to justify support for the campaign.[35] The Foundation agreed to provide $25 million if the university raised $75 million from nongovernment sources; that left an additional $60 million to meet their overall target. The entire administration and senior faculty were mobilized to meet the ambitious target.

The university formally launched its "Campaign for Chicago" in mid-October of 1965 to a gathering of the city's most important citizens, including Mayor Daley. Fairfax Cone, speaking for the trustees, began the meeting matter of factly with the assertion that "it is inconceivable that we will not raise the money."[36] He knew full well that the money would be forthcoming because the university was held in such high esteem by Chicago's citizenry and its alumni. With only three years to achieve their goal, the entire fund-raising apparatus shifted into high gear. Beadle foresaw that he would be heavily engaged in visiting foundations, corporations, and alumni clubs, selling the message that a gift to the university was a worthy investment in its future. He was also aware that wealthy individuals who already had a history of giving to academic endeavors were most likely to respond positively.

It did not take Beadle too long to recognize that he had to overcome his natural reticence of asking for contributions and that being subtle about his purpose rarely elicited a voluntary offer of money. A great deal of preparation preceded any appointment, and it was understood by the prospective donor that Beadle was there to present a specific proposal and was counting on a positive response. Occasionally, in spite of all the preparation, however, these overtures did not succeed, and these "failures" preyed on his mind and accentuated his growing fatigue.[37] When successful, it was his custom and preference to acknowledge gifts and pledges with handwritten notes, often with a reminder of some personal incident at their meeting. Levi was similarly engaged, guiding the development effort on the home front, and reminding his faculty colleagues of their responsibility to contribute to the campaign and its symbolic importance.

The trustees did their share by collectively pledging $12 million to the campaign, but their extensive contacts with wealthy friends and business colleagues were even more important. By October of 1967, two years after the campaign was launched, somewhat more than $104 million had been raised and nearly half of the Ford Foundation grant had been matched. About two-thirds of the total was obtained from fewer than a hundred individuals. He was optimistic, Beadle told the faculty, that the remaining $56 million would be raised by the three-year deadline.[38] But deep down, he was concerned that the campaign's goal would not be reached before his retirement from the presidency at age 65.

Students found a way to confound the campaign's momentum. Until the mid-1960s, Selective Service Board policy was to defer college students who remained in good academic standing. But in the spring of 1966, the criteria used to defer students from military service changed; the quality of student performance became the metric for determining who was called and who was permitted to continue studies. In practical terms, draft boards were instructed to use class rankings to set priorities for drafting students into the army. Many students and faculty at the University of Chicago recognized that a ranking system would inevitably place some students near the bottom even though they were likely to be superior to students in the top ranks of lesser institutions. Moreover, since women students generally outranked the men in academic standing, ranking alone might increase a student's chances of having a deferment revoked. In the ensuing two months before the semester was to end, students questioned the legitimacy of the practice. Many were outraged by the university's implicit agreement to be an active agent of the Selective Service system by providing data on ranking and, by extrapolation, its support of the war in Vietnam, which they considered immoral. Administrators and most of the faculty justified the university's action as similar to providing student grade records to potential employers, and graduate and professional schools. But that view was dismissed with the argument that placing students in line to engage in an immoral war could not be equated with helping them further their careers.

What had been a heated, albeit relatively peaceful, debate during early May turned militant when the faculty Senate refused to rescind the practice of class ranking and the provision of that information to draft boards upon a student's request.[39] A faculty council suggestion that the entire issue be put on hold pending more temperate discussions over the summer went unheeded. By mid-May, the students realized that they would not get their way, and

350–400 students, some of them experienced members of the activist Students for a Democratic Society, entered the administration building and occupied the president's office suite.[40] Up to this point, Beadle had had very little to do with the discussions, but he very quickly became engaged. As before, police were not asked to intervene; the prospect of having large numbers of police officers dragging kicking and screaming students from the building was to be avoided at all costs. The decision was to do nothing.

Beadle condemned the sit-in as attempted coercion and refused to discuss students' grievances until they left the premises. But he and Levi, along with other top administrators, chose to adopt the same strategy that had been successful in the previous sit-in and vanish so as not to be available for running debates with the students. They worked at home or at scattered places on campus. Police were allowed into the building to guard against out-and-out vandalism, internal battles, and any actions that might get out of hand. There were periodic meetings with the students, but they led nowhere because Beadle and Levi surmised that the purpose of the sit-in was more about students seeking power in university decision-making than about the issues that initiated the confrontation. The students' strategy was coercion, hoping the administration would blunder and tilt public sympathies to their cause.

There were shifting alliances, and students and faculty who saw the activities as "spiritually uplifting." Other students were completely unaware of what was happening. After four days, however, most of the students vacated the building, leaving a small force to "hold the fort" and threatening to reoccupy the building if police were brought in to arrest those who remained. Both Beadle and Levi made it known that they would not enter the building until all of the students left. Within a day, the small occupying force vacated the building and the process of restoring the academic setting began. Beadle and Levi kept the faculty informed and thereby gained their loyalty for administrative decisions and actions.

A student government assembly-sponsored referendum indicated that an overwhelming majority supported the university's position. To head off future sit-ins, Beadle and Levi offered a resolution to the faculty council of the university senate recommending that disruptive acts such as sit-ins be prohibited and that the consequence of participating in one would incur penalties up to expulsion. Only half the faculty senate attended the extraordinary meeting, but that proposal was approved overwhelmingly, although about 30% of those present abstained. Many of those who refused to support the faculty council's resolution felt that the students' sincerity justified their behavior. To Beadle, however, full professors who argued that "sincerity" was sufficient justification for the students' disruptive behavior had abdicated their responsibility to promote rational discourse as a means to settling a dispute. It was also left to Beadle to bear the brunt of "Monday morning quarterbacking" from parents, alumni, potential donors, and the press. Some callers implied that he was a coward for not being more forceful and throwing the "bums" out. In response, Beadle drafted a letter possibly to alumni and donors, explaining the origin of the sit-in and his handling of the matter.[41] The trustees, however, were wholly supportive of the way the administration had handled the crisis. Beadle's usual calm demeanor failed to reveal his inner turmoil, since he had lost 15 pounds during the ordeal.[42]

The issue that initiated the "Student Against the Rank" action resurfaced in early 1967 after the faculty senate agreed to continue creating class rankings. However, they decided not to make total class rankings available to draft boards, although this action soon became moot because the draft law was changed, making class ranking irrelevant. Nevertheless, students felt betrayed, and believing that any ranking was an affront, the student activists once again called for a takeover of the administration building for a nondisruptive "study-in." Having been warned that a sit-in is a sit-in, most of the students who proceeded with the plan were suspended. This action only poisoned the air, and as the term was coming to an end, destroyed the comity of the campus. Out of 1500 invitations sent to students for the Beadles' annual open house, only 150 came. Also, the convocation was marred by many students wearing white arm bands and refusing to shake Beadle's hand when he awarded them their diplomas. The experience left the Beadles hurt and discouraged but hopeful that things would improve in their last year.[43] In addition, they hoped that the respite afforded by their planned travel during the summer might heal the anguish.

Despite the gloom of the sit-in's aftermath, Beadle was able to enjoy a meaningful triumph. He had long argued that recruiting great scholars from institutions at which they are generally revered takes more than just attractive salaries. Depending on their specialties, except for the experimental sciences, outstanding library collections and facilities rank as top priorities. Having made do for years with antiquated and scattered libraries, the Chicago faculty was making its displeasure known, and one could assume that the inadequate facilities handicapped recruiting. Beadle believed a new library was "without doubt the greatest single need of the university." By the end of 1965, architectural plans for an elaborate structure to house all graduate libraries except the natural sciences were complete, and only the lack of $18 million was delaying its construction. Ironically, the new graduate library was to be sited on Stagg Field, where, before being banned, Chicago had been among the powerhouses in intercollegiate football.

In medical centers, grateful and wealthy patients often provide a steady flow of money to those physicians who provided them with attentive care or cured them of their infirmities. One such "grateful patient," Mr. Joseph Regenstein, had been a patient of Leon Jacobson. Beadle appealed to Jacobson for permission to approach Regenstein about making a gift toward the library. Regenstein made a modest lead gift, and after his death his widow pledged $10 million. These funds, along with support from the Schermerhorn Foundation, were sufficient to initiate construction of the graduate library. Presiding over the groundbreaking on October 23, 1967, Beadle ended his remarks by noting that no president in history ever received so heartwarming and significant a gift, and so well timed as a birthday present.[44] Today, the Regenstein Graduate Library stands as a hub of the intellectual life of the university.

Throughout Beadle's tenure as president, the university trustees were continuously preoccupied with the ongoing renewal of the Hyde Park–Kenwood areas, although Beadle himself was only tangentially involved. Much of the ongoing oversight and management continued to be the business of the South East Chicago Commission and the broader community coalition. But Muriel had a renewal project all her own: the Harper's Court Development.[45] She and a small group of veterans of the neighborhood's most intense renewal battles recognized that the focus on restoration of suitable housing ignored the fate of the community's merchants. With the encouragement of several members of the Hyde Park–Kenwood Community Planning Committee, Muriel agreed, with considerable trepidation but with Beadle's blessing, to lead the effort to construct a shopping center devoted exclusively to artists and craftsmen. When Muriel began the planning and fund-raising, she warned Beadle that they might be liable for paying any debts. To reassure her, he quipped that if the effort was to fail just be sure to have it happen when the stock market was up. Thus was born the Harper Court Foundation whose corporate charter was to promote "the continuation of small businesses of special cultural or community significance."[46] Harper Court survives and serves the university community and its immediate neighbors. Muriel's efforts on behalf of the university were rewarded with an honorary degree on March 21, 1969.

Amid all the turmoil on the campus, the announcement on June 27, 1967 that Beadle would retire on or about October 22, the following year, the date of his 65th birthday, caught his colleagues and the community by surprise. The date for his retirement had been negotiated with the trustees when he accepted the position nearly seven years earlier. In his announcement, Fairfax Cone, speaking for the trustees, lauded Beadle's role in restoring the university's faculty and its physical plant to the top rank of American universities. In his words "He's the plainest, nicest guy in the world who quite simply turned a very great university into one of the world's greatest."[47] The newspapers were replete with praise when they published that "The greatest thing about Dr. Beadle's tenure is that we made significant advancements with a minimum of internal controversy"; "Beadle was important to the University as an intellectual symbol. How many universities have Nobel Prize winners as Presidents?" "Beadle's chief virtue was that he maintained an open door policy. He was very direct, informally charming in dealing with you and always open to suggestions."[48]

Personal letters from faculty echoed the same sentiments. One in particular, from the director of the university library, paid special tribute to Beadle's contributions: "the Library probably has made more solid and enduring progress during your administration than at any other time in the history of the university," it continued, "your astute selection of some distinguished administrators, notably Edward; your recognition of the importance of the physical plant–including grass;... your combined patience and firmness in trying to open and maintain better channels of communication with students; your candor and honesty; and by no means least, Mrs. Beadle's hard work on community and related problems; along with other efforts and qualities that have made this a much stronger university than it was when you came."[49] Another added that the university's spirit had changed the day Beadle arrived, and added "the grass has become greener in many ways... it has become a place hard to leave."[50]

These accolades of Beadle's presidency by his colleagues contrasted with his own thoughts.[51] During gloomy periods, he would sometimes think "about the only thing I've accomplished around here is to get some good lawns started... That's a hell of a note, isn't it? To be remembered as the president who was psychopathic about grass." After being reminded that he had in fact raised $160 million, his rebuttal was "I mean something I accomplished by myself. Think of it: I couldn't even make a dent in the Biology Division. They're still fiddling around trying to get the building designed, and the course work isn't anywhere near as good as it could be." He was quite honest in admitting "that he would leave no specific program or institutional structure which could forever be linked with his name and personality." Nevertheless, from the beginning, he was impressed with having been president of the University of Chicago and, perhaps, he preferred being remembered for his years at the university than for receiving the Nobel Prize.[52]

The student newspaper, *The Chicago Maroon,* gave him mixed reviews, suggesting that "a definitive history of the University of Chicago's second half century...will likely devote more space to the Beadle years than to George Wells Beadle himself." To students, Beadle was a remote figure who smiled at them as he walked through the campus, avoiding direct dealings with them. He was, nevertheless, the first president to address the student body on the "State of the University." Students commented that while "working with an outspoken and powerful faculty, a limited budget, and massive, largely insoluble institutional problems, it can certainly be said of him that he did his best. And ...his best was enough to make the university a better school for his seven years here. Chicago will be lucky if it finds a successor of whom the same can be said."[53]

For Beadle, however, neither the generous praise nor the criticism of his accomplishments as president mitigated his concern that the Chicago campaign's three-year goal of $160 million would not be reached before his projected retirement date. Determined that it should succeed, he spent much of the summer of 1967 traveling and visiting potential donors no matter what their capacity to contribute. At the trustee–faculty dinner to begin 1968, Beadle literally enacted his fund-raising pitch to leading corporate executives and alumni groups to give the faculty some flavor for what he could say about the university's distinctions and needs. Some of these forays paid off with substantial contributions, often more than he expected, while others left him disappointed and dispirited. By the beginning of 1968, the total campaign receipts totaled about $112 million The challenge for the remaining 10 months was approximately $48 million.[54]

Beadle's old colleague James D. Watson, a graduate of the college and the discoverer of the DNA double helix for which he received the Nobel Prize, was one of Beadle's targets. Knowing Watson well enough to be quite direct in his purpose he wrote, "Now that you are in the best seller category," referring to Watson's much heralded book, *The Double Helix,* "and correspondingly wealthy, the powers that be in the Division of Biological Sciences and the Alumni Association tell me that they have promoted you from the list of minor donors to the list of major gift prospects—$10,000 and up for the Campaign for Chicago. Doing as I'm told," he added, "I'm asking whether you want to build the new Biology building, endow a professorship or do something a little closer to the $10,000 level?"[55] There is no record of how successful this "pitch" was. Evidently the final push to reach the campaign's goal was successful, for much to Beadle's relief the trustees announced in early November 1968 that the goal of $160 million was in hand.

When Beadle's retirement was announced, a trustee–faculty committee was created to identify and recruit his successor.[56] That search was short-lived, for it was evident to the trustees and the faculty that Edward Levi was the most appropriate candidate to assume the presidency, and by September he had accepted: "I welcome the opportunity to serve an institution which in many ways has been my life...I will do my best for this institution which I love."[57] Beadle was pleased by the trustees' choice, commenting that once again the trustees had demonstrated their "devotion, wisdom and good judgement" by appointing Levi as the university's eighth president. Over and above creating $70 million worth of new buildings and bringing 250 new faculty to the university, Beadle ventured that "getting Levi in the presidency line is my greatest contribution to higher education.[58] Muriel welcomed Kate Levi and helped her begin the renovations of the president's house to suit her family's taste and needs.

Levi's inauguration took place on November 14, 1968, at which time Beadle's retirement became official. Shortly after the ceremonies, Levi wrote to Beadle thanking him for a beautiful plant and added, "My chief virtue, if I have any at all, is that I have seen at closest hand a master on the job. I will try to emulate you but I know I will be disappointed with how far I will fall short of being able to do so. We reciprocate the love; we need the sympathy."[59] Freed from the everyday responsibilities of the president's office, Beadle could now contemplate a more relaxed lifestyle and a return to research.

CHAPTER 18

The Corn Wars

Retiring university presidents face a daunting personal challenge. Overnight they go from being the center of attention, the final stop for all university decisions, and influential figures in national and international arenas, to a private life. Left behind are the professional assistants and grand university homes. Some find it possible to return to teaching and research. Others discover that their field has moved forward in ways they cannot understand. The intellectual, political, and social voids can be overwhelming. Beadle had no problem. He almost seems to have breathed a big sigh of relief. Perhaps he realized that student activism would grow in fervor and complexity and that he was well to be out of it. Besides, he knew exactly what he wanted to do. He wanted to return to research and establish the origin of corn.

First, the Beadles organized their personal lives. Colleagues at Cal Tech urged them to return to Pasadena, but they felt at home in Hyde Park. They bought a "100 year old slum house near the University for a fantastic sum and are fixing it up."[1] Beadle promptly went to work hauling top soil, manure, and garden supplies from the front of the house into the garden in the rear and setting up the greenhouse the university's trustees had given him as a retirement gift. Eventually Muriel hung signs in front inviting neighbors to come and look at the garden where Beadle was experimenting with flower varieties. Muriel kept up her work with the artisan's shopping center and by 1969 her book about early childhood development was in press.[2] They met monthly with five other couples for scholarly presentations and dinner. Beadle usually talked about corn.[3] And they began traveling for pleasure. For the first time they attended the Nobel Conference in Lindau and toured Europe, stopping in England to visit David Beadle and his new family. He and his wife, Jacqueline Isitt, had two small children: Katherine was born in late 1966 and Maiwenn about a year later. Altogether, Beadle now had five grandchildren.

Not surprisingly, several Chicago institutions were anxious to bolster their enterprises through Beadle's prestige and fund-raising skills. In 1963, the American Medical Association created the AMA Institute for Biomedical Research with the intention of glossing its image as a political lobby with an effort to advance scientific knowledge. The institute's laboratories and 80 scientists, many recruited from leading research centers, were housed on the top floor of the AMA headquarters building in downtown Chicago. The effort, which was largely supported by the AMA's Education and Research Foundation, had modest success. When its director left in 1967, Beadle was invited to take over. He argued that the institute would remain viable only if it was considerably enlarged and relocated adjacent to the university campus. The AMA board accepted Beadle's demands and he was appointed director on December 1, 1968, barely a month after leaving the university presidency. A suitable site was identified, the selected architect produced a finished design, and Beadle started to raise the required $7 million. He was ambivalent about the fund-raising: "I thought after helping the Univ [*sic*] of Chicago raise $160,000,000 in three years I'd be out of fund-raising. But, alas, no...But raising money is fun in a way—especially when you get it."[4] In this case, it was not fun at all. The AMA proved indecisive and, in spite of Beadle's effort to convince its House of Delegates, "made up of pretty reactionary MDs," that it was a good plan, the rank-and-file physician members were unwilling to contribute.[5] Beadle knew that without their support there was no point in approaching other donors. In December of 1969, the AMA closed the institute. Beadle was out of a paying job, but it's doubtful that he had any regrets.

Even before Beadle handed the presidential responsibilities over to Edward Levi, the Chicago Horticultural Society announced his election as a trustee and its new president.[6] The society proudly broadcast Beadle's skill as a gardener and quoted his earlier proclamation to the university's students "that grass, like ideas, be allowed to grow freely, and without oppression on the campus of the University of Chicago." They expected that Beadle would take the lead in the planning, organizing, and fund-raising for a 300-acre Botanic Garden 20 miles north of Chicago. He did not disappoint them although he sensed that he had been "conned into becoming Pres [*sic*]."[7] The mammoth undertaking included reshaping the landscape by forming hills, lagoons, and islands on the Skokie marshes and moving over a million and a half tons of soil.[8] It was to be a place for the public to enjoy and for research. And it is. On a fine Sunday in the spring of 1998, the garden was full of visitors admiring the landscape and the profusion of blooms. The official guide emphasized the research plots and greenhouses. Beadle's work for the Garden wasn't entirely altruistic. His investigation of the origin of corn required more planting space than was allocated to him on the Chicago campus. While the Garden was taking shape, there was plenty of room for growing corn.[9]

Beadle wrote exuberantly to Delbrück, "And believe it or not, *I do research*—early each morning, evenings, and weekends."[10] Muriel relished his pleasure in the enterprise. "Without a grant, and without an assistant, but with a lot of energy and excitement, he is testing a hypothesis that domesticated corn is descended from the Mexican teosinte. I've never seen him so enthusiastic."[11] When he finally wrote to McClintock about his work, she wrote back, "Yes indeed, I have heard of your return to maize and, it seems, just where you left off. Of course, you never really turned-off the corn plant. It was always lurking somewhere—in the backyard, on the front lawn, or on any other piece of soil that was handy!"[12]

Corn has ancient origins in middle America perhaps as long ago as 7000 years.[13] Tiny ears, estimated to be more than 4000 years old by radioactive carbon dating, were discovered in southwestern U.S. and Mexican caves that had been occupied by humans. The grain was unknown to the rest of the world until the Spaniards arrived in the Western Hemisphere in 1492. By then, whole civilizations and millions of lives depended on it from the far north to the far south of the Americas. But where did it come from? This question has intrigued anthropologists as well as plant biologists for 150 years or more. In contrast to other major food plants, it was difficult to find a wild plant that looked like something akin to cultivated corn. There were several possible answers to the puzzling question of its origins. It might have been the outcome of purposeful crossbreeding by ancient humans or a chance cross between two closely related wild species that were quite different from their hybrid offspring and thus unrecognizable. Or, perhaps corn had a wild ancestor that was extinct or remained undiscovered. Or, as Beadle and a few botanists before him believed, a "wild" ancestor of corn actually existed and was even known but was not easily recognizable as the sought-after parent. Their hypothesis was that ancient Central American people had derived corn by conscious selection of advantageous mutants of the wild, indigenous plant called teosinte. The name itself suggested the hypothesis. Teosintl means God's corn in the Unto-Azteca language. The idea had been planted in Beadle's mind by R.A. Emerson and taken root during his student days at Cornell. He actually thought he had proved the hypothesis in the 1930s.[14]

Teosinte grows only about one-half to three-quarters as tall as a typical corn stalk and instead of having a single stalk, it is bushy. Corn has a single tassel, the "male" organ that produces the pollen, at the top of the stalk while tassels adorn each of the many teosinte branches. The teosinte "spike" (analogous to the corn cob) is about the width of a single row of corn kernels and only about three to four inches long. Teosinte spikes have only one row of small, hard kernels compared to the eight (or more) rows of plump kernels typical of the Indian corn that Beadle and others studied, which is more properly called maize. The leafy husk that covers corn cobs is missing in mature teosinte, where each kernel is surrounded by a hard, tough shell, or "glume."

The individual kernels readily separate from one another and fall off the teosinte plants. These seed-containing "fruits" can be dispersed by animals because their tough shells make them undigestable, an important advantage for a wild plant. In contrast, corn cannot grow in the wild, as Beadle continually reiterated. It needs people to cultivate it. Because the kernels are firmly fixed to the cob which is packaged tightly inside the husk, few kernels can be scattered to assure a new generation in the wild. Birds and small mammals do not spread the kernels because they eat and digest them. The differences between corn and teosinte appeared so great that they had been classified in different genuses: *Zea* for corn and *Euchlaena* for teosinte.

Emerson began working with teosinte long before Beadle arrived at Cornell. As his mentor's research assistant, Beadle was given the task of continuing the experimental work. Emerson had already figured out why it was that teosinte, which was brought to the United States from Mexico as early as the 1880s as a possible forage crop, grew but produced no seeds in Ithaca; the summer days were too long in the northern latitudes. To enable crossbreeding of the two plants, he artificially shortened the exposure of teosinte to daylight. Consequently, the teosinte flowered simultaneously with corn and breeding experiments could be carried out. Cross-species hybrids were produced in abundance and the hybrids themselves yielded fertile seed. This itself suggested a close kinship between the two plants (the results depend somewhat on the particular strain of teosinte used) because, by definition, organisms are considered to be the same species if they can breed and produce fertile offspring. Moreover, both plants had ten pairs of chromosomes. It had in fact been suggested more than a decade earlier that the two plants should at least be classified in the same genus, *Zea*.[15]

Beadle first analyzed the corn–teosinte hybrids that Emerson had already produced. He observed in the microscope that during meiosis in the hybrids, the ten teosinte chromosomes readily paired with the ten corn chromosomes, a surprising result if the two plants were really different species.[16,17] Not only did the two types of chromosomes pair properly during meiosis in the hybrid, but they engaged in recombination, that is, they exchanged chromosome segments, in the same way and at the same frequency as occurred in plants with pure corn lineage.[18] Thus, not only were corn and teosinte chromosomes and genes very similar, but the arrangement of the genes on the chromosomes of the two plants was also the same. Curiously, the paper does not state the obvious conclusion that teosinte and corn must in fact be the same species or that teosinte was likely the ancestor of domestic corn.

The first time Beadle actually proposed in print that teosinte was the ancient wild parent of domesticated corn was in 1939.[19] Although he was busy with the *Drosophila* work and teaching at Stanford, he felt compelled to respond to an alternative hypothesis proposed by Paul C. Mangelsdorf and R.G. Reeves, an idea he considered dead wrong. These scientists claimed that an extinct or undiscovered ancestor of corn existed or exists and that teosinte was of more recent origin than corn.[20] According to Mangelsdorf and Reeves, teosinte was the product of crossbreeding through which some four or five chromosome segments from a plant called tripsacum were inserted into primitive, wild corn about 1300 years ago. The idea was not original. As Mangelsdorf and Reeves noted, it had been suggested by Edgar Anderson, a highly respected scholar of plant variation at the Missouri Botanical Garden and one-time roommate of Mangelsdorf at Harvard (and no relation to E.G. Anderson but also called Andy).[21] Anderson took credit for the hypothesis,[22] although in later years Mangelsdorf broadcast the impression that this so-called "tripartite hypothesis" was his own idea.

Like corn and teosinte, tripsacum is native to Central America and Mexico. It has, to the eye, some similarities to both corn and teosinte, although it was classified in a third genus, *Tripsacum*. Beadle knew that tripsacum with 18 chromosome pairs was not very closely related to corn or teosinte, both of which had ten. More important, it was very difficult to hybridize tripsacum and corn. Various laboratory "tricks" were needed to have tripsacum pollen fertilize corn plants. Very few of the resulting seeds matured, the hybrids show little or no pairing between the two sets of chromosomes in meiosis, and, although the hybrid female germ cells could be fertile, the little pollen that was produced was sterile. Beadle thought it was unlikely that fertile, tripsacum–corn hybrids would form naturally given the difficulty of making them artificially. On the other hand, he found data supporting the close relation between teosinte and corn in Mangelsdorf and Reeves' own paper. Although not noted by the authors, their data indicated that there was only a small number of genetic differences between teosinte and corn and that these were localized in four or five chromosomal segments. According to Beadle, there was little justification for continuing to classify the two plants in different genuses. The Mangelsdorf and Reeves hypothesis was, in his view, untenable. He did, however, feel obliged to defend his own idea that teosinte was the wild parent of corn, against two criticisms.

Mangelsdorf and others had argued that it would have taken an impossibly long time for ancient people to develop corn by the selection of naturally occurring teosinte mutants. In rebuttal, Beadle pointed out that there were so few genetic differences between the two plants that selection could have taken place over a reasonable amount of time. The second objection was expressed as a question. Why would ancient people have thought that teosinte was worth cultivating in the first place? The covering on its kernels is so tough that getting at the edible part is very difficult. Beadle thought he had the answer. In February of 1939 he wrote enthusiastically to Emerson[23] about the "poppability [*sic*] of teosinte," enclosing a copy of the manuscript he was about to submit for publication. There was archeological evidence that pre-Columbian inhabitants of middle America popped corn. So Beadle put teosinte seeds in a corn popper, heated them, and produced what looked like ordinary popcorn. He argued that this food potential was reason enough for the ancients to cultivate teosinte and select desirable mutant traits as they appeared, ultimately obtaining plants and kernels that were like corn. Emerson wrote back saying that the idea was very interesting and had not occurred to him.[24] He told Beadle that others, namely, E.M. East and D.G. Langham, also had evidence that there are only five or six distinguishing characteristics between the two kinds of plants and that in those that were tested, the corn trait was dominant. He himself had noted three or four mutations to teosinte-like characteristics. Emerson had an additional possible explanation for the ancient interest in teosinte; he had obtained a wild teosinte with a mutation that made the seeds easy to thresh. Years later, Beadle showed that teosinte can actually be ground with primitive stone grinding tools and the grain separated from the hard parts by water flotation. To prove it was edible, he ate considerable quantities of ground teosinte over a period of four days with no untoward effects.[25] Colleagues, too, were forced to eat such things as poor-tasting cookies made from teosinte flour.[26]

For the 30 years following 1939, Beadle paid little attention to the question of the origin of corn. Meanwhile, Mangelsdorf pursued the matter relentlessly. Like Beadle, Mangelsdorf, born in 1899 in Kansas, was raised in corn country. By the time he went to Kansas State University to study agronomy he knew a great deal about plants, especially corn. After finishing college, he went east and obtained a Ph.D. at Harvard, where his mentors included the great corn breeders and geneticists East and Donald F. Jones at the Connecticut Agricultural Experiment Station. He took a position at the federally supported Agricultural Experiment Station at Texas A&M University where he accomplished enough by 1940 to be appointed professor of botany and, soon after, director of the botanical museum at Harvard. Mangelsdorf's whole life was dedicated to corn. Even his leisure time was spent collecting objects with corn motifs—everything from modern junk jewelry to a 16th-century Chinese ivory corn ear.

Mangelsdorf worked for decades to gather evidence in support of his tripartite hypothesis, and the idea gained ground in the 30 years Beadle was otherwise occupied. A large number of research papers and publications for the general public assured that Mangelsdorf's idea found its way into textbooks of biology, history, and archeology as presumably established fact. He frequently stated that "the hypothesis that teosinte is the progenitor of maize is definitely no longer tenable."[27] Although he eventually agreed that teosinte should be classified with corn in the genus *Zea*,[28] he stuck to the tripartite hypothesis. In the 1968 edition of the Encyclopedia Britannica, which was published by the University of Chicago while Beadle was still president, Mangelsdorf said bluntly that it had been "proved that the ancestor of corn was not teosinte." Dozens of others became involved in the controversy generating hundreds if not thousands of dense, printed pages and plenty of dispute. Experiments and polemics had to address not only the question of the origin of corn but also whether teosinte was a recent hybrid of corn and tripsacum as claimed by the tripartite hypothesis.[29]

When he caught up with what had happened, Beadle realized that "My opinion had zero influence—in part, because I had gone off onto other types of investigation."[30] A monograph by H. Garrison Wilkes[31] sent him "through the ceiling" because it said that the notion that teosinte is the ancestor of corn was "crude" and a "myth."[32] He "resolved to do something about 34 years of confusion that I attributed to the tripartite hypothesis."[33] He was ready to take on the competition, especially as it meant spending his days in the fields.

Beadle began with three goals. One was to determine the number of significant differences among the genes of corn and teosinte. The second was to determine what those genes did. Finally, he wanted to hybridize primitive corn and teosinte in the hope of generating ears like the ancient ears found in caves; Mangelsdorf had already tried and failed to work this last angle and Beadle did not succeed either.[34] His plan was to use purely Mendelian methods to determine the amount of genes that distinguished corn and teosinte. He did this by making hybrids and then determining how often the parental types showed up in the hybrid's offspring after self-pollination. He started with a primitive, teosinte-like corn called Chapalote and a corn-like teosinte, Chalco, hoping to avoid confusing results that could be introduced by the use of highly cultivated corn. Mendelian principles predict that if the hybrid's two parents differ in one gene, then each parental type should reemerge in one out of four new plants produced by self-pollination of the hybrids. Similar logic predicts one of each parental type out of 16 plants if the parents differ by two genes (see Chapter 2). Extending the same principle further, if the parents differ in no more than six genes, the parental types should reappear in one out of about 4000 plants. For statistically significant results, Beadle would eventually need to grow approximately 50,000 plants.[35]

By 1970, the project was in full swing. Beadle had corn growing at the

Botanical Garden and the small plots on the University of Chicago campus, and he traveled frequently to Mexico. He was ready to plant thousands of hybrid kernels he had painstakingly cut out of cobs with a nail clipper as well as additional backcrosses of hybrids to the parental plants.[36] The 20,000 third-generation plants were grown at El Batan, an experimental station 20 miles north of Mexico City set up by the Rockefeller Foundation's International Maize and Wheat Improvement Center (Centro International de Mejoramiento de Maiz y Trigo or CIMMYT). Several Mexican plant scientists collaborated with him, and his prestige galvanized like-minded colleagues in the United States who had despaired at the acceptance of Mangelsdorf's tripartite hypothesis.

Many of the participants in the dispute over the origin of corn met at a conference in Urbana, Illinois, in September of 1969, including two, H. Garrison Wilkes and Hugh H. Iltis, who had not been invited.[37] Wilkes, a student of Mangelsdorf, disagreed with Beadle. Like some of the other corn warriors, Wilkes was a near mystic on the subject: "When you're working with a wheat plant, who cares? But a corn plant, it's different. It's of human heights, and you can look it in the eye."[38] Iltis, a morphologist and taxonomist who is typically outspoken, agreed with Beadle. He was born in Mendel's city, Brno, Moravia, in 1925 and received a Ph.D. from Washington University in 1952. Before the family emigrated to the United States, his father, Hugo Iltis, had taught natural science in Brno and wrote a biography of Mendel.[39] Although Beadle had been absent from the fray for 30 years, his stature allowed him to influence the meeting's outcome by inviting Iltis to give the summary at the end. Later, Iltis recollected his talk. "By all taxonomic and genetic standards Teosinte cannot be a hybrid between *Tripsacum* and a mystical and now reputedly extinct [*sic*] 'wild corn,' whose morphologically ridiculous reconstructions would seem to place it in any case outside of the Tripsaceae into fantasy."[40] No wonder that "Mangelsdorf left in a huff."[41]

Walton C. Galinat, another member of the corn crowd, also had an unusual approach to his research. He told a 1983 visitor to his laboratory that "Corn is my religion and this lab is my church."[42] Besides his serious work, the flamboyant Galinat spent years trying to breed corn with kernels that were square, or displayed the stars and stripes of the American flag, or came on three foot-long cobs. He had started out convinced that Mangelsdorf, his professor, was correct. By 1970, however, his own data and Beadle's arguments convinced him to switch sides.[43] Galinat observed that vestiges of the hard, cuplike structures called cupules that hold teosinte kernels occur between the kernels in corn. This observation, which was also made by Iltis at about the same time, was another blow to the tripartite hypothesis which assumed that corn had no cupules.[44]

In October of 1970, Beadle and Galinat examined the architecture of the spikelets on the corn–teosinte hybrids growing at El Batan. They worked 14-hour days and Muriel worried that "all that standing didn't do his knee any good."[45] Spikelets are the structures that grow into cobs and kernels on the female organs or pollen-producing structures on the male organs or tassels, respectively. Corn normally has paired spikelets on both tassels and the female pistils while teosinte has paired spikelets in the tassels but single spikelets on the pistils; one of the embryonic spikelets does not fully develop. Their data suggested that two different genes were needed to determine the single-spikelet trait in teosinte.

Beadle planned to plant the 25,000 third-generation plants at the Mexican site, and by the end of summer 1973, 50,000 plants had been grown in all. The results were clear-cut. Parental types appeared in about one of every 500 plants, consistent with there being a total of four or five genetic differences between corn and teosinte.

For the second of his aims, to document the essential differences between teosinte and corn, Beadle wanted to look for mutant teosinte plants that had one or more corn-like properties. This would require looking at thousands of wild plants and collecting their seeds. To find such plants, Beadle organized "teosinte mutation hunts." In November of 1971, 18 people representing seven universities and volunteers from the Chicago Horticultural Society, and including T.A. Kato, a young Mexican cytologist trained by McClintock, and L.F. Randolph (who was at Cornell in the 1920s and was now 77 years old), spent a week south of Mexico City on the first hunt.[46] A grant of $6582 from the National Science Foundation covered most of the expenses. Beadle made a separate trip a few weeks earlier to arrange the logistics and harvest some corn. He organized everything from food and lodging to transportation both to and within Mexico and provided detailed instructions to everyone. The group collected seeds from 75,000 teosinte plants, demonstrating that it would not have been difficult for primitive societies to harvest large crops. In particular, Beadle was looking for signs of the natural occurrence of a mutation called *tunicate*. When the corn *tunicate* gene was bred into teosinte experimentally, the normally hard teosinte kernel covering became soft. Beadle thought that a natural mutation in that gene might have been an important step in the domestication of teosinte by ancient people.[47] Disappointingly, no such mutant turned up in the hunt.

When the Chicago Horicultural Society recruited a full-time president, Beadle had no commitments except for his research and lecturing. His part-time appointment at Chicago gave him access to greenhouses. He and Muriel escaped part of the Chicago winter by spending two weeks lecturing at the

University of California at Davis. They made a joint presentation about urban renewal and spoke separately about their respective interests. Then, in November of 1972 he organized another teosinte mutation hunt. With the help of villagers from Mazatlan, in the State of Guererro about 170 miles south of Mexico City, the group collected 73 kilos of ripe teosinte seeds in two days. The technique was to beat the plants so that the seeds fell onto the plastic sheets and blankets they had laid out on the ground. Beadle took the seeds back to El Batan where, with the help of Dr. Mario Gutierrez, he sorted, cleaned, and dried the seeds for study and determined that he could look at 2 kilos (about 60,000 seeds) in an 8-hour work day. When he returned home he examined each of the 2 million seeds individually but once again found none with the soft covering typical of the corn tunicate allele.

Wilkes, who was neither rigid nor immune to Beadle's arguments, joined the team for the second teosinte mutant hunt. Beadle valued his collaboration because he was a knowledgeable collector and they got along well in spite of their disagreement about the origin of corn. He also thought that if he worked hard, all the corn warriors would come around to his view. He and Wilkes together critiqued Galinat's manuscript for a major review article.[48] Beadle boasted that they had helped Galinat get up the courage to challenge Mangelsdorf and believed that both Galinat and Wilkes "see the light."[49] But the war went on, and years later he worried that Wilkes had not "abandoned his former heretical views."[50]

After more than 40 years of going their separate ways, Beadle realized that McClintock, too, was now interested in the origin of corn and might have relevant data or be able to obtain them. She had officially "retired" by 1972 but had an active laboratory at Cold Spring Harbor. Her 25 years of trying to convince biologists that corn chromosomes contained segments that moved around in the genome were vindicated.[51] Others had demonstrated that these transposable elements—what she called "controlling elements"—occur not only in corn but in virtually all genomes. She was now caught up in efforts to trace the origins of the various strains of corn that were growing all over the world because an encounter with Mangelsdorf had left her in "high dudgeon."[52] Whatever else might be said about Mangelsdorf, he gets a lot of credit for motivating other scientists. McClintock was especially interested in figuring out which of the different teosinte genomes had contributed to which strains of corn. She and her colleagues in Mexico were using the unusual knob structures that decorate corn and teosinte chromosomes as a key to the relationships.

Visible knobs are a notable feature of corn, teosinte, and tripsacum chromosomes. The DNA sequences in the knobs are known, but the knobs' functions, if any, are still unknown. Knobs are useful in trying to understand the history of cultivated corn because various strains differ as to the size, location, and number (from none to 20) of knobs, depending on the geographic origin of the plant. Beadle argued that corn and teosinte, growing in proximity over long periods of time and exchanging genes by cross-pollination, might share knob characteristics. He hoped McClintock already had the data and, if not, he suggested that T.A. Kato, with whom both collaborated in Mexico, might do the work. A bonus from this would be that Kato could use the project as the basis for a Ph.D. thesis and stepping-stone to a good job at CIMMYT.[53] McClintock, too, was concerned about Kato's future, and she helped Kato obtain his degree, with Galinat as his research professor. Beadle and McClintock tried to convince the Rockefeller Foundation to provide Kato with a fellowship but, because he was over 40, he did not qualify. Beadle was sufficiently concerned to offer him some of his own money.[54]

McClintock had already collected a lot of knob data with her Mexican collaborators. They had characterized the knobs (and other chromosome features) in thousands of plants from all over Central, South, and North America. As to Beadle's question, the data varied from one place to another. Teosinte and corn from Guatemala did not have similar knobs, but in at least some parts of Mexico, they did.[55] Beadle was not surprised. He and Emerson had long ago noted that Guatemalan teosinte was sterile when crosses were attempted with corn or even Mexican teosinte. He concluded that Guatemalan teosinte was a distinct, ancient strain that separated from the teosinte parent of corn well before corn evolved and had not contributed at all to the origin of corn.[56] This conclusion was consistent with McClintock's proposal that the Rio Balsas region south of Mexico City in the State of Guerrero was the area from which corn had emerged.

The evidence favoring teosinte as the parent of corn was mounting, but Mangelsdorf remained skeptical. To try to convince him, Beadle asked him to identify the true corn in a collection of coded cobs, some of which were "real" corn and others the corn-like second-generation products of the hybrid breeding. Mangelsdorf named some of the hybrid products "good corn." It was hard to argue his way out of that admission and he weakened. When the corn warriors gathered at Harvard in June 1972 there was "some three hours or so, of heated, loud arguments, of even a certain amount of name-calling and innuendo," after which Mangelsdorf grudgingly conceded that the evidence did not support the idea that teosinte was a recent hybrid between tripsacum and corn, and the tripartite hypothesis should be abandoned.[57] Soon after, however, he claimed that an ancient, unknown, and probably extinct wild corn was the parent of teosinte and presented this hypothesis in scathing commentaries on Beadle's writings.[58] Reviewing Mangelsdorf's book, Beadle called it authoritative and acknowledged his adversary's great influence on the field of corn evolution.[59] But he also wrote, "Now, as a rather sad anticlimax, we find...an admission that the [tripartite] hypothesis is no longer tenable." Moreover, Beadle was "not persuaded" that Mangelsdorf's basis for withdrawal was even sound. Beadle was confounded. If corn could be the ancestor of teosinte, why couldn't Mangelsdorf admit that teosinte, a successful wild plant, could be the ancestor of corn?[60] And this was not to be the end of it.

All through the 1970s, Beadle was a man with a mission. Most of his scientific dealings were with the corn warriors, his Mexican colleagues, and McClintock. Elsewhere, genetics was rapidly becoming a molecular science and his correspondence with Delbrück kept him aware of new developments. But there is no indication that he considered applying the new methods to the question of the origin of corn in spite of sharp questions from Delbrück.[61] Delbrück wanted to know what kind of alleles were important for the differences between teosinte and corn. Were they structural changes in the genes that would give altered proteins? Or were they regulatory differences that affect the cell type in which the genes are expressed or the level or timing of gene expression? Delbrück reported that Allan C. Wilson had suggested that "most evolutionary major mutations are regulatory." And if they were regulatory, were they changes in one or a few DNA base pairs or mutations caused by insertion of one of McClintock's controlling elements? Did teosinte even have such transposable elements? (Teosinte has the same transposable elements as corn, but that was not known at the time.) Beadle replied, saying he "broods" over Delbrück's questions and his own "rather naïve and primitive" approach. He summarized more sophisticated data being obtained by others and asked Delbrück if it would be useful to gather a small group of people to discuss new approaches.[62] Brooding or not, he seemed contented with being out of the mainstream of genetics and boasted that after eight years of trying he had finally succeeded in breeding the photoperiod genes from corn into teosinte, which allowed the two plants to be grown side by side in northern latitudes.

In the same letter he admitted that he and Muriel were thinking about moving to a warmer climate if they could find a place where corn colleagues and facilities were available. The University of Hawaii was attractive but isolated. Would Cal Tech have greenhouse space available? Between concern about their physical surroundings and Beadle's detachment from current science, the letter seems almost despondent. Enthusiasm emerged in describing Muriel's latest book about the history of cats, their biology, their role as pets, and their behavior.[63] The past was also on his mind. He set up a foundation named in memory of Bess McDonald Higgins, the Wahoo teacher who had convinced him to go to the university. He, his sister, Ruth, and residents of Wahoo contributed the funds to provide a scholarship each year for a student graduating from Wahoo High School and going to the University of Nebraska.[64]

Delbrück sensed that his old friend and colleague needed a boost and promptly talked with Robert Sinsheimer who then held Beadle's old position as chairman of the biology division. Sinsheimer immediately wrote encouraging Beadle to think about coming to Pasadena. But Beadle was realistic and recognized that by the time such a move could be organized it would be the following year and "by then I'll be 75+, and though I'd wish it were not so, I'm slowing down physically and mentally....I'm sad to conclude that it's a beautiful dream."[65] Almost a year later he wrote to Delbrück and his wife about trying to derive primitive-like corn from backcrosses of teosinte/corn hybrids.[66] His corn work was old-fashioned, but it was providing purpose and occupation for his aging hands and mind.

Except for minor illnesses or discomforts, Beadle's health had been robust and he, like many active people, probably brushed aside the usual discomforts of aging. His arthritic knees were troubling but they didn't keep him from his work. He did stay out of the fields for a while in the summer of 1977 when he suffered severe headaches, but neurological tests at the University Medical Center revealed nothing and in time the headaches disappeared.[67] Muriel as well as friends and colleagues had, like Beadle himself, noted occasional lapses in Beadle's memory and behavior. A few months short of his 75th birthday, he noted that memory blackouts had "developed rather rapidly, over the past 2–3 years...I have to stop and think about a person I know perfectly well, and often I can't get the name."[68] Muriel, too, had been reasonably healthy and as untiring as ever in her writing and community activities.[69] Between 1970 and 1977, while her husband was occupied by his research on the origin of corn and frequent trips to Mexico, she published three books.[70–72] While *The Cat* was still in press she began her eighth book with a tentative title of *Aging*; Beadle quipped, "maybe as a result of observing me!"[73] Then, in the spring of 1978, all her plans went to pieces and the Beadles' lives changed. Muriel suffered a debilitating stroke followed soon after by a heart attack.[74] For a while it was doubtful if she would survive, and she remained in the hospital for more than six weeks. In time she rallied and began to recover, although she would need a leg brace and a cane for the rest of her life.

Neither his worries about Muriel, new responsibilities at home, or his own memory problems kept Beadle from completing an arduous Sigma Xi-sponsored lecture tour during which he visited campuses all over the United States. In November, delighted with the invitation, he traveled to Pasadena to present a lecture on his ideas about the origin of corn at the Cal Tech biology division's 50th anniversary celebration.[75] If he was slowing down, he was not about to give up. However, the Beadles soon realized that the house on Dorchester was too inconvenient, and they rented an apartment a few blocks away on E. 56th Street. Sadly, it meant giving up the flower garden and greenhouse that Beadle had nurtured since retiring as president. For the first time in his life, he could not open his door directly to the out-of-doors.

As time went on, Beadle continued to admit to all kinds of health problems, and his difficulties became more obvious to others. At a meeting of the Welch Foundation's scientific advisory board, he was found wandering in the hotel's hallways unable to remember where he was or how to find his room.[76] Traveling alone to the annual meetings in Houston, Texas was no longer safe for him and he resigned from the board.[77] At about the same time he began a letter to his sister "Dear Ruth," and forgetfully signed it "Ruth" at the end.[78] At the same time his research went on. "I really can't get away from a greenhouse full of corn I'm working on. I'm now ready to do in for good the Mangelsdorf-Barghoorn et al wild corn hypothesis—other than teosinte."[79] At 77, age and infirmity had not squelched Beadle's competitive fervor. Besides, the work was important because wild teosinte represented the only straightforward source of new genetic traits that could be used to improve cultivated corn.[80] Mangelsdorf, too, was still ready to fight on.

Despite the deteriorating mental condition that Muriel reported in her letters to family, Beadle's scientific correspondence with Norman Horowitz, David Perkins, and Hugh Iltis over the succeeding eight months was coherent and purposeful. He published a review in *Scientific American*[81] and worked on a paper for *Science* magazine. Amazingly, he had no reservations about agreeing to give the Proctor Prize lecture at Sigma Xi's annual meeting almost a year later in October 1981. He couldn't resist gloating that he had finally done in the Mangelsdorf–Barghoorn wild-corn hypothesis.[82] His proposals for new experiments that could nail down additional aspects of the teosinte story bear little evidence of his mental or physical problems.[83] It was as though his scientific mind and his social mind were under different controls.

Perhaps his satisfaction with the scientific work explains why Muriel admitted to the burden of her own and Beadle's infirmities more openly than he did. She was "very despondent at times,"[84] although eventually she learned to manage the depression and resumed her long letters to family and friends.[85] Muriel worried about their security and considered moving to a retirement community, although admitting that life "would be more circumscribed and lacking in diversity" than what they then enjoyed. Beadle shared neither her concerns nor her sense of urgency and doubted the necessity of the solution she was contemplating. His reticence, Muriel thought, was because "he is afraid he will languish in purposeless leisure if he enters one [a retirement home]."[86] Perhaps of greater concern to him was the loss of contact with other scientists and whether he could continue his corn experiments.

In summer of 1981, Mangelsdorf, then 82 years old, sent Beadle a copy of the conclusions of two papers about to be published by Harvard's Bussey Institute. He now had another scheme for the origin of corn.[87] Soon, Mangelsdorf wrote again, commenting on Beadle's paper in *Science*, the last he published.[88] Beadle had succinctly reviewed all the old biological, genetic, archeological, and anthropological evidence about the relation between teosinte and corn. He was clearly exasperated with the whole body of Mangelsdorf's work and concentrated on new experiments designed to deal with longstanding claims by Mangelsdorf and his colleagues concerning the ancient pollen obtained from a drill core sample made in Mexico City. The pollen was estimated to be more than 25,000 years old, much older than any documented human remains in the region. The Mangelsdorf faction had based much of their observation on the fact that some of the grains were the size of modern corn pollen. Beadle disputed their conclusion on various grounds, including the likelihood that the grains are more recent contaminants of the drill core sample and his own observations that the size range of teosinte and corn pollen overlap. He argued, as he had done earlier, that pollen size is not reliably diagnostic of whether the source is teosinte or corn. Mangelsdorf's long letter argues that Beadle is wrong and concludes with this parting shot: "I am sending Walt [Galinat] and Hugh [Iltis] copies of this letter since you are all in the same boat and it seems to be leaking." The argument as well as the chief protagonists had clearly reached a point of diminishing strengths if not enthusiasms.

These scientists were civil, if tough, antagonists. But their real thoughts about each other were not always so polite. Paul C. Mangelsdorf, Jr. says that "Dad was not impressed with Beadle's credentials, particularly since he had had to work closely with George Wald and Jim Watson in the Biology Department at Harvard, and was of the general impression that Nobel laureates were inclined to throw their weight around in areas in which they had too little expertise."[89] Further, he reports "I fear that Dad really considered Beadle more of a nuisance than a serious scholar whose opinions needed to be carefully considered." It was not too different for Beadle, who confided in McClintock. She had written to him that "Paul Mangelsdorf lived in a dream-world and with total conviction of its reality."[90] He responded "Paul has now been pushed back to the fossil pollen evidence which I'm convinced is no good...I don't believe one word of the 'fossil' [*sic*] corn pollen and Walt [Galinat] doesn't either."[91]

Beginning in the mid-1970s, biologists learned how to isolate genes as pure segments of DNA, determine the sequence of bases in DNA, and transfer genes from one organism to another. This made it possible to identify the chromosomal regions that account for the differences between teosinte and primitive corn. Among the scientists who applied the new methods to corn genetics is John Doebley at the University of Wisconsin, a student of Iltis. This work speaks as well to broader, long-standing questions, including Morgan's concerning the relation between genes and development, the nature of the quantitative trait loci (or blended characteristics) described by Emerson and East in 1914, those that Delbrück posed to Beadle, and evolutionary mechanisms.

The series of concise, definitive papers published by John Doebley and his colleagues starting in the early 1990s make a remarkable statement about the way in which science progresses. Decades of argument, tens of thousands of plants, and months of measuring and cataloguing plant properties fall by the wayside after a decade of modern experiments on a laboratory bench and in a greenhouse. The key is combining classical genetic breeding experiments with insights from molecular genetic analysis. Yet Doebley recognizes that among the corn warriors, Beadle used a unique approach that allowed him to reach correct conclusions.[92] Unlike the others, Beadle asked a relatively simple question: How many genes or DNA regions distinguish teosinte from corn? Also, Beadle avoided many complications by growing the 50,000 second-generation teosinte–corn hybrids in Mexico where the short day allowed teosinte to flower normally and by choosing a primitive, teosinte-like corn rather than the highly selected modern corn typically grown in the United States. Nevertheless, Beadle's efforts were complicated by two factors. He had to depend on chromosome maps that were derived from modern corn varieties, not from the primitive corns that he used in his experiments. Also, the phenotypes he analyzed depended on more than one gene, that is, they were "blended" characteristics (or quantitative trait loci). Modern methods make both of these problems tractable.

Molecular tools allow variations in the DNA sequence to replace mutant genes as markers for defining chromosome maps. Therefore, a difference in the DNA sequence at a particular chromosomal location defines an allele, just as a mutation does. Sequences in large numbers of individual samples from plants or seeds are readily determined. The problem of blended characteristics is handled by sophisticated statistical methods that allow definition of the multiple chromosomal sites that contribute to these quantitative traits. Using these tools, Doebley confirmed Beadle's conclusion: Five chromosomal sites are responsible for most of the characteristic differences between corn and teosinte.[93]

According to Doebley's experiments, DNA variations (alleles) in only two genes account for several of the major differences in traits. The corn allele of each of these genes is dominant. One of the genes, called *teosinte branched 1*, controls whether the plant will have the single main stem and tassel of corn or the multiple branches and tassels of teosinte. The *teosinte branched 1* mutant was described as early as 1959 by the corn geneticist Charles Burnham, another of those who had spent time in Emerson's sphere around 1930. Burnham offered the seeds to Doebley and when Doebley was asked to speculate why none of the corn warriors had picked up on Burnham's discovery, he responded that they had concentrated on the morphology of the ears, not of the plant itself.[94] In the early years of the corn war, many of the warriors would not have realized that Burnham's corn mutant looked like a real teosinte plant because they had never seen one. Beadle was partly at fault for their misconceptions. He had published a misleading drawing of a teosinte plant in 1972 and was using the same illustration as late as 1980.[95] Iltis published a correct drawing in 1983,[96] but Beadle's sketch was still occasionally used in recent years.[97]

The *teosinte branched 1* mutation also affects other traits including the sex of the organs at the ends of side branches and the length of the side branches. When the DNA region containing the *teosinte branched 1* allele on the teosinte chromosome is replaced with the corresponding region from corn DNA, teosinte becomes corn-like and grows short side branches bearing ears instead of long branches with tassels. When the *teosinte branched 1* gene was purified and its DNA sequenced, it turned out to be related to genes known to affect the organization of side branches and flowers in other plants as well. The most striking discovery is that there is virtually no difference in the proteins encoded by the corn and teosinte alleles. Rather, the mutation affects the level of expression of the gene and thus the amount of the corresponding protein that is produced in the plant.[98] Delbrück had certainly posed the right question when he asked about the regulation of gene expression. If the normal action of this gene is to modulate the growth of certain organs, then the increased expression in corn might explain the decreased length of the lateral branches as well as other differences between corn and teosinte.[99] The multiple effects of the different alleles could be explained if the protein encoded by *teosinte branched 1* gene regulates the activity of multiple genes.

Major differences are also associated with a gene called *teosinte glume architecture*.[100] Alleles of this gene affect whether or not the kernels will have the hard seed covering typical of teosinte. The corn allele changes the teosinte seed covers so that the seed is partly exposed. It is easy to imagine how important it would have been to ancient Central Americans to find a mutant teosinte with soft seeds. The degree of hardness depends, at least partly, on changes in the distribution and concentration of silica grains and the rate of growth of the cells of the seed covering.[101] Besides these two major genetic differences, regions on other chromosomes contribute to the occurrence of many double rows of kernels on modern corn compared to the two rows of single kernels on teosinte and influence whether the kernels fall from the plant as they do in teosinte or stay bound to the cob as with corn.[102]

With the benefit of hindsight, it seems that Beadle's views derived from his genetic approach while Mangelsdorf emphasized morphology. Each of their views also reflected divergent understandings of evolution. Manglesdorf was troubled by the rapid change implied by the evolution of corn from teosinte in less than 7000 years. Iltis points out that Mangelsdorf adhered to the original Darwinian notion of gradualism, the slow accumulation of many, small changes leading eventually to a new species. Beadle, on the other hand, recognized that a few genetic changes and strong selective pressure applied by human intervention could rapidly effect profound changes. Doebley has now settled the argument on Beadle's side. Even small changes in the regulation of a gene's activity can produce extraordinary phenotypic alterations. Also, silent genetic differences in different strains of teosinte can, when combined, rapidly yield distinct new phenotypes.[103]

In 1980, while he was a graduate student, Doebley saw firsthand Beadle's struggles with his failing memory. Seeking to talk with Beadle, he telephoned their home and spoke with Muriel about an appointment. He was astonished when, without consulting her husband, she set the day, time, and place they would go for lunch and told him that Beadle would be waiting in front of their apartment house in a gray suit and red tie. Doebley better understood the reasons for her careful instructions after what turned out to be a deeply disappointing and difficult visit. The legendary great scientist was no longer consistently lucid.[104]

Epilogue

Over the last half-century, biology has advanced at a swift pace. Even five-year-old discoveries sometimes seem like ancient history as they are quickly incorporated into our understanding of the basic frameworks of living systems. With few exceptions, the legendary scientists who made great advances also fade from view. Although geneticists and molecular biologists operate daily on the premise that one gene encodes one protein—more or less—few know George Beadle's name.

Although imprecise in detail, Beadle's enunciation that each gene specifies a unique enzyme (protein) was, in its time, a "bombshell." Many biologists could not accept the proposition that each gene has only one function, preferring instead the long-held view that a gene contributes to an organism's traits in multiple ways. Beadle understood the revolutionary nature of his proposition and was well aware of the biological community's initial skepticism about his hypothesis. Taking on the skeptics in his usual forthright way, Beadle engaged the debate head on. The pace and scale of his lectures to scientific audiences throughout the country was breathtaking.[1] Doubts about the validity of the methodology and the unrepresentative nature of *Neurospora* as a model biological system fed the skepticism. Occasionally he wondered whether there were any more supporters than the fingers on his hand. His attempt to clarify the situation by saying that each gene specifies a single function rather than a single enzyme did not still the clamor. Finally, resistance lessened as more evidence was assembled to establish the fundamental correctness of his ideas.

As with most concepts in biology, time and the advance of knowledge have modified the original one gene–one enzyme proposal in many ways. First came the change to one gene–one protein. Then, one gene–one protein gave way to one gene–one polypeptide as it became clear that many proteins contain more than one polypeptide chain, each specified by a different gene. Then we learned that, particularly in eukaryotes, one gene often gives rise to more than one polypeptide as a consequence of the organization of genes into polypeptide-encoding exons and noncoding introns and the phenomenon of alternative splicing. Some genes do not encode a polypeptide at al, but rather encode functional RNA molecules. Even Beadle's own revision—that each gene specifies a single function—belies our current understanding that a gene product may have multiple functions. Today, the notion that each gene specifies a single polypeptide is appropriately replaced by the statement that a single gene specifies one or more macromolecules.[2]

Until Beadle and Tatum's experiments with *Neurospora*, mutations were used primarily to establish the existence of a gene and as genetic markers to study the mechanisms of inheritance. Geneticists relied on spontaneous, random events that altered an observable or measurable phenotypic property of an organism; the normal or wild-type form of the gene was more an inference than a measurable entity. Also, there was no need to know the function of the mutated gene or its corresponding normal allele to map it to a specific chromosomal locus, to determine its linkage to other genes, or to follow its inheritance from one generation to the next.

The one gene–one polypeptide paradigm proposed that the normal form of a gene directed the formation of the functional form of a single polypeptide chain. The mutant form of the gene could then be understood as directing the formation of an impaired or inactive form of that polypeptide, or perhaps no polypeptide at all. Geneticists were then obliged to view genes from three perspectives; as before, genes were units of mutation and units of recombination (crossing-over), and they were now also units of function. It was not until genes were known to be segments of DNA that these three descriptions of genes could be reconciled as fundamental features of the DNA structure. Thus, mutations result from one or more changes in the DNA sequence, genetic recombinations result from exchange of segments between two DNA chains, and gene function results from translation of DNA segments into a specific polypeptide.

In ascribing an instructional role to genes, Beadle and Tatum implicitly acknowledged a fourth perspective on a gene; namely, that genes are also units of information. Importantly, genes acquired a physical reality from this new perspective. This initiated the transformation of genetics into a molecular science. Consequently, discovering the physical and chemical nature of the gene became the highest priority for understanding the gene–protein connection. Beadle knew full well that the solution to understanding gene action lay in solving the problem of the physical and chemical nature of genes, but he chose to leave that task to others. That was surprising, for he had long since proved his versatility by venturing into fields with which he had only cursory experience. His ready adoption of disparate organisms to explore intriguing genetic questions was a measure of his self-confidence and aggressive eagerness to get on with new challenges.

Although it was widely acknowledged that chromosomes were the repository of the genes, disputes persisted over whether genes resided in the chromosomal proteins or DNA. During the 1940s and despite strongly suggestive experimental evidence that DNA itself possessed genetic properties, many argued that DNA could not account for the gene's biological specificity or its capacity for autocatalytic replication. Beadle was among many who were inclined to overlook the data favoring DNA as the genetic material, preferring instead to assume that DNA was merely the architectural support for the genetic proteins. That uncharacteristically blurred vision was one of the few times his remarkable scientific intuition failed him.

With the confirmation that genes were made of DNA, new players were energized to explore the use of X-ray diffraction methods coupled with model building to solve its structure. They were aided by new data showing that the base composition of DNA was indeed sufficiently complex to serve as the repository of biological information. The success that followed in 1953 was an intellectual tour de force if less than an exemplary scientific story. Most importantly, DNA's double-helical structure was more informative than even the most optimistic expectations. It revealed the essence of the gene's functions: how genetic information is stored, how it is replicated, and how it is used. Emerging from that picture was the realization that genes consist of discrete segments of paired bases along the length of the DNA molecule. The one gene–one polypeptide dictum together with the demonstration that a mutant form of human hemoglobin was associated with a change in the protein's amino acids pointed directly to the existence of a relationship between the gene's base sequence and the corresponding protein's amino acid sequence. A colinear "genetic code" provided a molecular explanation for the consequences of mutations: A change in the base sequence of a gene changes the amino acid sequence of the encoded protein and hence its biological property.

Breaking the code and revealing how the base sequence is translated into discrete polypeptides occupied center stage for the next decade and produced its own set of "heroes." Beadle's role in revealing the gene–protein paradigm was past history, and he was never a direct participant in the "rough and tumble" of explosive advances that followed from solving the DNA structure. His modest demeanor and "nice guy, straight arrow" manner would likely have put him at odds with the flamboyant and quirky behavior of several of the major players of the new biology. They likely viewed him as "a noncombatant" figure from the past.

The overall trajectory of Beadle's career was similar to that of many scientists who switch gears after years devoted to demanding, highly original research. Very few people who have made what Thomas Kuhn called a "paradigm-shifting" discovery within their lifetimes make another. Some continue to do what Kuhn called "ordinary" research.[3] By the time Morgan went to Cal Tech in 1928, for example, he was no longer in the mainstream of genetics, and while he wrote extensively about the need to find the genetic basis of embryological development, he left that for others to explore. Emerson became the dean of the Cornell graduate school while keeping up his own teaching and research. Sturtevant and Bridges were steady in their devotion to research, but the paradigm-shifting discoveries of their youth were not repeated. Delbrück, always a fount of new ideas for others, spent his last decades in a fruitless pursuit. Others, like Beadle, turn their energy, expertise, and status to the support of other scientists or become engaged in science policy. Beadle moved among the many possible paths for a scientist with a self-confident ease. And no matter what activity engaged him, Beadle the farmer was a consistent presence. Neither family, nor administrative or advising commitments kept him from his cornfield. He never returned to Nebraska or a farm, but they never left him. Everywhere he went he grew corn, if not for research then for the pleasure of growing, eating, and sharing.

Beadle was an early, untiring, and articulate spokesman for the integration of biochemistry and genetics. Although not a biochemist himself, he saw clearly that the future of genetics was in its interactions with biochemistry. It was, at the time, not a common position. Many of the phage group and their acolytes considered chemistry irrelevant for genetics. Also, many biochemists were skeptical of the interpretations of purely genetic experiments.[4] But Beadle had learned the relevance of enzymes and biochemistry to biology from his earliest university studies and, as a graduate student, knew of James Sumner's crystallization of the enzyme urease. Indeed, Beadle and Ephrussi's explanation of their experiments on *Drosophila* eye color was fundamentally a chemical one. And, of course, the collaboration between Beadle and Tatum embodied much of what is today widely called molecular biology. Beadle's visionary belief in the importance of interactions between genetics and biochemistry contributed significantly to his decision to join Pauling in a bold proposal responding to Warren Weaver's eagerness to support molecular biology. Curiously, Beadle used the term "biochemical genetics" rather than "molecular biology" in his many lectures and published reviews, even as late as the 1970s.[5] The term molecular biology does not appear in the book *The Language of Life* that he and Muriel published in 1966 for a general audience, although by then it was in widespread use.[6]

In rebuilding Cal Tech's biology division, Beadle ensured the flourishing of biochemistry and genetics by recruiting people without much regard for their subfields within biology. The wonderful science and brilliant, productive students that emerged from that environment speak for themselves as do the four Nobel Prizes to people he recruited—Max Delbrück, Edward Lewis, Renato Dulbecco, and Roger Sperry. The extraordinary advances that emerged from the marriage of chemistry and biology in the second half of the 20th century affirm Beadle's view. Perhaps the most dramatic evidence is how advances in the chemistry and ease of manipulation of DNA made sequencing of the human and other genomes possible, thereby providing enormously productive new approaches to genetics.

Besides assigning a physiological role to genes, the Beadle–Tatum experiments introduced a powerful experimental approach for the analysis of biological systems. Most often, geneticists had relied on spontaneous, random events that altered an observable or measurable phenotypic property of an organism to establish the existence of a gene. That strategy could identify mutants associated with a particular biological process. For example, mutations affecting *Drosophila* eye color, wing structure, or mating behavior could logically be assumed to affect some step or process leading to those characteristics. Similarly, mutations in corn that altered seed color or the ability to perform meiosis were understood to impair some aspect of the relevant processes. Because Beadle and Tatum produced mutants at high efficiency with radiation rather than relying on spontaneous mutations, they obtained many different mutants, including groups of mutants that affected the same property. The existence of the groups, they realized, indicated that the affected property was the consequence of many consecutive reactions, each catalyzed by, facilitated by, or structurally dependent on one or more proteins. For example, the biosynthesis of cellular constituents such as amino acids, purines, pyrimidines, vitamins, sugars, and fats results from sequential enzymatically catalyzed chemical reactions. Similarly, the assembly of amino acids into proteins, of purines and pyrimidines into DNA and RNA, and of single sugars into complex polymeric carbohydrates, and the regulation of all of these processes, are dependent on the action of a series of proteins and therefore, multiple, specific genes.

Almost immediately following publication of Beadle and Tatum's *Neurospora* reports, mutational analysis became the predominant approach to the analysis of metabolic pathways, not least in Beadle's lab itself. For example, mutational analysis became the basis for studying the mechanisms organisms use to derive energy. Mutations that alter the normal course of events are, today, the dominant strategy for identifying discrete steps in such complex processes as memory, learning, vision, and smell, and even how the shape of an organism is patterned during embryonic and fetal development, and more recently, the process of aging. Generally, many genetically different mutants affecting the same process are isolated and are then studied to determine the order in which the various mutants are required for the normal process. Thereafter, the effort focuses on isolating the affected gene, examining its entire base sequence, identifying the macromolecule that the gene specifies, and attempting to deduce the function it performs. It is likely, for example, that a mutational analysis of yeast could have identified the 12 discrete proteins and associated individual enzymatic reactions responsible for conversion of sugar to alcohol with less time and effort than the nearly 40 years of painstaking biochemical work it took by the classical route of enzyme fractionation.

The connection between genes and proteins has also paved the way for understanding the genetic basis of many human diseases. Mutations cause alterations in the abundance or structure—and thus the function—of proteins, and such changes in proteins are implicated as the causes of many inherited diseases as well as many cancers that develop spontaneously during a human life. Ironically, more than 40 years earlier, Beadle's discovery that each gene informs the production of a particular protein laid the ground for a molecular understanding of the disease that was destroying his mind. A month before his 80th birthday, observation of his behavior and performance on standardized neuropsychological tests led to a diagnosis of Alzheimer's disease, senile onset.[7]

More than 95% of Alzheimer's disease cases are associated with aging, the frequency increasing strikingly as people approach their 80s. Most often, there is no apparent cause. But some individuals with Alzheimer's have a familial history of the disease, and, even as Beadle's mind was deteriorating, evidence was accumulating that certain alleles of particular genes can predispose individuals to Alzheimer's. There is no way to know now whether the genes he inherited constituted a recipe for Alzheimer's, or if mutations or injury acquired during his lifetime played a part in its emergence. None of Beadle's ancestors is recorded as having become senile in old age, but some died well before the age of 70. The culprit in both the early- and late-onset forms of Alzheimer's is the accumulation of an aberrant protein which forms a highly insoluble material referred to as amyloid. The amyloid is deposited as plaque around blood vessels and around dying brain cells. Considerable progress has been made in elucidating the molecular nature of amyloid, the genetic control of its formation, and the properties of the accumulated amyloid in plaque. Presently, however, the mechanism of the toxicity or even if the amyloid itself is the cause of the toxicity is not known.

When the diagnosis of Alzheimer's was made, the Beadles had been living for 18 months at Mt. San Antonio Gardens, a well-regarded retirement community straddling the boundary of Claremont and Pomona in southern California. Aware of her own frailty and Beadle's diminished capacities, Muriel decided against staying in Chicago. Squarely facing their prospects she wrote, "If I have another heart attack, or another stroke, I can't count on dying; I might survive in so debilitated a condition that I would need longtime care beyond George's or the Barnett's ability to provide." Although she admitted that life in a retirement community "would be more circumscribed and lacking in diversity" than what they then enjoyed in Chicago, she felt that security was essential. Beadle shared neither her concerns nor her sense of urgency and doubted the necessity of the solution she was contemplating. His reticence, Muriel thought, was because he probably equated "modern retirement communities with the old people's homes he may have known in his youth, and he is afraid he will languish in purposeless leisure if he enters one."[8] Perhaps of greater concern to him was the loss of contact with other scientists and whether he could continue his corn experiments.

The milder weather and familiarity of southern California appealed to Muriel. She selected Mt. San Antonio Gardens after a visit with old Cal Tech friends who had already settled there. The community occupied a four-block, park-like area, with a mix of cottages and apartment complexes, and there were medical and recreational facilities on the grounds and readily available transportation to local communities and nearby Los Angeles. The financial commitment was considerable but manageable. A more serious issue was an anticipated two-year delay before a cottage would be available.

Waiting in Chicago, Beadle was fading right before Muriel's eyes. "Remembering him at Stanford or even much more recently at Cal Tech," Muriel wrote, "you would probably be saddened to see him today. He has become an old man, snowy of hair, creaky of joints, lacking in vigor and very forgetful." But perhaps "because it's the thing that matters most to him, he still seems able to do productive research."[9] Then, in September of 1981, just when a place unexpectedly became available in Pomona, doctors discovered that Beadle had a malignant but not yet metastasized prostate tumor requiring surgery and several months of radiotherapy. The move to California had to be postponed. The physical illness seemed to exacerbate his mental problems and Muriel told the family that his dementia "concerns me much more than the cancer itself." Moreover, she was certain that "he is quite well aware of these lapses and disorientations and inabilities to make sense, and he must be scared silly."[10] In spite of this, Beadle could write to his sister thanking her for sending her story about their father,[11] to David Perkins,[12] and to McClintock to congratulate her on receiving the prestigious Wolf Prize.[13]

Finally, in February of 1982, the Beadles were able to move to California.

Initially disoriented by the new surroundings, Beadle eventually adjusted, especially as he had a sizeable plot for corn and teosinte right on the grounds and could help others with the communal flower garden.[14] Muriel, however, sorely missed her Chicago friends and uninterrupted time for writing. Confessing that she missed the racial and ethnic diversity of her Chicago circle, she found the "Waspish" surroundings appalling. Noting that "Nobody swears. Nobody tells dirty jokes," she wondered what they did in private.[15] After 30 years of marriage, she missed his intelligent companionship and worried deeply over how they would fare during their remaining years. "The part that's hard on me is the feeling that I am completely alone."[16] Undaunted, Muriel continued to invite friends to join them on special occasions. Whether he understood the purpose of the occasions and appreciated the many cards and letters wishing him well was problematic. "Sometimes, though, he does comment intelligently on people or events, past or current. So, communication is still possible."[17] As time went by, such periods of apparent lucidity all but disappeared. Earlier "when he couldn't order his thoughts into a comprehensible sentence, he would become so frustrated and angry he would weep and pound on the table with a fist. Now he just smiles, shrugs, and abandons the effort."[18]

By the middle of 1985, Beadle had deteriorated to the point where Muriel could no long take care of him and he was moved to "The Lodge" at Mt. San Antonio Gardens. There, along with others similarly afflicted, he received close supervision virtually around the clock. He still went to his garden and Muriel visited him once a week, occasionally sharing a meal in the lodge dining room. Somewhat oblivious to his new situation, he remained "a sweet-tempered, cooperative man, not happy about this move but making the best of it." She was no longer certain that he recognized her and regarded their separation "like a death in the family," one "that has been and will be drawn out over such a long period of time that it makes a mockery of real death." Watching "George's mind die in small increments, as it's doing, is almost more than I can bear." She likened Alzheimer's disease to "a never-ending funeral."[19] Fearing senility with advancing age, John Quincy Adams referred to it as "dying at the top."[20]

Suspecting that his father might not live much longer, David came to Pomona from England. He recalled that his father had not "recognized me specifically, but he knew somebody was there, he knew somebody cared about him, and as I left I put my arms around him and gave him a great big hug which he did not resist...always before he would have stiffened up...I went back with him to his room and I just let him talk about whatever was both-ering him which was where he'd been two minutes ago" and how "he just hated being this burden on people."[21]

By the fall of 1988, Beadle was moved to Mt. San Antonio Garden's medical unit where, after nearly nine months of round-the-clock care, he died on June 9, 1989. Muriel's health deteriorated and although previously fiercely independent, she now welcomed her dependency on others and became a recluse.[22] She lingered on for almost five years and died on February 13, 1994, of heart failure stemming from the heart attack sixteen years earlier.[23]

As requested in their wills, the Beadles' ashes rest in urns in the Rockefeller Chapel at the University of Chicago along with those of several other former presidents. Both wills provided modest sums for David, Redmond, Ruth, and Muriel's brother, and Beadle left smaller amounts to David's legitimate children. He provided for Muriel's support and also left a substantial sum for Marion, who died two years later. When Muriel died, the remaining funds went to David, Redmond, and the University of Chicago.[24]

Within months of Beadle's death, friends and former colleagues convened a memorial service at the Athenaeum on the Cal Tech campus.[25] Muriel was too ill to attend but David, Redmond, and Ruth were present and the two sons delivered eulogies. They each talked of Beadle as their "Pop" and agreed that he was not a person who was easy with the language of emotion. But, "If he wasn't too good with words," David said, "he sure knew the music." Several of Beadle's closest colleagues recalled that he had brought the division to world leadership in biology and in doing so had become, in James Bonner's words, "the central figure in the dawn of this new golden age of biology." Many spoke of how he had touched their lives and careers personally. Shortly thereafter, friends at Cal Tech and Chicago established a George W. Beadle professorship in the Cal Tech biology division.

Years later, James D. Watson echoed the thoughts about Beadle's personal attentions to colleagues in his keynote address at the dedication of the University of Nebraska's George Beadle Center for Genetics and Biomaterials Research in Lincoln.[26] He recalled how Beadle unhesitatingly invited him home for dinner on a summer visit to Cal Tech. This was well before Watson's discovery of the double-helix structure of DNA had made him a celebrity. The young man was well aware of Beadle's towering scientific reputation and impressed by his host's unpretentious, unself-conscious warmth and his genuine excitement to be with someone who shared his passion for science. Many who met and knew George Beadle remember him that way.

Notes*

CHAPTER 1

1. Ruth Beadle, "An uncommon farmer." Unpublished memoir, October, 1981. RB.
2. W. Cather. *My Antonia*. 1995. Paperback edition. Houghton Mifflin Company, New York, 1918.
3. Contemporary newspaper clippings. Box 36.2, George Wells Beadle Papers. Archives, CIT.
4. R. Beadle, "An uncommon farmer."
5. Record of Deeds, 1896. Wahoo, Nebraska.
6. Walter J. Beadle. "SAMUEL BEADLE FAMILY, History and Genealogy of Descendants of Samuel Beadle, Planter, Who lived in Charlestown, Massachusetts in 1656 and died in Salem, Massachusetts in 1664." Privately printed, 1970. Deposited at the New England Historic Genealogical Society, 101 Newbury Street, Boston, MA 02116.
7. R. Beadle, "An uncommon farmer."
8. Cora Babbitt to R. Beadle, June 3, 1961.
9. J.A. Warren. 1908. Small farms in the corn belt. *USDA Farmers Bull.* 325 (May 9): 5–17. US Government Printing Office. George W. Beadle Papers, CHG.
10. Yields on the Beadle farm were almost double those of its neighbors although most of the other farms in the vicinity were at least four times as large because the Homestead Act of 1862 provided settlers with 160 free acres (Plat of Stocking Township, 14 North, Range 7 East, Saunders County, Nebraska, 1907). Storing as much as 1400 bushels of potatoes was a challenge CE met by building a model potato cellar to replace the old cave he had been using. Half a silo was laid on the ground over a deep trench and covered with dirt. The front end had a conveyer belt to help move the potatoes. Almost a century later in 1997, the sturdy structure was still reasonably intact although overgrown by grass and weeds.
11. The Populist Party was, by then, little more than a decade old and on its way to oblivion. It had been organized by a Farmers' Alliance and flourished after the extensive crop failures brought on by the droughts of 1893 and 1894 as a dominant element of the State Democratic Party led by William Jennings Bryan.
12. Record of Deeds, 1903. Wahoo, Nebraska.

*The following abbreviations of archival sources are used in these Notes. (APS) American Philosphical Society; (CIT) California Institute of Technology; (CHG) University of Chicago; (COL) Columbia University; (COR) Cornell University; (GWB) George W. Beadle; (NRC) National Research Council; (RFA) The Rockefeller Foundation; (RB) Ruth Beadle; (WAH) Saunders County Historical Society.

13. Laura Mote and Ken Carlson, interview with authors, Wahoo, Nebraska, July 10, 1997; Photograph at WAH.
14. *Wahoo's Century Round-up: 1870–1970* at Nebraska State Historical Society Museum and Library, Lincoln, Nebraska.
15. W. Cather. *My Antonia; O, pioneers*. 1994. Penguin paperback edition. Houghton Mifflin, New York, 1913.
16. *Wahoo's Century Round-Up*.
17. Ibid.
18. Hattie Albro to GWB. 1908. Box 36.3, George Wells Beadle Papers, CIT.
19. GWB. 1975. Biochemical genetics: Reflections. In *Three Lectures. January 15–17, 1975.* The Edna H. Drane Visiting Lectureship, University of Southern California, School of Medicine, p. 11, CHG.
20. R. Beadle, "An uncommon farmer."
21. Ibid.
22. Laura Motes, Ken Carlson, and Dorothy Miller, interview, July 10, 1997, Wahoo, Nebraska.
23. R. Beadle, interview, August 14, 1997.
24. Redmond Barnett, interview, July 22, 1997.
25. Dorothy Miller (daughter-in-law of Lulu Stapleford Miller), interview, Wahoo, Nebraska, July 10, 1997; R. Beadle, interview.
26. R. Beadle, "An uncommon farmer."
27. David Beadle, interview, August 13, 1997.
28. B. Fussell. *The story of corn*. A.A. Knopf, New York, 1992.
29. E.W. Irish. *Sioux City's corn palaces: 1890, 1889, 1888, 1887.* Pinckney Book and Stationery, Chicago, 1890.
30. E.J. Kahn. 1984. The staffs of life, I. The golden thread. *New Yorker* June 18: 46–88.
31. Fussell. *Story of corn*.
32. D. Beadle, interview, June 27, 1997.
33. *The Wahoo Wasp*. April 17, 1913:2, WAH.
34. Several different Protestant churches were established in the first decade of Wahoo's existence. A Catholic priest and church arrived in 1877 with Czech settlers. In 1883, the growing Swedish immigrant community founded a Lutheran congregation and services in Swedish continued until after World War II (*Wahoo's Century Roundup*). When George was born, nearly half of Nebraska's inhabitants had been born abroad. While we know from Willa Cather's novels that these communities, Anglo Saxon, Czech, and Swedish, tended to remain cohesive and separate, the books also show the growing interdependence among the three, especially in the towns. Public schools like the single high school in Wahoo brought the young people together. A photograph of an August 1900 swimming party shows a large group of wet young men and women in typical bathing costumes of the time, their diverse origins revealed by their names.
35. R. Smith. 1998. Nebraska standing tall again. *Natl. Geograph. Mag.* November: 118–139.
36. *Wahoo's Century Roundup*.
37. R. Beadle, "An uncommon farmer."
38. Laura Motes, interview, July 10, 1997.
39. GWB, "Biochemical genetics: Reflections."
40. Ibid; Quoted in Bryce Nelson. Corn Patch Laureate. *Bulletin of the Atomic Scientists*, October 1977, pp. 48–50.

41. George S. Round. Unpublished interview with GWB. March 7, 1975. Institute of Agriculture and Natural Resources, The Beadle Center, University of Nebraska, Lincoln.
42. *New York Times*, October 8, 1997.
43. GWB, "Biochemical genetics: Reflections."
44. Round, interview.
45. GWB, "Biochemical genetics: Reflections."

CHAPTER 2

1. R.N. Copple. *Tower on the plains: Lincoln's Centennial history, 1859–1958.* Lincoln Centennial Commission, 1959.
2. R.E. Knoll. *Prairie University: A history of the University of Nebraska.* University of Nebraska Press, 1995, p. 70.
3. Knoll, *Prairie University*, p. 75.
4. Knoll, *Prairie University*, p. 72.
5. E.F. Frolik and R.J. Graham. *The University of Nebraska-Lincoln College of Agriculture: The first century.* Board of Regents of the University of Nebraska, 1987.
6. Knoll, *Prairie University*, p. 87.
7. R.C. Tobey. *Saving the prairies: The life cycle of the founding school of American plant ecology*. University of California Press, Berkeley, 1981; R. Overfield. *Science with practice.* Iowa State University Press, Ames, Iowa, 1993.
8. M.M. Rhoades. 1984. The early years of maize genetics. *Ann. Rev. Genet.* 18: 1–29. Reprinted in *The dynamic genome* (ed. N. Fedoroff and D. Botstein). Cold Spring Harbor Laboratory Press, Cold Spring Harbor, New York, 1992, pp. 45–69; R. Morris. "Rollin Adams Emerson (1873–1947): Horticulturalist, pioneer in plant genetics, administrator, inspiring student advisor." A paper presented to the Nebraska Academy of Sciences, April 25, 1969, personal communication, www.cornell.edu/ Brutnell_lab2/ Projects/ESGP.
9. D.B. Paul and B.A. Kimmelman. 1988. Mendel in America: Theory and practice, 1900–1919. In *The American development of biology* (ed. R. Rainger, K.R. Benson, and J. Maienschein). University of Pennsylvania Press, pp. 281–310 and at http://www.netspace.org/MendelWeb/Mwpaul.html
10. A.H. Sturtevant. *A history of genetics.* Harper & Row, New York, 1965. (Republished by Cold Spring Harbor Laboratory Press, New York, 2001.)
11. V. Orel. *Gregor Mendel: The first geneticist.* Oxford University Press, Oxford, 1996.
12. Paul and Kimmelman, "Mendel in America."
13. Wayne Keim. Personal communication.
14. R.A. Emerson. 1902. Preliminary account of variation in bean hybrids. *Nebraska Agric. Exp. Station Rep.* 15: 30–43.
15. Overfield, *Science with practice.*
16. G. Mendel. 1865. Versuche uber Pflanzen-Hybriden. *Verhandlungen des naturforschenden Vereines in Brunn* 4: 3–47. First English translation. 1901. *J. R. Horticult. Soc.* 26: 1–32; C. Stern and E.R. Sherwood, eds. *The origin of genetics: A Mendel sourcebook.* W.H. Freeman, San Francisco, 1966.
17. P. Berg and M. Singer. *Dealing with genes: The language of heredity.* University Science Books, Mill Valley, California, 1992.

18. W. Johanssen. *Elemente der Exacten Erbkichkeitslehre.* Fischer, Jena, 1909.
19. R.A. Emerson. 1904. Heredity in bean hybrds (*Phaseolus vulgaris*). *Nebraska Agric. Exp. Station Rep.* 17: 33–68.
20. G. Mendel to Carl Nageli, April 18, 1867 and July 3, 1870. (Reprinted in Stern and Sherwood "*Origin of genetics*.")
21. M.M. Rhoades, "Early years of maize genetics."
22. It turned out that a second gene, one that controls the formation of pollen tubes, is linked to *sugary* on chromosome 4. The second gene was also altered in the *sugary* mutants and they therefore produced fewer kernels of the sugary type than expected, lowering the ratio of sugary to starchy kernels recovered. This discovery was made in 1921; in 1900 the linkage of different genes on a single chromosome was unknown.
23. GWB. *Genetics and modern biology.* Jayne Lectures for 1962, Lecture 1. American Philosophical Society, Philadelphia, 1963; GWB. 1971. Genes, intelligence, and education. *Stadler Genetics Symp.* 1 and 2: 111–122, University of Missouri, Agricultural Experiment Station, Columbia, Missouri; T.H. Morgan. 1909. What are "factors" in Mendelian explanations? *Am. Breeders Assoc.* 5: 365–368.
24. The American Breeders' Association became the Genetics Society of America. W.E. Castle reports that Morgan made a similar criticism of Mendelism at the 1909 meeting. Beadle may have given the wrong year or perhaps Morgan repeated himself; W.E. Castle. 1951. The beginning of Mendelism in America. In *Genetics in the 20th century* (ed. L.C. Dunn). Macmillan, New York, p. 63.
25. M.M. Rhoades, "Early years of maize genetics"; R.A. Emerson and E.M. East. 1913. The inheritance of quantitative characters in maize. *Bull. Agricult. Exp. Station Nebraska* 2: 1–120.
26. Francis Haskins, personal communication.
27. GWB, official transcript, University of Nebraska College of Agriculture.
28. Farm House Newsletters and House Records, published by Farm House, University of Nebraska and obtained through the courtesy of Professor Francis Haskins, Emeritus Professor of Agronomy, College of Agriculture, U. Nebraska.
29. Francis Haskins, personal communciation.
30. Wayne Keim, personal communication.
31. GWB, official transcript.
32. Francis Haskins to Don Weeks, September 5, 1995.
33. Knoll, *Prairie university*, p. 87.
34. GWB. 1959. Genetics at Nebraska. *Nebraska Agric. Exp. Station Quart.*, Winter: 8 and 16; GWB, interview by Mark Herbst. Pasadena, California, April, 1981. Oral History Project, CIT Archives.
35. GWB. 1966. Biochemical genetics: Some recollections. In *Phage and the origins of molecular biology* (ed. J. Cairns, G.S. Stent, and J.D. Watson). Cold Spring Harbor Laboratory of Quantitative Biology, Cold Spring Harbor, New York, pp. 23–32.
36. F. Keim to H.H. Love, January 14, 1925; Love to Keim, January 23, 1925. Box 4, Department of Plant Breeding Records, #21/28/889, COR.
37. Keim to Love, March 13, 1925, Box 4, Department of Plant Breeding Records, #21/28/889, COR.
38. Unnamed instructor to Keim, April 7, 1925, Box 4, Department of Plant Breeding Records, #21/28/889, COR.
39. GWB. 1975. Biochemical genetics: Reflections. In *Three lectures. January 15–17, 1975.*

The Edna H. Drane Visiting Lectureship, University of Southern California, School of Medicine, p. 12, CHG.

40. GWB, official transcript.
41. GWB, "Biochemical genetics: Reflections," p. 12.
42. F.D. Keim. "Inheritance studies of a cross between T. compactum and T. spelta." Ph.D. thesis, Cornell University, 1927.
43. GWB, "Biochemical genetics: Some recollections."
44. GWB, official transcript. GWB, "Genetics at Nebraska."
45. Keim to Love, February 3, 1925. Box 4, Department of Plant Breeding Records, #21/28/889, COR.

CHAPTER 3

1. W.E. Castle. *Genetics and eugenics*, 3rd edition. Harvard University Press, 1924. Castle was Professor of Zoology at Harvard. A copy of this book bearing Beadle's name in ink, his address at 2545 O Street, and the initials F.H. for Farm House, is in the library of the Beadle Center of the University of Nebraska.
2. GWB. 1966. Biochemical genetics: Some recollections. In *Phage and the origins of molecular biology* (ed. J. Cairns, G.S. Stent, and J.D. Watson). Cold Spring Harbor Laboratory of Quantitative Biology, Cold Spring Harbor, New York, pp. 23–32.
3. H.E. Walter. *Genetics: An introduction to the study of heredity*. Macmillan, New York, 1914 and subsequent editions. Walter was professor at Brown University. The preface to his book says that it is based on a course he gave at Brown in the 1911–1912 academic year and again as a summer course at the Cold Spring Harbor Laboratory in the summer of 1912.
4. W.E. Castle. 1951. The beginning of Mendelism in America. In *Genetics in the 20th century* (ed. L.C. Dunn), Macmillan, New York, p. 63.
5. W.S. Sutton. 1902. On the morphology of the chromosome group in *Brachystola magna. Biol. Bull. Marine Biol.* Labor IV: 1.
6. T.H. Montgomery. 1901. A study of the chromosomes of the germ cells of metazoa. *Transact. Amer. Philos. Soc.* XX (New series): 154.
7. W.S. Sutton. 1902. The chromosomes in heredity. *Biol. Bull. Marine Biol.* Labor IV: 231–251.
8. G.E. Allen. *Thomas Hunt Morgan: The man and his science*. Princeton University Press, 1978.
9. R.E. Kohler. *Lords of the fly*. University of Chicago Press, 1994, p. 23.
10. Castle, "Beginnings of Mendelism."
11. Wild type refers to the standard or normal phenotype; Drosophila's wild-type (normal) eye color is deep reddish-brown.
12. S.F. Gilbert. *Developmental biology*, 5th edition. Sinauer Associates, Sunderland, Massachusetts, 1997, p. 37.
13. T.H. Morgan. 1910. Sex-limited inheritance in *Drosophila. Science* 32: 120–122.
14. T.H. Morgan. 1911. Random segregation versus coupling in Mendelian inheritance. *Science* 34: 384.
15. F.A. Jannsens. 1909. La theorie de la chiasmatype. *La Cellule* 25: 389–411.
16. T.H. Morgan, "Random segregation versus coupling."

17. I. Shine and S. Wrobel. *Thomas Hunt Morgan: Pioneer in genetics.* University of Kentucky Press, Lexington, Kentucky, 1976, p. 92.
18. A.H. Sturtevant. 1913. The linear arrangement of six sex-linked factors in *Drosophila* as shown by their mode of association. *J. Exp. Zool.* 14: 43–59; A.H. Sturtevant. 1965. "The fly room." *Am. Sci.* 53: 303–307; A.H. Sturtevant. *A history of genetics.* Harper & Row, New York, 1965. (Reprinted by Cold Spring Harbor Laboratory Press, New York, 2001, p. 47.)
19. Today, map distances along a chromosome are expressed in Morgans where one Morgan is equivalent to a 1% frequency of crossing-over.
20. A.H. Sturtevant and GWB. *An introduction to genetics.* W.B. Saunders Co., New York, 1939. (Reprinted by Dover Publications, Inc., New York, 1962.)
21. J.S. Fruton. *Proteins, enzymes, genes.* Yale University Press, 1999.
22. N.W. Gillham. *A life of Sir Francis Galton: From African exploration to the birth of eugenics.* Oxford University Press, England, 2001.
23. Thomas Hunt Morgan initially kept his objections to eugenics private. Eventually, he resigned from the American Breeders Association because of its tolerance for eugenics as a science and clearly stated his objections in the 1925 edition of his book, *Evolution and genetics.*
24. D.J. Kevles. *In the name of eugenics.* University of California Press, Berkeley and Los Angeles, 1985.
25. George S. Round, unpublished interview with GWB. March 7, 1975. Institute of Agriculture and Natural resources, The Beadle Center, University of Nebraska, Lincoln.
26. GWB. 1959. Genetics at Nebraska. *Nebraska Agricult. Exp. Station Quart.,* Winter: 8 and 16.
27. B. Nelson. 1977. Corn patch Nobel Laureate. *Bull. Atomic Sci.* October: 48–50.
28. GWB, interview by Mark Herbst, Pasadena, California, April, 1981. Oral History Project, CIT Archives.
29. F.D. Keim and GWB. *Prairie hay in Nebraska: A progress report of prairied hay investigations in the upper Elkhorn Valley region.* Lincoln, Nebraska, September 13, 1926. Box 32, Beadle Collection, CHG.
30. J.A. Warren. 1908. Small farms in the corn belt. *USDA Farmers Bull.* 325 (May 9): 5–17. U.S. Government Printing Office. George W. Beadle Papers, CHG.
31. GWB, official transcript, University of Nebraska.
32. GWB. "Identification of the more important graminaceous constituents of the prairie hays of Nebraska by means of their vegetative characteristics." Master's thesis. University of Nebraska. 1927. Beadle Collection, CHG.
33. F.D. Keim, A.L. Frolik, and GWB. 1932. Studies of prairie hay in north central Nebraska. *Univ. Nebraska Agricult. Exp. Station Res. Bull.* 60: 5–54; GWB., F.D. Keim, and A.L. Frolik. 1932. The identification of the more important prairie hay grasses of Nebraska by their vegetative characteristics. *Univ. Nebraska Agricult. Exp. Station Res. Bull.* 65: 5–40.
34. Ruth Beadle, interview, August 14, 1997.
35. Round interview.
36. R. Beadle, interview.
37. Nelson, "Corn patch Nobel Laureate."
38. Keim to Emerson, June 29, 1925, Box 4 and January 26, 1926, Box 13, Department of Plant Breeding Records, #21/28/889, COR.

CHAPTER 4

1. GWB. 1959. Genetics at Nebraska. *Nebraska Agric. Exp. Station Quart.*, Winter: 8 and 16.
2. G.S. Round, unpublished interview with GWB, March 7, 1975. Institute of Agriculture and Natural Resources, The Beadle Center, U. Nebraska, Lincoln.
3. GWB, interview by Mark Herbst, Pasadena, California, April, 1981. Oral History Project, CIT Archives.
4. Jo Srb, personal communication.
5. GWB, interview, p. 30.
6. Ernest W. Lindstrom papers. Special Collections Department, College of Agriculture, Department of Genetics, Iowa State University.
7. R.A. Emerson to Secretary, John Simon Guggenheim Memorial Foundation, November 17, 1927. Box 8, Department of Plant Breeding Records, #21/28/889, COR.
8. R.A. Emerson to National Research Council Fellowship Committee April 12, 1929. Box 15, Department of Plant Breeding Records, #21/28/889, COR.
9. B. Glass. 1971. Milislav Demerec: January 11, 1895–April 12, 1966. *Biographical Memoirs* 42: 1–27. National Academy of Sciences, Washington, D.C. Published by Columbia University Press.
10. Synapsis Club Records, Box 49, Department of Plant Breeding Records, #21/28/889, COR.
11. GWB. 1966. Biochemical genetics: Some recollections. In *Phage and the origins of molecular biology* (ed. J. Cairns, G.S. Stent, and J.D. Watson). Cold Spring Harbor Laboratory of Quantitative Biology, Cold Spring Harbor, New York, pp. 23–32.
12. Synapsis Club Records.
13. GWB, "Biochemical Genetics: Some recollections."
14. R.A. Emerson to F.R. Lillie, January 14, 1930. Cornell Graduate School Records, Beadle File, COR.
15. Harriet Zuckerman, interview with GWB, December 12, 1963, COL.
16. GWB. 1974. "Biochemical genetics: Recollections." *Ann. Rev. Biochem.* 43: 1–13.
17. David Beadle, interview, August 13, 1997.
18. GWB, Biochemical genetics: Some recollections, 1966.
19. GWB. 1956. Rollins Adams Emerson, 1873–1947. *Genetics* 35: 1–3.
20. Emerson's judgment was not perfect. He refused to take Lewis J. Stadler as a doctoral student when he applied in 1918. The young man returned to the University of Missouri where he received his Ph.D. in 1922 for work in corn breeding. Later, when Stadler visited Ithaca in the summer of 1926, Emerson changed his mind about his talents. Stadler became an outstanding corn geneticist and discovered the mutagenic effect of x-rays at the same time as H.J. Muller.
21. E. Fox Keller. *A feeling for the organism: The life and work of Barbara McClintock*. W.H. Freeman, San Francisco, 1983; N.C. Comfort. *The tangled field: Barbara McClintock's search for the patterns of genetic control.* Harvard University Press, Cambridge, Massachusetts, 2001; M.M. Rhoades. 1992. The early years of maize genetics. In *The dynamic genome: Barbara McClintock's ideas in the century of genetics* (ed. N. Fedoroff and D. Botstein). Cold Spring Harbor Laboratory Press, Cold Spring Harbor, New York, pp. 45–69.
22. N. Fedoroff. 1995. Barbara McClintock: June 16, 1902–September 2, 1992. *Biographical Memoirs* 68: 210–235. National Academy Press, Washington, D.C.

23. M.M. Rhoades, "The early years of maize genetics."
24. L.F. Randolph and B. McClintock. 1926. Polyploidy in *Zea mays* L. *Am. Nat.* 60: 99–102.
25. B. McClintock. 1929. Chromosome Morphology in *Zea mays. Science* 69: 629.
26. B. McClintock. 1929. A cytological and genetical study of triploid maize. *Genetics* 14: 180–222.
27. M.M. Rhoades, "The early years of maize genetics."
28. H.B. Creighton. 1992. Recollections of Barbara McClintock's Cornell years. In *The dynamic genome: Barbara McClintock's ideas in the century of genetics* (ed. N. Fedoroff and D. Botstein). Cold Spring Harbor Laboratory Press, Cold Spring Harbor, New York, pp. 13–18.
29. Zuckerman, interview, 1963; GWB, "Biochemical genetics: Some recollections," 1966.
30. C. Burnham. 1992. Barbara McClintock: Reminiscences. In *The dynamic genome: Barbara McClintock's ideas in the century of genetics* (ed. N. Fedoroff and D. Botstein). Cold Spring Harbor Laboratory Press, Cold Spring Harbor, New York, pp. 19–24.
31. E.G. Anderson to R.A. Emerson, January 15, 1927. Department of Plant Biology Records, Box 8, Anderson file. 21/28/889, COR.
32. M.M. Rhoades, "The early years of maize genetics."
33. GWB and B. McClintock. 1928. A genetic disturbance of meiosis in *Zea mays. Science* 68: 433.

34. The name *asynaptic* illustrates an aspect of genetic terminology that is often confusing to people who are not in the field. Like many other genes, *asynaptic* was named after the apparent abnormality in a visible trait (that is, the phenotype) caused by a mutant allele. The alleles in normal corn plants permit the developing germ cells to undergo proper synapsis, yet the normal allele is also called *asynaptic*. There is an advantage to this seemingly peculiar convention; the name reveals something about the function of the normal gene.

35. GWB, "Biochemical genetics: Some recollections," 1966.
36. N.C. Comfort, *The tangled field,* p. 52.
37. Harriet B. Creighton, personal communication, April 13, 1998.
38. GWB. "Genetic and cytological studies of Mendelian asynapsis in *Zea mays.*" Ph.D. thesis, Cornell University, 1930. SB124, 1930, B365, COR.
39. GWB. 1930. "Genetical and cytological studies of Mendelian asynapsis in *Zea mays. Cornell Univ. Agric. Exp. Station Memoir* 129: 1–22.
40. GWB to Adrian Srb, November 9, 1958. Personal communication from Jo Srb.
41. David Beadle, personal communication, June 14, 1999.
42. R. Beadle, interview, August 1, 1997.
43. Wedding announcement from Mrs. Fred Hill. Personal communication from R. Beadle; David Beadle, interview, June 14, 1999.
44. David Beadle, interview, June 27, 1997; Ed Lewis, interview, October 16, 1996.
45. Keller, *A feeling for the organism,* p. 32.
46. Jo Srb, personal communication.
47. GWB. 1929. A gene for supernumerary mitoses during spore development in *Zea mays. Science* 70: 406–407; GWB. 1930. A gene for supernumerary cell divisions following meiosis. *Cornell Univ. Agric. Exp. Station Memoir* 135: 1–11.
48. L.C. Strong. 1929. Transplantation studies on tumors arising spontaneously in heterozygous individuals. *J. Cancer Res.* 13: 103–115.

49. GWB, op. cit., Collected papers 1928–1950. SB 43 B36, Mann Library, Cornell University.
50. GWB, Ph.D. thesis.
51. M.M. Rhoades, "The early years of maize genetics."
52. GWB. 1929. Yellow Stripe—A factor for chlorophyll deficiency in maize located in the Pr pr chromosome. *Am. Nat.* 63: 189–192.
53. C. Curie, Z. Panaviene, C. Loulergue, S.L. Dellaporta, J-F. Briat, and E.L. Walker. 2001. Maize *yellow stripe1* encodes a membrane protein directly involved in Fe(III) uptake. *Nature* 409: 346–349.
54. R. Beadle, August 14, 1997.
55. GWB. Application to National Research Council for a National Research Fellowship in the Biological Sciences, December 26, 1929. National Research Council Fellowships, Fellows Roster Files, NRC.
56. R.A. Emerson to E.G. Anderson, February 7, 1930. Box 15, Department of Plant Breeding Records, #21/28/889, COR.
57. Ibid.
58. Frank R. Lillie to R.A. Emerson, January 2, 1930. Box 15, Department of Plant Breeding Records, #21/28/889, COR.
59. Lester W. Sharp to Frank R. Lillie, January 6, 1930. National Research Council Fellowships, Fellows Roster Files, NAS; Lewis Knudson to Frank R. Lillie, January 8, 1930. National Research Council Fellowships, Fellows Roster Files, NAS.
60. R.A. Emerson to Frank R. Lillie, January 14, 1930. Box 15, Department of Plant Breeding Records, #21/28/889 COR and National Research Council Fellowships, Fellows Roster Files, NAS.
61. R.A. Emerson to Edith E. Conger, January 6, 1930. National Research Council Fellowships, Fellows Roster Files, NAS.
62. C.E. Allen to R.A. Emerson, February 12, 1930. Box 15, Department of Plant Breeding Records, #21/28/889 COR and National Research Council Fellowships, Fellows Roster Files, NAS.
63. GWB, "Genetics at Nebraska."
64. R.A. Emerson to C.E. Allen, February 18, 1930. Box 15, Department of Plant Breeding Records, #21/28/889, COR.
65. C.E. Allen to R.A. Emerson, February 21, 1930. Box 15, Department of Plant Breeding Records, #21/28/889, COR; GWB Graduate Student File, COR.
66. C.E. Allen to GWB, February 13, 1930. National Research Council Fellowships, Fellows Roster Files, NAS.
67. GWB to G.B. Rigg, June 18, 1930. Box 8, Department of Plant Breeding Records, #21/28/889, COR.
68. Zuckerman, 1963; GWB, "Biochemical genetics: Some recollections," 1966, p. 25.

CHAPTER 5

1. J.R. Goodstein. *Millikan's School: A history of the California Institute of Technology*. W.W. Norton, New York, 1991.
2. T.H. Morgan. *Experimental embryology*. Columbia University Press, New York, 1927.
3. R.A. Emerson to Secretary, John Simon Guggenheim Memorial Foundation,

November 17, 1927. Box 8, Department of Plant Breeding Records, #21/28/889, COR.

4. R.A. Emerson to National Research Council Fellowship Board, April 12, 1929. Box 15, Department of Plant Breeding Records, 1926–1934 series; Wayne F. Keim, personal communication.
5. R.A. Emerson to E.G. Anderson, May 18, 1928. Box 8, Department of Plant Breeding Records, #21/28/889, COR.
6. GWB. 1970. Alfred Henry Sturtevant (1891–1970). *American Philosophical Society Yearbook 1970*, pp. 166–171; E.B. Lewis. 1976. Alfred Henry Sturtevant. In *Dictionary of scientific biography*, vol. 13, pp. 133–138. Scribner's, New York; E.B. Lewis. 1997. Alfred H. Sturtevant, November 21, 1891–April 6, 1970. *Biographical Memoirs* 73: 348–363, National Academy Press, Washington, D.C. At http://www.nap.edu/reading room/ books/biomems/asturtevant.html).
7. Ibid.
8. R.E. Kohler. *Lords of the fly*. University of Chicago Press, 1994; G.E. Allen. *Thomas Hunt Morgan: The man and his science*. Princeton University Press, 1978; I. Shine and S. Wrobel. *Thomas Hunt Morgan: Pioneer in genetics.* University of Kentucky Press, 1976.
9. Kohler, *Lords of the fly*, p. 115.
10. Eggs lacking an X chromosome fail to develop if fertilized by sperm bearing a Y chromosome but fertilization with sperm carrying an X chromosome results in male offspring; these are said to be XO males.
11. C.B. Bridges. 1914. Direct proof through non-disjunction that the sex-linked genes of *Drosophila* are borne by the X chromosome. *Science* 40: 107–109; C.B. Bridges. 1916. Non-disjunction as proof of the chromosome theory of heredity. *Genetics* 1: 1–52; 107–163.
12. E. Heitz. 1933. Uber totale und partielle somatische Heteropycnose, sowie strukturelle Geschlechtschromosomen bei *Drosophila* funebris Zeits. *Zellf. Mikr. Anat.* 19: 720–742; T.S. Painter. 1933. A new method for the study of chromosome rearrangements and plotting of chromosome maps. *Science* 78: 585–586.
13. Alfred H. Sturtevant. Memoir of C.B. Bridges. Folder 1.1, Sturtevant Collection, CIT.
14. Th. Dobzhansky, oral history, COL.
15. Max Delbrück, oral history, CIT.
16. GWB to Jack Schultz, August 4, 1970. Schultz Collection, APS.
17. T.H. Morgan. 1941. Calvin Blackman Bridges: 1889–1938. *Biographical Memoirs* 22: 31–40. National Academy of Sciences, Washington, D.C.
18. R.H. Lock. *Recent progress in the study of variation, heredity and evolution*. E.P. Dutton, New York, 1906.
19. Allen, *Morgan*, pp. 206–207.
20. H.J. Muller. 1927. Artificial transmutation of the gene. *Science* 66: 84–87.
21. L.J. Stadler. 1928. Genetic effects of X-rays in maize. *Proc. Natl. Acad. Sci.* 14: 69–75; L.J. Stadler. 1928. Mutations in barley induced by X-rays and radium. *Science* 68: 186–187.
22. Both are quoted in I. Shine and S. Wrobel. *Thomas Hunt Morgan: Pioneer in genetics.* University of Kentucky Press, Lexington, Kentucky, 1976, p. 92.
23. Kohler, *Lords of the fly*, p.139.
24. Dobzhansky, oral history, p. 240.
25. T. Mohr. Personal recollections of T. H. Morgan presented at the centenary celebra-

tion of Morgan's birth, Lexington, Kentucky, 1966.
26. Ibid.
27. Kohler, *Lords of the fly*, 1994; Morgan, "Bridges Biographical Memoir," 1941.
28. A.H. Sturtevant. 1959. Thomas Hunt Morgan, September 25, 1866–December 4, 1945. *Biographical Memoirs* 33: 283–325, National Academy of Sciences. Published by Columbia University Press, New York.
29. A.H. Sturtevant. 1965. The "Fly Room." *Am. Sci.* 53: 303–307.
30. A.H. Sturtevant. *A history of genetics.* Harper & Row, New York, 1965, pp. 49–50. (Reprinted by Cold Spring Harbor Laboratory Press, 2001.)
31. T.H. Morgan, A.H. Sturtevant, H.J. Muller, and C.B. Bridges. *The mechanism of Mendelian inheritance.* Henry Holt, New York, 1915.
32. GWB. Biochemical genetics: Some recollections. In *Phage and the origins of molecular biology* (ed. J. Cairns, G.S. Stent, and J.D. Watson). Cold Spring Harbor Laboratory of Quantitative Biology, Cold Spring Harbor, New York, 1966, p. 26.
33. Most *minute* alleles cause the embryo to develop more slowly so that the flies reach maturity about two days later than normal. See P.A. Lawrence. *The Making of a fly.* Blackwell Scientific Publications, Oxford, 1992, p. 149.
34. J. Schultz. 1935. Aspects of the relation between genes and development in *Drosophila. Am. Nat.* 69: 30–54.
35. F.J. Ayala. 1985. Theodosius Dobzhansky, January 25, 1900–December 19, 1975. *Biographical Memoirs* 55: 163–213. National Academy of Sciences, National Academy Press, Washington, D.C.
36. Dobzhansky, oral history, pp. 221–238.
37. Such chromosomes are referred to as having translocations.
38. The spindle-fiber attachment site, now referred to as the centromere, is the region of the chromosome at which the "apparatus" (microtubules) for separating pairs of chromosomes is attached during meiosis and mitosis.
39. Kohler, *Lords of the fly*, p. 256.
40. Provine W.B. *Sewall Wright and evolutionary biology.* University of Chicago Press, 1986, pp. 334-340; W.B. Provine. In *Dobzhansky's genetics of natural populations I-XLIII* (ed. R.C. Lewontin, J.A. Moore, W.B. Provine, and B. Wallace). Columbia University Press, 1987, pp. 1–76.
41. Provine, 1987; Donald Poulson. Joint Oral History with J. Bonner, S. Emerson, and N. Horowitz. Interviewed by Judith Goodstein, Harriet Lyle, and Mary Terrall, 1981, CIT.
42. Th. Dobzhansky. *Genetics and the origin of the species.* Columbia University Press, 1937.
43. A.H. Sturtevant. T.H. Morgan: The Biologist and the Man. Centennial Celebration for T.H. Morgan at University of Kentucky, Lexington, Kentucky, 1966.
44. Dobzhansky, oral history, COL.
45. Francis Haskins to Maxine Singer, Feb. 26, 1998, personal communication.
46. H.J. Muller. 1935. A viable two-gene deficiency phenotypically resembling the corresponding hypomorphic mutations. *J. Heredity* 26: 469–478; H.J. Muller. The development of the gene theory. In *Genetics in the 20th century: Essays on the progress of genetics in its first fifty years* (ed. L.C. Dunn). MacMillan, New York, 1951, pp. 93–94.
47. Understanding between Cal Tech and E.G. Anderson. Undated but prior to September 25, 1929. Folder 1/2, Biology Division Collection, CIT.
48. E.B. Lewis, personal communication.

CHAPTER 6

1. Francis Haskins, personal communication.
2. C. Burnham. 1992. Barbara McClintock: Reminiscences. In *The dynamic genome: Barbara McClintock's ideas in the century of genetics* (ed. N. Fedoroff and D. Botstein). Cold Spring Harbor Laboratory Press, Cold Spring Harbor, New York, pp. 19–24.
3. G.E. Allen. Thomas Hunt Morgan: The man and his science. Princeton University Press, 1978.
4. Norman Horowitz, oral history, 1987, CIT.
5. N. Horowitz. 1998. T.H. Morgan at Cal Tech: A reminiscence. *Genetics* 149: 1629–1632.
6. A.H. Sturtevant. *Thomas Hunt Morgan: The biologist and the man.* Centennial of Morgan's Birth. University of Kentucky, Lexington, Kentucky, 1966. Sturtevant Collection, CIT.
7. Jack Schultz to GWB, July 31, 1970. Schultz Collection, APS.
8. GWB to Jack Schultz, July 2, 1970. Schultz Collection, APS.
9. GWB, interview with Harriet Zuckerman, 1966. Oral History Collection, COL.
10. F.D. Keim, A.L. Frolik, and GWB. 1932. Studies of prairie hay in north central Nebraska. *Univ. Nebraska Res. Bull.* 60: 5–54; F.D. Keim., GWB, and A.L. Frolik. 1932. The identification of the more important prairie hay grasses of Nebraska by their vegetative characters. *Univ. Nebraska. Res. Bull.* 65: 5–40.
11. R.A. Emerson to GWB, March 6, 1931; R.A. Emerson to GWB, March 21, 1932. Box 8, Department of Plant Breeding Records #21/28/889, COR.
12. GWB. 1932. A gene in *Zea mays* for failure of cytokinesis during meiosis. *Cytologia* 3: 142–155.
13. GWB. Thomas Hunt Morgan: Some personal recollections. Presented at the Centennial Celebration of Morgan's birth, University of Kentucky, 1966. Folder 5.34, Beadle Collection, CIT.
14. L. Kay. *The molecular vision of life.* Oxford University Press, New York, 1993, p. 147.
15. L.V. Morgan. 1922. Non criss-cross inheritance in *Drosophila melanogaster. Biol. Bull.* 42: 267–274.
16. A.H. Sturtevant and GWB. *An introduction to genetics.* W.B. Saunders, Philadelphia, 1939. (Reprinted by Dover Publications, New York, 1967); GWB and S. Emerson. 1935. Further studies of crossing over in attached-X chromosomes. *Genetics* 20: 192–206.
17. N.H. Horowitz, "T.H. Morgan at Cal Tech."
18. GWB, "Gene in *Zea mays*"; GWB, A.C. Fraser, and R.A. Emerson to Maize Geneticists, November 28, 1930, Folder 1, Demerec Papers, APS.
19. H.B. Creighton and B. McClintock. 1931. A correlation of cytological and genetical crossing-over in *Zea mays. Proc. Natl. Acad. Sci.* 17: 492–497.
20. C. Stern. 1931. Zytologische-genetische Untersuchungen als Beweise für die Morgansche Theorie des Faktorenaustausches. *Biol. Zeit.* 51: 547–587.
21. GWB. 1932. Genes in maize for pollen sterility. *Genetics* 17: 413–431.
22. E.G. Anderson to Fellowship Board, April 4, 1931, NRC.
23. R.A. Emerson to E.E. Conger, Secretary of the Board of Fellowships, April 3, 1931, NRC.
24. E.J. Kraus to E.E. Conger, undated, NRC.
25. GWB. 1932. A gene for sticky chromosome in *Zea mays. Zeitschrift fur Induktive Abstammungs und Vererbungslehre* 63: 195–217; GWB. 1937. Chromosome aberra-

tion and gene mutation in sticky chromosome plants of *Zea mays*. *Cytologia Fujii Jubilee Volume:* 43–56.

26. GWB. 1966. Biochemical genetics: Some recollections. In *Phage and the origins of molecular biology* (ed. J. Cairns, G.S. Stent, and J.D. Watson). Cold Spring Harbor Laboratory of Quantitative Biology, Cold Spring Harbor, New York, p. 26.
27. Ibid; GWB. Foreword to Selected Papers of A.H. Sturtevant. In *Genetics and evolution* (ed. E.B. Lewis). W.H. Freeman, San Francisco and London, 1961.
28. B. McClintock. 1941. Spontaneous alterations in chromosome size and form in *Zea mays*. *Cold Spring Harbor Symp. Quant. Biol.* 9: 72–80.
29. B. McClintock, interview by William.B. Provine and P. Sisco, August 18, 1980, Cold Spring Harbor Laboratory, New York, COR.
30. Burnham, "Barbara McClintock."
31. Provine and Sisco, interview, 1980, COR.
32. Harriet B. Creighton, personal communication.
33. R.A. Emerson to GWB, Sept 19, 1931. Box 8, 1926–1934 series, Department of Plant Breeding Records, #21/28/889, COR.
34. GWB to M. Demerec, October 12, 1931. Folder 2, Demerec Papers, APS.
35. M. Demerec to GWB, December 28, 1931. Folder 2, Demerec Papers, APS.
36. GWB to M. Demerec, January 9, 1932. Folder 2, Demerec Papers, APS.
37. R.A. Emerson to GWB, January 5, 1932. Box 8, Department of Plant Breeding Records, #21/28/889, COR.
38. R.A. Emerson to GWB, March 21, 1932. Box 8, Department of Plant Breeding Records, #21/28/889, COR.
39. M. Demerec to GWB, January 16, 1932. Folder 2, Demerec Papers, APS.
40. W.J. Robbins, Chair of NRC Fellowship Board, February 8, 1932, NRC.
41. R.A. Emerson to GWB, May 25, 1932. Box 8, Department of Plant Breeding Records, #21/28/889, COR.
42. Ibid.
43. T.H. Morgan to Max Mason, May 15, 1933. RG 1.1 Series 205, Box 5, Folder 71, RFA.
44. T.H. Morgan to W.W. Weaver, November 9, 1933. RG 1.1 Series 205, Box 5, Folder 71, RFA.
45. Rockefeller Foundation to T.H. Morgan, December 19, 1933. RG 1.1 Series 205, Box 5, Folder 71, RFA.
46. Warren Weaver Diaries, April 16 and 17, 1934, RFA.
47. T.H. Morgan to W. Weaver, Jan. 11, 1934. RG 1.1 Series 205, Box 5, Folder 72, RFA.
48. H.M. Miller, Diary, June 7, 1934, RG 1.1 Series 205, Box 5, Folder 71, RFA.
49. W.E. Tisdale, Diary, May 9, 1934. RG 1.1 Series 205, Box 5, Folder 71, RFA.
50. H.M. Miller. Report, September 25–27, 1935. RG1.1 Series 205D, Box 6.74, quoted in Kay, 1993.
51. Letters between M. Demerec and GWB, 1929–1930. Folder 1, M. Demerec Papers, APS.
52. GWB to M. Demerec, November 23, 1929. Folder 1, M. Demerec Papers, APS.
53. GWB, Fraser, and Emerson to Maize Geneticists, November 28, 1930. Folder 1, M. Demerec Papers, APS.
54. Letters between R.A. Emerson and GWB, November 1930 through 1935, Boxes 8, 10, and 20, Beadle Files, COR.
55. R.A. Emerson to GWB, March 4, 1932. Box 8, Department of Plant Breeding Records, #21/28/889, COR.
56. Maize (or Corn) Genetics Cooperation. microfilm 1417, APS.

57. http://www.w3.ag.uiuc.edu/maize-coop/wwwlist.html
58. R.A. Emerson, GWB, and A.C. Fraser. 1935. A summary of linkage studies in maize. *Cornell Univ. Agric. Exp. Station Memoirs* 180: 1–83.
59. Agricultural Statistics. 1962. United States Department of Agriculture. The United States Government Printing Office, Washington, D.C.
60. M.M. Rhoades. 1992. The early years of maize genetics. In *The dynamic genome: Barbara McClintock's ideas in the century of genetics* (ed. N. Fedoroff and D. Botstein). Cold Spring Harbor Laboratory Press, Cold Spring Harbor, New York, pp. 45–69.
61. GWB. 1933. Polymitotic maize and the precocity hypothesis of chromosome conjugation. *Cytologia* 5: 118–121.
62. Provine and Sisco, interview, 1980, COR.
63. GWB, "Biochemical genetics: Some recollections," pp. 23–32.
64. A.H. Sturtevant. 1913. The linear arrangement of six sex-linked factors in *Drosophila* as shown by their association. *J. Exp. Zool.* 13: 43–59.
65. Th. Dobzhansky. 1930. Translocations involving the third and fourth chromosomes of *Drosophila melanogaster. Genetics* 15: 347–399; Th. Dobzhansky. 1931. The decrease in crossing-over observed in translocations and its probable explanation. *Am. Nat.* 65: 214–232.
66. GWB. 1932. A possible influence of the spindle fiber on crossing-over in *Drosophila. Proc. Natl. Acad. Sci.* 18: 160–165.
67. L.V. Morgan, "Non criss-cross inheritance."
68. GWB. 1935. Crossing-over in attached X triploids of *Drosophila melanogaster. J. Genet.* 29: 277–309; GWB and S. Emerson. 1935. Further studies of crossing-over in attached chromosomes of *Drosophila melanogaster. Genetics* 20: 192–206.
69. GWB and A.H. Sturtevant. 1935. X chromosome inversions and meiosis in *Drosophila melanogaster. Proc. Natl. Acad. Sci.* 21: 384–390.
70. A.H. Sturtevant and GWB. 1936. The relations of inversions in the X chromosome of *Drosophila melanogaster* to crossing-over and disjunction. *Genetics* 21: 544–604.
71. GWB and S. Emerson. Studies on the mechanism of crossing over in *Drosophila* II. Experiments with certain translocations. In *Proceedings of the Sixth International Congress of Genetics.* Ithaca, New York. Published by the Brooklyn Botanical Garden, 1932. Vol. 1–2, p. 7.
72. GWB, "Thomas Hunt Morgan."
73. D. Joravsky. *The Lysenko affair.* The University of Chicago Press, 1970; Z.A. Medvedev. *The Medvedev papers: The plight of Soviet science today.* Macmillan, St Martins Press, London, 1971.
74. Kay, *Molecular Vision,* 1993, p. 125 suggests that according to Hiram Bentley Glass this visit took place during Beadle's first trip from Ithaca to Pasadena in 1930. A personal communication from Glass confirmed the story but indicated that the year of the visit was more likely 1932.

CHAPTER 7

1. E.B. Lewis. 1995. Remembering Sturtevant. *Genetics* 141: 1227–1230.
2. Jack Schultz to GWB, July 31, 1970. APS Schultz Collection.

3. Lewis, "Remembering Sturtevant."
4. GWB. In *Genetics and evolution; Selected papers of A.H. Sturtevant* (ed. E.B. Lewis). W.H. Freeman, San Francisco, 1961.
5. A. Roe. *The making of a scientist.* Dodd, Mead, New York, 1953.
6. C.M. Child. *The origin and development of the nervous system from a physiological viewpoint.* University of Chicago Press, 1921.
7. S. Wright. 1921. Review of the origin and development of the nervous system. *J. Hered.* 12: 72–75.
8. R.E. Kohler. *Lords of the fly.* University of Chicago Press, 1994, p. 178.
9. T.H. Morgan. 1926. Genetics and the physiology of development. *Am. Nat.* 60: 489–515.
10. G.E. Allen. *Thomas Hunt Morgan: The man and his science.* Princeton University Press, 1978, p. 301.
11. T.H. Morgan. *The rise of genetics.* Proceedings of the Sixth International Congress of Genetics, 1932, pp. 87–103.
12. Allen. *Thomas Hunt Morgan*, p. 374.
13. Ibid. pp. 374–375.
14. Morgan to Sturtevant June 17, 1934. Sturtevant Collection, Box 3.17, CIT.
15. I. Shine and S. Wrobel. *Thomas Hunt Morgan.* University Press of Kentucky, 1976, p. 120.
16. T.H. Morgan. 1935. The relation of genetics to physiology and medicine. *Sci. Monthly* 41: 5–18. Nobel Lecture, June 4, 1934, Stockholm, Sweden.
17. O.T. Avery, C.M. Macleod, and M. McCarty. 1944. Studies on the chemical nature of the substance inducing transformation of pneumococcal types. I. Induction of transformation by a deoxyribonucleic acid fraction isolated from *Pneumococcus* type III. *J. Exp. Med.* 79: 137–158.
18. T.H. Morgan, "The relation of genetics to physiology."
19. T.H. Morgan. *Embryology and genetics.* Columbia University Press, New York, 1934.
20. S. Wright. 1934. Genetics of abnormal growth in the guinea pig. *Cold Spring Harbor Symp. Quant. Biol.* 2: 137–147.
21. R. Goldschmidt. *Physiological genetics.* New York, McGraw-Hill, 1938.
22. J. Schultz. 1935. Aspects of the relation between genes and development in *Drosophila. Am. Nat.* 69: 30–54.
23. A.H. Sturtevant. *The use of mosaics in the study of the developmental effects of genes.* Proceedings of the Sixth International Congress of Genetics 1, 1932, pp. 304–307.
24. GWB. 1974. Biochemical genetics: Recollections. *Annu. Rev. Biochem.* 43: 1–13.
25. T.H. Morgan and C.B. Bridges. 1916. Sex-linked inheritance in *Drosophila. Carnegie Inst. Wash. Rep.* pp. 1–87.
26. While not easy, male and female cells can be distinguished by the character of the bristles or by their colors.
27. A.H. Sturtevant. 1920. The *vermilion* gene and gynandromorphism. *Proc. Soc. Exp. Biol. Med.* 17: 70–71.
28. Sturtevant, "Use of mosaics."
29. J.B.S. Haldane. 1920. Some recent work on heredity. *Trans. Oxford Univ. Junior Sci. Club* 1: 3–11; R. Scott-Moncrieff. *The classical period in chemical genetics: Recollections of Muriel Wheldale Onslow, Robert and Gertrude Robinson and J.B.S. Haldane.* Notes and Records of the Royal Society of London 36 (1981), pp. 126–154.

30. T.H. Morgan. 1926. Genetics and the physiology of development. *Am. Nat.* 60: 489–515.
31. Irène Ephrussi Barluet, interview November 11, 1997 and letter December 17, 1997.
32. B. Ephrussi. *Contribution a l'Analyse des Premier Stades du Development de l'Oeuf; Action de la Temperature.* Paris, Imprimerie de l'Academie, 1932; B. Ephrussi. Croissance et Regeneration Dans les Cultures des Tissus. Paris, Masson, 1932.
33. R.M. Burian, J. Gayon, and D. Zallen 1988. The singular fate of genetics in the history of French biology 1900–1940. *J. Hist. Biol.* 21: 357–402.
34. B. Ephrussi. 1933. Sur le Facteur Lethal des Souris Brachyures. *Compte Rendus Acad. Sci.* 197: 96–98.
35. B. Ephrussi. 1934. The absence of autonomy in the development of the effects of certain deficiencies in *Drosophila melanogaster. Proc. Natl. Acad. Sci.* 20: 420–422.
36. GWB, "Biochemical genetics: Recollections."
37. Morgan, *Embryology and genetics.*
38. B. Ephrussi. 1958. The cytoplasm and somatic cell variation. *J. Cell. Comp. Physiol.* (suppl. 1) 32: 35–53.
39. GWB, "Biochemical genetics: Recollections."
40. G.E. Allen. *Thomas Hunt Morgan,* pp. 6–18.
41. Barluet, November 11, 1997, personal communication.
42. Barluet, February 27, 1998, personal communication.
43. R.E. Kohler, *Lords of the fly,* p. 215.
44. Although the isolated granules appear as purplish red and ocher yellow, they impart a deep reddish or brown color, respectively, in the eyes.
45. E. Caspari. 1933. Uber die Wirkung eines pleiotropen gens beider Melmotte *Ephestia kuhniella* Zeller. *Arch. Entw. Mech.* 130: 353–381; A. Kuhn et al. 1935. Uber hormonale genwirkung bei Ephestia kuhniella Zeller. *Nachr. Ges. Wiss. Gottingen* 2: 1–29.
46. GWB to Milislav Demerec June 5, 1935. APS Demerec Collection, Folder 2, 1931–1949.
47. GWB and B. Ephrussi. 1935. Transplantation in *Drosophila. Proc. Natl. Acad. Sci.* 21: 642–646.
48. GWB. 1975. Biochemical genetics: Reflections. In *Three lectures January 15–17, 1975.* The Edna H. Drane Lectureship at the University of Southern California, School of Medicine. CHG
49. Ibid.
50. GWB to Demerec, September 1, 1935, APS Demerec Collection, Folder 2.
51. Demerec to GWB, September 13, 1935, APS Demerec Collection, Folder 2.
52. GWB to Demerec, October 17, 1935, APS Demerec Collection, Folder 2.
53. GWB and B. Ephrussi. 1936. The differentiation of eye pigments in *Drosophila* as studied by transplantation. *Genetics* 21: 225–247.
54. A.H. Sturtevant. 1920. The *vermilion* gene and gynandromorphism. *Proc. Soc. Exp. Biol. Med.* 17: 70–71.
55. GWB and Ephrussi, "The differentiation of eye pigments."
56. B. Ephrussi and GWB. 1935. La Transplantation des Disques Imaginaux chez la *Drosophile. Comptes Rendu Acad. Sci. Paris* 201: 98–99; GWB and B. Ephrussi. 1935. Differenciation de la Couleur de l'Oeil *cinnabar* chez la *Drosophile. Comptes rendu Acad. Sci. Paris* 201: 620–622.
57. GWB and Ephrussi, "The differentiation of eye pigments."
58. F.B. Hanson's diary 14, August 27 1935. 1.1 205F 7.87, RFA.
59. George C. Streeter to GWB. January 8, 1936. APS Demerec Collection, Folder 2.

60. F.B. Hanson 1935 Diaries. 1.1 205F 7.87, p. 142, RFA.
61. GWB, 1966 interview with Harriet Zuckerman. Oral History Collection, COL.
62. B. Ephrussi and M.H. Harnley. 1936. Sur la Presence Chez Differents Insectes des Substances Intervenant Dans la Pigmentation des Yeux de *Drosophila melanogaster. Comptes Rendu Acad. Sci. Paris* 202: 1028–1029; GWB, R.L. Anderson, and J. Maxwell. 1937. A comparison of the diffusible substances concerned with eye color development in *Drosophila, Ephestia* and *Habrobracon. Proc. Natl. Acad. Sci.* 24: 80–85.
63. GWB. 1937. Development of eye colors in *Drosophila:* Fat bodies and Malpighian tubes as sources of diffusible substances. *Proc. Natl. Acad. Sci.* 23: 146–152.
64. GWB and B. Ephrussi. 1937. Development of eye colors in *Drosophila:* Diffusible substances and their interrelations. *Genetics* 22: 76–86.
65. GWB, C.W. Clancy, and B. Ephrussi. 1937. Development of eye colours in *Drosophila:* Pupal transplants and the influence of body fluid on *vermilion. Proc. R. Soc. Lond. Ser. B* 826: 98–105; M.H. Harnley and B. Ephrussi. 1937. Development of eye colors in *Drosophila:* Time of action of body fluid on *cinnabar. Genetics* 22: 393–401.
66. R.M. Burian, J. Gayon, and D. Zallen. 1988. The singular fate of genetics in the history of French biology 1900–1940. *J. Hist. Biol.* 21: 357–402.
67. A. Lwoff. 1979. Recollections of Boris Ephrussi. *Somatic Cell Genet.* 5: 677.
68. H.F. Judson. *The eighth day of creation.* Simon and Schuster, New York, 1979, pp. 356–357.

CHAPTER 8

1. Lloyd Law, interview, January 30, 1997.
2. David Beadle, interview, August 13, 1997.
3. Frank B. Hanson diary, September 4 and 5, 1937. RFA 205D 7.88.
4. K.V. Thimann and GWB. 1937. Development of eye colors in *Drosophila:* Extraction of the diffusible substances concerned. *Proc. Natl. Acad. Sci.* 23: 143–146.
5. Norman H. Horowitz, personal communication.
6. Law, interview.
7. David Regnery, interview, July 3, 1996.
8. GWB. 1974. Biochemical genetics: Recollections. *Annu. Rev. Biochem.* 43: 1–13.
9. GWB. 1975. Biochemical genetics: Reflections. In *Three lectures. January 15–17, 1975.* The Edna H. Drane Visiting Lectureship, University of Southern California, School of Medicine.
10. Charles V. Taylor to Warren Weaver, April 5, 1937. 205D 10. 135, RFA.
11. Law, interview.
12. A.H. Sturtevant and GWB. *An introduction to genetics.* W.B. Saunders, Philadelphia and London, 1940.
13. GWB to Edward L. Tatum, July 31, 1937; provided by Joshua Lederberg.
14. Law, interview.
15. Norman Horowitz, interview, October 1, 1996.
16. GWB to Tatum, June 23, 1937; provided by Lederberg.
17. W.H. Peterson and E.B. Fred to Tatum, July 2, 1937; provided by Lederberg.
18. Tatum to GWB, July 21, 1937, acknowledged in letter from GWB to Tatum July 31, 1937.
19. GWB to Tatum July 31, 1937.

20. B. Ephrussi. 1946. Analysis of eye color differentiation in *Drosophila. Cold Spring Harbor Symp. Quant. Biol.* 11: 40–48.
21. Ephrussi to GWB, January 9, 1937. Box 1.26, CIT.
22. Ibid.
23. Ephrussi to GWB, November 8, 1937. Box 1.26, CIT.
24. Ephrussi to GWB, November 29, 1937. Box 1.26, CIT.
25. Ephrussi to GWB, December 26, 1937. Box 1.26, CIT.
26. Thimann and GWB. "Eye colors in *Drosophila*."
27. GWB and L.W. Law. 1938. Influence on eye color of feeding diffusible substances to *Drosophila melanogaster. Proc. Soc. Exp. Biol. Med.* 37: 621–623.
28. Y. Khouvine, B. Ephrussi, and S. Chevais. 1938. Development of eye colors in *Drosophila:* Nature of the diffusible substances; effects of yeast, peptones and starvation on their production. *Biol Bull.* 75: 425–446; GWB, E.L. Tatum, and C.W. Clancy. 1938. Food level in relation to rate of development and eye pigment formation in *Drosophila melanogaster. Biol. Bull.* 75: 447–462.
29. E.L. Tatum and GWB. 1938. Development of eye colors in *Drosophila:* Some properties of the hormones involved. *J. Gen. Physiol.* 22: 239–253.
30. E.L. Tatum. 1939. Development of eye colors in *Drosophila:* Bacterial synthesis of v^+ hormone. *Proc. Natl. Acad. Sci.* 25: 486–490.
31. E.L. Tatum and GWB. 1940. Crystalline *Drosophila* eye color hormone. *Science* 91: 458.
32. A.H. Sturtevant and GWB. *An introduction to genetics.* W.B. Saunders, Philadelphia and London, 1940.
33. Ephrussi to GWB, March 17, 1939. Box 1.26, CIT.
34. Ephrussi to GWB, April 17, 1939. Box 1.26, CIT.
35. David Beadle, interview, June 27, 1997.
36. E.A. Carlson. *Genes, radiation and society: The life and work of H.J. Muller.* Cornell University Press, Ithaca, 1981, pp. 231–232.
37. The Genetic Manifesto was published in the *Journal of Heredity*, September 1939; a copy exists in the Muller archive in the Lilly Library at Indiana University.
38. Remarks by Otto L. Mohr at the opening plenary session of the 7th International Congress of Genetics August 23, 1939.
39. *Proceedings of the 7th International Congress of Genetics,* August 23, 1939.
40. Ibid.
41. J.B.S. Haldane. *New paths in genetics.* Harper and Brothers, New York and London, 1942.
42. The German army invaded Holland in May 1940, two months after Haldane's lectures.
43. W.S. Churchill. *The gathering storm*, Vol. 1. Houghten Mifflin, Boston, 1948, p. 423.
44. Ibid., p. 484.
45. David Beadle, interview, June 27, 1997; as a memento of Beadle's experience, the blanket is now on display in the Saunders County Historical Museum.
46. GWB to Chauncey Beadle September 18, 1939; reprinted in the *Wahoo Wasp,* September 28, 1939, WAH.

CHAPTER 9

1. E.L. Tatum. 1939. Development of eye-colors in *Drosophila:* Bacterial synthesis of v hormone. *Proc. Natl. Acad. Sci.* 25: 486–490; E.L. Tatum and GWB. 1940. Crystalline

Drosophila eye-color hormone. *Science* 91: 458.

2. A. Butenandt, W. Weidel, and E. Becker. 1940. Kynurenine als Augenpigmentbildung Auslosendes Agens bei Insekten. *Naturwissenschafen* 28: 63–64.
3. E.L. Tatum and A. Haagen-Smit. 1940. Identification of *Drosophila* V hormone of bacterial origin. *J. Biol. Chem.* 140: 575–580.
4. Clarence P. Berg to GWB, January 14, 1947. Biological Division Records, Box 12.24, CIT.
5. H. Kikkawa. 1941. Mechanism of pigment formation in *Bombyx* and *Drosophila. Genetics* 26: 587–607.
6. A. Butenandt, W. Weidel, and H. Schlossberger. 1949. 3-oxy-kynurenine als cn-Gen Abhangiges Gleid im Intermediaren Tryptophan Stoffwechsel. *Z. Naturforsch* 4B: 242–244.
7. D. Ghosh and H.S. Forrest. 1967. Enzymatic studies on the hydroxylation of kynurenine in *Drosophila melanogaster. Genetics* 55: 423–431.
8. J.P. Phillips and H.S. Forrest. 1980. Ommachromes and pteridines. In *The genetics and biology of* Drosophila, Vol. 2D, Chap. 38 (ed. M. Ashburner and T.R.F. Wright). Academic Press, New York.
9. Ephrussi to GWB, October 23, 1939. Beadle Collection, Box 1.26, CIT.
10. Ibid.
11. Ephrussi to GWB, February 19, 1940. Beadle Archive, Box 1.26, CIT.
12. Louis Rapkine, a distinguished biologist at the Pasteur Institute in Paris, played a huge role in plotting the escape and settlement of Jewish scientists in America and other neutral countries.
13. Irène Barluet to P. Berg dated December 14, 1997 and records of the Rockefeller Foundation RG 10 F/S Recorder Cards Room 103 Red Drawers Draw #2.
14. GWB. *Genetic control of the production and utilization of hormones. Proceedings of the 7th International Genetics Congress* (suppl.). Cambridge University Press, 1941, pp. 58–62.
15. R. Scott-Moncrieff. *The classical period in chemical genetics recollections of Muriel Wheldale Onslow, Robert and Gertrude Robinson and J.B.S. Haldane.* Notes and Records of the Royal Society of London, 36, 1981, pp. 126–154.
16. A.E. Garrod. *Inborn errors of metabolism.* 2nd edition, Frowde and Hodder and Stoughton, 1923.
17. Because of the considerable delay in receiving, redacting and publishing the papers presented during August 1939 at the 7th International Genetics Congress, it is uncertain if the published text (1941) is the same as what Beadle presented at the time of the Congress.
18. GWB and E.L. Tatum. 1941. Experimental control of development and differentiation. *Am. Nat.* 75: 107–116.
19. GWB, *Genetic control of production.*
20. GWB. 1975. Biochemical genetics: Reflections. In *Three lectures, January 15–17, 1975.* The Edna H. Drane Visiting Lectureship, University of Southern California, School of Medicine. Norris Medical Library Archives W9B365t.
21. Lecture notes were provided by Joshua Lederberg; original in National Library of Medicine Archives.
22. A. Lwoff. 1938. Les Facteurs de Croissance Pour les Microorganismes. *Ann. Inst. Pasteur* 61: 580–617; B.C.J.G. Knight. 1936. Bacterial nutrition. *His Majesty's*

Stationary Office, pp. 1–182; R.J. Williams. 1941. Growth-promoting nutrilites for yeasts. *Biol. Rev.* 16: 49–80.

23. Sydney Raffel, interview, August 31, 1999.
24. GWB. 1966. Biochemical genetics: Some recollections. In *Phage and the origins of molecular biology* (ed. J. Cairns, G.S. Stent, and J.D. Watson), Cold Spring Harbor Laboratory of Quantitative Biology, Cold Spring Harbor, New York, pp. 23–32; Beadle often recalled that he and the other graduate students had to correct Dodge's misinterpretation of results he obtained in certain genetic crosses. Dodge seemed unaware of the new ideas that explained how the genetic character of spores produced in a given mating varied depending on when crossing-over took place during meiosis.
25. At the time of Dodge's retirement, Beadle commented that "the ancestors of the Dodge and Beadle lines may have been associated in a grotesque manner some three centuries before." It seems that one of Dodge's relatives was burned as a witch at Salem and Beadle's cousin, eight times removed, owned Beadle's Tavern in Salem where it is documented that alleged witches were taken for interrogation. Beadle wondered if "the Dodge relative was taken there?" Beadle Collection, Box 3.9, CIT.
26. Garland E. Allen to Joshua Lederberg September 6, 1979 provided by J. Lederberg; GWB. 1959. Genes and chemical reactions in *Neurospora. Science* 129: 1715–1719; C.C. Lindegren. 1942. The use of fungi in modern genetical analysis. *Iowa State Coll. J. Sci.* 16: 271–290; C.C. Lindegren and G. Lindegren. 1941. X-ray and ultra-violet mutations induced in *Neurospora. J. Hered.* 32: 405–412.
27. A.H. Sturtevant and GWB. *An introduction to genetics.* W.B. Saunders, Philadelphia and London, 1940.
28. GWB to Bernard O. Dodge, February 27, 1941. New York Botanical Garden Library. Dodge Collection.
29. Ibid. GWB to Dodge, March 13, 1941.
30. F. Kögl and N. Fries. 1937. Uber den Einfluss von Biotin, Aneurin und Meso-Inositol Wachstum Verschiedener Pilzarten. *Hoppe-Seyl. Z.* 249: 93–106.
31. Biotin's scarcity ended soon after W.J. Robbins, Dodge's colleague at the Botanical Garden, discovered that biotin could be readily obtained from cotton batting by extraction with hot water.
32. H.J. Muller. 1927. Quantitative methods in genetic research. *Am. Nat.* 61: 407–419.
33. David Regnery, interview, July 3, 1996.
34. GWB and E.L. Tatum. 1945. *Neurospora* ll. Methods of producing and detecting mutations concerned with nutritional requirements. *Am. J. Bot.* 32: 678–686.
35. GWB to Carl C. Lindegren, July 25, 1941. Beadle Collection, Box 5.10, CIT.
36. GWB to Dodge, August 11, 1941. New York Botanical Garden, Dodge Collection.
37. GWB and E.L. Tatum. 1941. Genetic control of biochemical reactions in *Neurospora. Proc. Natl. Acad. Sci.* 27: 499–506.
38. N.H. Horowitz. *George Wells Beadle: Biographical memoirs of the American Philosophical Society Year Book,* American Philosophical Society, 1995.
39. Sydney Raffel, interview, August 31, 1999.
40. B. Ephrussi to GWB, August 22. 1941. Beadle papers, Box 1.26, CIT.
41. GWB to Ephrussi, November 17. 1941. Beadle papers, Box 1.26, CIT.
42. GWB to Warren Weaver, November 28, 1941. 205D RG 1.1 Box 10, Folder 141, RFA.
43. Ibid.
44. Frank B. Hanson diary entry on his meetings with Beadle during Dec.15, 17, and 18,

1941. 205D, Box 10, Folder 141, RFA collection.
45. Hanson diary September 4 and 5, 1937. 205D, RG1.1, Box 7, Folder 88, RFA.
46. Telegram from Hanson to GWB, 1942. 205D, RG1.1, Box 10, Folder 141, RFA.
47. GWB. 1978. Recollections. *Annu. Rev. Biochem.* 43: 1–13.
48. A.M. Srb. 1973. Beadle and *Neurospora*. Some recollections. *Neurospora Newslett.* pp. 8–9.

CHAPTER 10

1. David Beadle, interview, June 27, 1997.
2. Ibid.
3. Norman Horowitz, interview, Oct. 16, 1996.
4. David Beadle, interview, June 27, 1997.
5. Jo Srb, interview, April 1998.
6. D. Beadle, interview, June 27, 1997.
7. D. Beadle, interview, August 13, 1997.
8. D. Beadle, interviews, June 27 and August 13, 1997.
9. D. Beadle, interview, August 13, 1997.
10. D. Beadle, interview, May 14, 2001.
11. Ibid.
12. Ephrussi to GWB, January 5, 1942. CIT, Beadle collection, box 1.26.
13. GWB. 1966. Biochemical genetics: Some recollections. In *Phage and the origins of molecular biology* (ed. J. Cairns, G.S. Stent, and J.D. Watson), pp. 23–32. Cold Spring Harbor Laboratory of Quantitative Biology, Cold Spring Harbor, New York.
14. GWB to Sterling Emerson, March 14, 1942. CIT, Beadle collection, box 3.15.
15. N.H. Horowitz and GWB. 1943. A microbiological method for the determination of choline by use of a mutant of *Neurospora*. *J. Biol. Chem.* 150: 325–333.
16. GWB and E.L. Tatum. 1941. Genetic control of biochemical reactions in *Neurospora*. *Proc. Natl. Acad. Sci.* 27: 499–506.
17. F.J. Ryan, GWB, and E.L. Tatum. 1942. The tube method of measuring the growth rate of *Neurospora*. *Am. J. Botany* 30: 784–799.
18. Ryan to GWB, June 27, 1942. CIT, Beadle collection, box 2.23.
19. GWB to Ephrussi, Aug 1, 1942. CIT, Beadle collection, box 1.26.
20. GWB to Ryan, January 4, 1943. CIT Beadle collection, box 6.22.
21. Norman H. Horowitz. George Wells Beadle. In *Biographical Memoirs of the National Academy of Sciences*, 1990, pp. 45–54.
22. A.M. Srb and N. Horowitz. 1944. The ornithine cycle in *Neurospora* and its genetic control. *J. Biol. Chem.* 154: 129–139.
23. GWB. 1945. The genetic control of biochemical reactions. *Harvey Lect.* 40: 179–194.
24. GWB and E.L. Tatum. 1941. "Genetic control of biochemical reactions"; E.L. Tatum, D. Bonner, and GWB. 1944. Anthranilic acid and the biosynthesis of indole and tryptophan by *Neurospora*. *Arch. Biochem.* 3: 477–478; E.L. Tatum and D. Bonner. 1944. Indole and serine in the biosynthesis and breakdown of tryptophan. *Proc. Natl. Acad. Sci.* 30: 30–37.
25. D. Bonner, E.L. Tatum, and GWB. 1943. The genetic control of biochemical reactions in *Neurospora:* A mutant strain requiring isoleucine and valine. *Arch. Biochem.* 3: 71–91.

26. D. Bonner. 1946. Further studies of mutant strains of Neurospora requiring isoleucine and valine. *J. Biol. Chem.* 166: 545–554.
27. R.E. Kohler. *Lords of the fly*. University of Chicago Press, 1994, p. 240.
28. GWB interview by Harriet Zuckerman in Chicago, Illinois on December 12, 1963, COL.
29. Ibid.
30. David Regnery, interview, July 3, 1996.
31. D. Regnery, interview, July 3, 1996; Norman Horowitz, interview, October 16, 1996; Esther Zimmer Lederberg, interview, October 19, 1996. (After receiving a masters degree with Beadle, she joined Tatum at Yale in 1946 and worked with Joshua Lederberg, whom she married soon after.)
32. D. Regnery, interview, July 3, 1996.
33. D. Beadle, interview, August 3, 1997; Jo Srb, interview, April 17, 1998; Esther Zimmer Lederberg, interview, October 19, 1996.
34. Jo Srb, interview, April 17, 1998.
35. D. Beadle, interview, June 27, 1997.
36. Esther Zimmer Lederberg, interview, October 19, 1996.
37. Telephone conversation with David Beadle, June 14, 1999.
38. GWB to Bonner's draft board. CIT, Beadle collection, box 1.19.
39. GWB to F.B. Hanson July 3, 1942. RFA 205D, record group 1.1, box 10, folder 142.
40. Hanson to GWB July 8, 1942, ibid.
41. Beadle to Hanson July 13, 1942, ibid.
42. A.N. Richards quoted in GWB's letter to Hanson November 6, 1942, ibid.
43. Letter from GWB to Hanson January 17, 1944. RFA 205D, record group 1.1, box 10, folder 144 and memo entitled, A Proposed Study of the Production of Antibiotic Substances by Neurospora. CIT, Beadle collection box 34 folder 4.
44. GWB to A.D. Welch dated February 18, 1944, ibid.
45. D. Bonner. 1946. Biochemical mutations in *Neurospora. Cold Spring Harbor Symp. Quant. Biol.* 11: 14–24.
46. R.D. Coghill and R.S. Koch. 1945. Penicillin: A wartime accomplishment. *Chem. Eng. News* 23: 2310–2316.
47. B. McClintock. 1945. *Neurospora*. I. Preliminary observations of the chromosomes of *Neurospora crassa. Am. J. Bot.* 32: 671–678; D.D. Perkins. 1992. *Neurospora* chromosomes. In *The dynamic genome: Barbara McClintock's ideas in the century of genetics* (ed. N. Federoff and D. Botstein). Cold Spring Harbor Laboratory Press, Cold Spring Harbor, New York; see also E.F. Keller. *A feeling for the organism: The life and work of Barbara McClintock.* W.H. Freeman, San Francisco, 1983.
48. GWB. 1945. Do bacteria have genes? *J. Hered.* 36: 23–24; J. Lederberg. 1987. Genetic recombination in bacteria: A discovery account. *Annu. Rev. Genet.* 21: 31–33.
49. GWB to J. Lederberg circa 1978–1979.
50. C.H. Gray and E.L. Tatum. 1944. X-ray induced growth factor requirements in bacteria. *Proc. Natl. Acad. Sci.* 30: 404–419.
51. E.L. Tatum. 1944. X-ray induced mutant strains of *Escherichia coli. Proc. Natl. Acad. Sci.* 31: 215–219.
52. GWB. 1978. Recollections. *Annu. Rev. Biochem.* 43: 1–13.
53. J. Lederberg, interview, September 26, 1996.
54. N. Horowitz, interview, October 27, 1996.

55. E.L. Tatum and J. Lederberg. 1947. Gene recombination in the bacterium *Escherichia coli*. *J. Bacteriol.* 53: 673–684; J. Lederberg "Genetic recombination in bacteria," see pp. 31–33.
56. GWB to Tatum, September 9, 1946. CIT, Beadle collection, box 7.24.

CHAPTER 11

1. Alfred H. Sturtevant to GWB, May 13, 1945. CIT, Beadle collection, box 7.19.
2. GWB. 1945. Biochemical genetics. *Chem. Rev.* 37: 15–96.
3. GWB. 1945. Genetics and metabolism in *Neurospora*. *Physiol. Rev.* 25: 643–663.
4. GWB. 1946. Genes and the chemistry of the organism. *Am. Sci.* 34: 31–53.
5. GWB. 1966. Biochemical genetics: Some recollections. In *Phage and the origins of molecular biology* (ed. J. Cairns, G.S. Stent, and J.D. Watson). Cold Spring Harbor Laboratory of Quantitative Biology, Cold Spring Harbor, New York, pp. 23–32.
6. GWB. 1945. The genetic control of biochemical reactions. *Harvey Lect.* 40: 192; 179–194.
7. By function, Beadle must have meant one type of enzymatic reaction. Today, however, we know that a given enzyme can interact with more than one substrate thereby affecting more than one function; in addition, many enzymes comprise more than one type of protein and are necessarily specified by more than one gene.
8. GWB. 1945. "Genetics and metabolism."
9. GWB, "Genetic control of biochemical reactions," pp. 192–193.
10. Sewall Wright quoted by Francis Bello. Great American Scientists: The Biologists. Fortune Magazine, 1960, p. 229.
11. F.B. Hanson memo February 14, 1945. RFA RG 1.1, box 10, folder 143.
12. R. Goldschmidt. *Physiological genetics*. McGraw-Hill Book, New York, 1938, pp. 309–312.
13. Ibid; R. Goldschmidt. 1938. The theory of the gene. *Sci. Monthly* 46: 268–273.
14. GWB, "Biochemical genetics," p. 18.
15. S. Wright. 1941. The physiology of the gene. *Physiol. Rev.* 21: 489–490.
16. J. Schultz. 1935. Aspects of the relation between genes and development in *Drosophila*. *Am. Nat.* 69: 30–54.
17. GWB, "Biochemical genetics."
18. GWB. "Genetics and metabolism"; GWB. 1946. The gene. The R.A.F. Penrose, Jr. Memorial Lecture April 18. *Proc. Am. Philos. Soc.* 90: 422–431; GWB. 1946. The gene and biochemistry. In *Currents in biochemical research*, pp. 1–12. Interscience, New York; GWB. 1947. Genes and the chemistry of the organism. In *Science in progress*, Fifth Series (ed. G.A. Baitsell). Yale University Press, pp. 166–196.
19. P.A. Levine. 1921. On the structure of thymus nucleic acid and on its possible bearing on the structure of plant nucleic acid. *J. Biol. Chem.* 48: 119–125.
20. GWB, "Biochemical genetics," p. 71.
21. E.L. Ellis and M. Delbrück. 1939. The growth of bacteriophage. *J. Gen. Physiol.* 22: 365–384.
22. GWB. 1948. Some recent developments in chemical genetics. Fortschritte der Chemie Organischer Naturstoffe, V. Wien, Springer-Verlag, pp. 300–330.
23. Ibid; GWB. Cal Tech Engineering and Science Monthly, May 1947, pp. 2 and 27.

24. GWB, "Biochemical genetics," p. 82.
25. E. Chargaff. 1950. Chemical specificity of nucleic acids and mechanism of their enzymatic degradation. *Experentia* 6: 201–209; E. Chargaff. 1955. Isolation and composition of the deoxypentose nucleic acids and of the corresponding nucleoproteins. In *The nucleic acids,* Vol. 1. (ed. E. Chargaff and J.N. Davidson). Academic Press, New York, pp. 307–407.
26. A. Hollaender and C.W. Emmons. 1941. Wavelength dependence of mutation production in the ultraviolet with special emphasis on fungi. *Cold Spring Harbor Symp. Quant. Biol.* 9: 179–186; L.J. Stadler and F.M. Huber. 1942. Genetic effects of ultraviolet radiation in maize. IV. Comparisons of monochromatic radiation. *Genetics* 27: 84.
27. GWB, "Biochemical genetics," p. 71.
28. O.T. Avery, C.M. MacLeod, and M. McCarty. 1944. Studies on the chemical nature of the substance inducing transformation of pneumococcal types. I. Induction of transformation by a desoxyribonucleic acid fraction isolated from pneumococcus type III. *J. Exp. Med.* 79: 137–158.
29. A. Boivin and discussion by A. Mirsky. 1947. *Cold Spring Harbor Symp. Quant. Biol.* 12: 7–17.
30. GWB, "Biochemical genetics," p. 76.
31. GWB, "Genetic control of biochemical reactions," p. 190; GWB, "Biochemical genetics," pp. 15–96.
32. GWB, "Biochemical genetics," p. 18.
33. E.L. Tatum. 1944. X-ray induced mutant strains of *Escherichia coli. Proc. Natl. Acad. Sci.* 31: 215–219.
34. S.E. Luria and M. Delbrück. 1943. Mutations of bacteria from virus sensitivity to virus resistance. *Genetics* 28: 491–511.
35. G. Knaysi. 1945. Elements of bacterial cytology. *J. Hered.* 36: 23–24.
36. GWB. 1945. Do bacteria have genes? *J. Hered.* 36: 23–24.
37. Tatum, "X-ray induced mutant strains."
38. J. Lederberg. 1947. Gene recombination and linked gene segregations in *Escherichia coli. Genetics* 32: 505–525; GWB. 1948. Some recent developments in chemical genetics. Fortschritte der Chemie Organischer Naturstoffe V, Springer-Verlag, Wien, pp. 300–330.
39. D. Bonner. 1946. Biochemical mutations in *Neurospora. Cold Spring Harbor Symp. Quant. Biol.* 11: 14–24; Max Delbrück in the discussion following Bonner's paper, pp. 22–23.
40. Many enzymes consist of more than one kind of polypeptide chain and, therefore, require more than one gene for their formation; in such cases, nonallelic mutations could affect either of the protein products causing the same deficiency.
41. J. Lederberg. 1956. Comments on gene-enzyme relationship. In *Enzymes: Units of biological structure and function* (ed. O.H. Gabler). Academic Press, New York.
42. GWB, "Biochemical genetics: Some recollections."
43. GWB. 1950. Chemical genetics. In *Genetics in the twentieth century; Essays on the progress of genetics in its first fifty years* (ed. L.C. Dunn). Macmillan, New York, pp. 221–239.
44. N. Horowitz and U. Leupold. 1951. Some recent studies bearing on the one gene–one enzyme hypothesis. *Cold Spring Harbor Symp. Quant. Biol.* 16: 65–74.
45. Such mutants are said to be thermosensitive.

46. Horowitz and Leupold, "One gene–one enzyme hypothesis."
47. Max Delbrück. 1966. A physicist looks at biology. In *Phage and the origins of molecular biology* (ed. J. Cairns, G.S. Stent, and J.D. Watson). Cold Spring Harbor Laboratory of Quantitative Biology, Cold Spring Harbor, New York, pp. 9–22.
48. H.K. Mitchell and J. Lien. 1946. A *Neurospora* mutant deficient in the enzymatic synthesis of tryptophan. *J. Biol. Chem.* 175: 481–482.
49. L. Pauling, H.A. Itano, S.J. Singer, and I.C. Wells. 1949. Sickle cell anemia: A molecular disease. *Science* 110: 543–548.
50. V.M. Ingram. 1957. Gene mutation in human hemoglobin: The chemical difference between normal and sickle cell hemoglobin. *Nature* 180: 326–328.
51. C. Yanofsky. 1956. Gene interactions in enzyme synthesis. In *Enzymes: Units of biological structure and function* (ed. O.H. Gabler). Academic Press, New York, pp. 147–160.
52. W.K. Maas and B. Davis. 1952. Production of an altered pantothenate-synthesizing enzyme by a temperature-sensitive mutant of *Escherichia coli*. *Proc. Natl. Acad. Sci.* 38: 785–797; N.H. Horowitz and M. Fling. 1953. Genetic determination of tyrosinase thermostability in *Neurospora*. *Genetics* 38: 360–374.
53. C. Yanofsky, B.C. Carlton, J.R. Guest, D.R. Helinski, and U. Henning. 1964. On the colinearity of gene structure and protein structure. *Proc. Natl. Acad. Sci.* 51: 266–272; A.S. Sarabhai, A.O. Stretton, S. Brenner, and A. Bolle. 1964. Colinearity of the gene with the polypeptide chain. *Nature* 201: 13–17.
54. F.H.C. Crick. 1966. The genetic code—Yesterday, today, and tomorrow. *Cold Spring Harbor Symp. Quant. Biol.* 31: 1–9.
55. Cited in letter from Horowitz to GWB, Jan. 21, 1963. CIT, Horowitz collection.
56. GWB to Horowitz Jan. 23, 1963. CIT, Horowitz collection.
57. GWB. 1941. Genetic control of the production and utilization of hormones. Proceedings of the Seventh International Genetical Congress in Edinburgh. *J. Genet.* (suppl.), p. 58.
58. GWB, "Genetic control of biochemical reactions," pp. 192–193; GWB, "Biochemical genetics," pp. 15–96.
59. R. Goldschmidt. *Physiological genetics*. McGraw-Hill Book, New York, 1938, pp. 309–312; S. Wright. 1941. The physiology of the gene. *Physiol. Rev.* 21: 487–527.
60. R.C. Olby. 1974. The physiology of the gene. In *The path to the double helix*, Chap. 8. University of Washington Press, Seattle, pp. 123–152; H.F. Judson. *The eighth day of creation*. Simon and Schuster, New York, 1979, pp. 1–686.
61. E.B. Wilson. *The cell in development and heredity*. New York, Macmillan. Four editions from 1896–1928.
62. Wright, "Physiology of the gene"; Goldschmidt, "Physiological genetics," pp. 309–312.
63. Wright, "Physiology of the gene."
64. GWB, "Chemical genetics," pp. 221–239; GWB, "Genes and chemical reactions in *Neurospora*." Le Prix Nobel, Stockholm, 1958.
65. A.G. Bearn. *Archibald Garrod and the individuality of man*. Clarendon Press, Oxford, 1993.
66. A.E. Garrod. 1902. The incidence of alcaptonuria: A study in chemical individuality. *Lancet* ii: 653–656.
67. W. Bateson. *Mendel's principles of heredity*. Cambridge University Press, London, 1909.
68. A.E. Garrod. Comments at the Proceedings of the Royal Society of Medicine 8 (I), 1908.

69. A.E. Garrod and W.H. Hurtley. 1906. Concerning cystinuria. *J. Physiol.* 34: 217–223.
70. A.E. Garrod. *The inborn factors in disease.* Clarendon Press, Oxford, 1931. [Reprinted as *Garrod's inborn factors in disease. Oxford monographs on medical genetics 16*, (ed. C.R. Scriver and B. Childs), Oxford University Press, 1989.]
71. O. Gross. 1914. Uber den Einfluss des Blutserums des normalen und des Alcaptonurikens auf Homogentisinsaure. *Biochemische Zeitschrift* 61: 165–170.
72. B.N. LaDu, V.G. Zannoni, L. Laster, and J.E. Seegmiller. 1958. The nature of the defect in tyrosine metabolism in alcaptonuria. *J. Biol. Chem.* 230: 251–260.
73. M.R. Pollack, Y.-H. Chou, J.J. Cerda, B. Steinmann, B.N. LaDu, J.G. Seidman, and C.E. Seidman. 1993. Homozygosity mapping of the gene for alcaptonuria to chromosome 3q2. *Nat. Genet.* 5: 201–204.
74. GWB, "Genes and chemical reactions."
75. J.B.S. Haldane, *New paths in genetics.* Harper Brothers, New York and London, 1942.
76. Bearn, *Archibald Garrod and the individuality of man*, p. 162.
77. J. Lederberg, interview, September 24, 1996.
78. Olby, *The path to the double helix*, pp. 132–133.
79. GWB to Horowitz Nov. 23, 1974. CIT, Horowitz collection.
80. H.W. Davenport to Norman Horowitz, Dec. 5, 1974. CIT, Horowitz collection.
81. Bateson, *Mendel's principles of heredity.*
82. Goldschmidt, *Physiological genetics.*
83. A.H. Sturtevant and GWB. *An introduction to genetics.* W.B. Saunders, New York, 1939.
84. S. Benzer. 1961. Genetic fine structure. *Harvey Lect.* 56: 1–21.

CHAPTER 12

1. Harriet Zuckerman, interview with GWB, December 12, 1963. COL.
2. GWB. 1945. Biochemical genetics. *Chem. Rev.* 37: 15–96.
3. GWB to H.M. Miller, August 25, 1945 and September 6, 1945. Box 11/1, Beadle collection, CIT.
4. F.B. Hanson. Memorandum, December 3, 1945. Folder 143, box 10, series 205D, RG1.1 RFA; A.H. Sturtevant to GWB, June 28, 1945. Box 7/19, Beadle collection, CIT; L. Kay. *The molecular vision of life.* Oxford University Press, New York, 1993, p. 212.
5. Sturtevant to GWB, May 13, 1945. Box 7/19, Beadle collection, CIT.
6. James Bonner, Sterling Emerson, Norman Horowitz, and Donald Poulson. Joint interview by J. Goodstein, H. Lyle, and M. Terrall, November 6, 1978, Pasadena, California, pp. 36–37, Oral History, CIT.
7. Ibid.
8. Warren Weaver's diary, January 28 and February 1, 1937, RFA.
9. Sturtevant to GWB, May 13, 1945.
10. Sterling Emerson to GWB, May 21, 1945. Box 3/15 Beadle collection, CIT.
11. F.B. Hanson's diary, 1942, p. 137. Folder 73, box 6, series 205D, RG1.1, RFA.
12. Sterling Emerson to GWB, June 20, 1945. Box 3/15, Beadle collection, CIT.
13. Sturtevant to Robert Millikan, June 21, 1945. Box 18/22, Millikan collection; Sturtevant to GWB, June 28, 1945. Box 7/19, Beadle collection, CIT.
14. GWB to Millikan, July 16, 1945. Box 18/22 Millikan collection, CIT.
15. GWB to Hanson, July 20, 1945. Folder 143, box 10, series 205D, RG1.1, RFA.

16. Sterling Emerson to GWB, July 25, 1945. Box 3/15, Beadle collection, CIT.
17. Sturtevant to GWB, August 1, 1945. Box 7/19, Beadle collection, CIT.
18. Kay, *Molecular vision,* pp. 210–214; T. Hager. *Force of nature: The life of Linus Pauling.* Simon and Schuster, New York, 1995, p. 277.
19. Interview by Barbara Land. 1961. *The reminiscence of Warren Weaver.* Oral History, Volume 2, pp. 337–340, 1961. Category IA, no. 434, COL.
20. Ibid.
21. Hager, *Force of nature,* p. 276.
22. Sturtevant to GWB, October 13, 1945. Box 7/19, Beadle collection, CIT.
23. Hager, *Force of nature,* p. 277.
24. C.V. Taylor to Warren Weaver, October 22, 1945. Folder 143, box 10, series 205D, RG1.1, RFA.
25. J.R. Page to GWB. November 5, 1945. Box 1/1, chemistry division collection, CIT.
26. GWB to Warren Weaver, October 26, 1945. Folder 143, box 10, series 205D, RG1.1, RFA; Rockefeller Foundation annual report, 1946, RFA.
27. Hanson, Memorandum, June 23, 1945. Folder 143, box 10, series 205D, RG1.1, RFA.
28. Warren Weaver to GWB, November 2, 1945. Folder 143, box 10, series 205D, RG1.1, RFA.
29. Bonner, Emerson, Horowitz, and Poulson, joint interview, 1981, p. 36.
30. Joint interview, 1981, p. 37, CIT.
31. R. Millikan to GWB, November 6, 1945. Box 18/22, Millikan collection, CIT.
32. J.R. Page to GWB, November 5, 1945; Linus Pauling to GWB, October 31, 1945. Box 1/1, chemistry division collection, CIT.
33. Sturtevant to GWB. November 1, 1945. Box 7/19, Beadle collection, CIT.
34. V. Bush, 1945. *Science the endless frontier.* (Reprinted by the National Science Foundation, 1990, NSF 90-8.)
35. Pauling to GWB, November 13, 1945. Box 1/1, chemistry division collection, CIT.
36. Barabara McClintock to Boris Ephrussi, December 29, 1945. Ephrussi correspondence collection, Gif-Sur-Yvette, France.
37. GWB to Pauling, October 28, 1945. Box 1/1, chemistry division collection, CIT; David Beadle, personal communication, suggests that his father may have thought that the change would improve family relations.
38. GWB to Pauling, October 28, 1945.
39. Pauling to GWB, November 6, 1945. Box 1.1, chemistry division collection, CIT.
40. GWB to Pauling, November 10, 1945. Box 1.1, chemistry division collection, CIT.
41. Sturtevant to GWB, November 1, 1945. Box 7/19, Beadle collection, CIT.
42. GWB to Pauling, November 10, 1945.
43. Pauling to GWB, November 19, 1945. Box 1.1, chemistry division collection, CIT.
44. GWB to Pauling, December 12, 1945. Box 1.1, chemistry division collection, CIT.
45. GWB to Pauling from Stanford, no day/month, 1946. Box 1.2, chemistry division collection, CIT.
46. GWB to Pauling, March 7, 1946. Box 1.2, chemistry division collection, CIT.
47. GWB to Barbara McClintock, September 9, 1946. Box 5/26, Beadle collection, CIT.
48. L.A. DuBridge to GWB, May 23, 1946. Box 26/29, biology division records, CIT.
49. Kay, *Molecular vision,* 1993, p. 236.
50. Mason recruited Weaver as the director of the division of natural sciences soon after he arrived at the Foundation in 1929. Earlier, he was responsible for Weaver's appointment to the Cal Tech faculty, an appointment that was maintained, in absen-

tia, when Weaver returned to Wisconsin and later moved to the Rockefeller Foundation.

51. Robert F. Bacher, interview by Mary Terrall, June 9, 1981, Pasadena, California, Oral History, CIT. p. 138.
52. GWB to B. McClintock, August 13, 1946. Box 5/26. Beadle collection, CIT; GWB, personnel security questionnaire, April 22, 1947, FBI file 116-HQ-6619, Washington, DC.
53. Edward B. Lewis, personal communication; Wayne Keim, personal communication; Francis Haskins, personal communication.
54. Lewis, personal communication; GWB to Dael Wolfle, May 11, 1954, AAAS archives.
55. Lewis, personal communication.
56. A.W. Galston and T.W. Sharkey. 1998. Frits W. Went. *Biographical Memoirs* 74: 3–17. National Academy of Sciences.
57. Bonner, Emerson, Horowitz, and Poulson, Joint Interview, 1981, p. 36.
58. Sturtevant to GWB, Nov. 1, 1945. Box 7/19 Beadle collection, CIT.
59. Pam and Ed Lewis, personal communication.
60. Box 65/36, biology division collection, CIT.
61. GWB to Pauling, February 27, 1946. Box 1/2, chemistry division collection, CIT.
62. Ibid.
63. William Hayes. 1982. Max Ludwig Henning Delbrück, *Biographical Memoirs* 62: 67–117. National Academy of Sciences.
64. Ibid; S.W. Golomb. 1982. Max Delbrück, An appreciation. *Am. Schol.* 51: 351–367.
65. R. Olby. *The path to the double helix*. University of Washington Press, Seattle, 1974, p. 231; Kay, *Molecular vision,* p. 133.
66. Max Delbrück, interviews by Carolyn Harding, July 14 and September 11, 1978, Pasadena, California, p. 39. Oral History, CIT.
67. Max Delbrück. *How it was.* 1980. Engineering and Science. March–April 21–26 and May–June 21–27, 1980, CIT; Hayes, *Delbrück,* 1982.
68. Delbrück, interview, 1978.
69. Delbrück, *How it was*; Hayes, *Delbrück*, 1982.
70. Golomb, "Max Delbrück."
71. N.W. Timoféeff-Ressovsky, K.G. Zimmer, and M. Delbrück. 1935. Ueber die Natur der Genmutation und der Genstruktur. *Nachr. Ges. Wiss. Göttingen, math. phys.* Kl., Fachgr. 6, 1: 189–245.
72. Delbrück, interview, 1978, p. 73.
73. Olby, *Path to the double helix,* p. 236.
74. Hayes, "Delbrück."
75. Ibid.
76. E.L. Ellis. 1966. Bacteriophage: One-step growth. In *Phage and the origins of molecular biology* (ed. J. Cairns, G.S. Stent, and J.D. Watson), pp. 53–62. Cold Spring Harbor Laboratory of Quantitative Biology, Cold Spring Harbor, New York.
77. Olby, *Path to the double helix*, 1974, p. 238.
78. E.L. Ellis and M. Delbrück. 1939. The growth of bacteriophage. *J. Gen. Physiol.* 22: 365–384.
79. E.L. Ellis, "Bacteriophage: One-step growth"; H.F. Judson. *The eighth day of creation,* Simon and Schuster, New York, 1979 and expanded edition, Cold Spring Harbor Laboratory Press, 1996.
80. Delbrück, interview, 1978, p. 73.

81. Manny Delbrück, interview by Carolyn Harding, August 24, 1978. Oral History, CIT.
82. GWB to McClintock, August 13, 1946.
83. GWB to Delbrück, telegram, December 10, 1946. Box 3/4, Delbrück collection, CIT.
84. GWB to Dean E.C. Watson, December 4, 1946. Box 1/2, chemistry division collection, CIT.
85. M. Delbrück, 1949. A physicist looks at biology. Reprinted in *Phage and the origins of molecular biology,* pp. 9–22.
86. G.S. Stent. 1966. In *Phage and the origins of molecular biology, Introduction: Waiting for the paradox.* (ed. J. Cairns, G.S. Stent, and J.D. Watson), pp. 3–8. Cold Spring Harbor Laboratory of Quantitative Biology, Cold Spring Harbor, New York.
87. Olby, *Path to the double helix,* p. 238.
88. Kay, *Molecular vision,* 1993, p. 255.
89. Ibid, p. 256.
90. J. Goodstein. *Millikan's school.* W.W. Norton Co., New York, 1991.
91. Pauling to DuBridge, June 28, 1946; Pauling to H.M. Weaver, June 27, 1946. Box 50/5, biology division collection, CIT.
92. W. Weaver to Pauling, March 12, 1947. Box 62/23, biology division collection, CIT.
93. N. Thompson to DuBridge, May 16, 1947. Box 62/63, biology division collection, CIT.
94. F.M. Rhind to DuBridge, April 8, 1948. Box 62/62, biology division collection, CIT.
95. Lewis, personal communication.
96. McClintock to GWB, October 6, 1946. Box 5/26, Beadle collection, CIT.
97. GWB to Dean Watson, November 26, 1946. Box 1/18, biology division collection, CIT; GWB to H.M. Lende and H.G. Betka at American Optical Co., fall, 1946. Box 10/4, biology division collection, CIT.
98. Ray Owen, talk at GWB Memorial at the Athenaeum. Tape recording, courtesy of Professor Norman Horowitz, original in CIT archive.
99. Academic catalogs, CIT.
100. Herschel K. Mitchell, interview by Shirley K. Cohen, Pasadena, California, December 3 and 22, 1997, CIT.
101. N.H. Horowitz. 1990. George W. Beadle, October 22, 1903–June 9, 1989. *Biographical Memoirs* 59: 27–51. National Academy of Sciences.
102. David and Judy Hogness, personal communication.
103. GWB. The genes of men and molds. *Scientific American,* September, 1948. [Reprinted in *Facets of genetics* (ed. A.M. Srb, R.D. Owen, and R.S. Edgar). W.H. Freeman, San Francisco, 1970.]
104. GWB. 1951. Chemical genetics. In *Genetics in the 20th century: Essays on the progress of genetics during its first 50 years* (ed. L.C. Dunn). Macmillan, New York.
105. D. Bonner and GWB. 1946. Mutant strains of Neurospora requiring nicotinamide or related compounds for growth. *Arch. Biochem.* 11: 319–328.
106. D. Beadle, personal communication.
107. C.C. Lindegren to GWB, November 8, 1950. Box 5/10, Beadle collection, CIT.

CHAPTER 13

1. T.A. Appel, *Shaping biology: The National Science Foundation and American biological research, 1945–1975.* Johns Hopkins Press, Baltimore, 2000.

2. E.G. Anderson to GWB, Research Report, November 24, 1946. Box 1/2, Beadle collection, CIT.
3. Now an obsolete unit, a roentgen was defined as the quantity of ionizing radiation that results in the formation of 2 billion ion pairs (one electrostatic unit of charge) per cubic centimeter of dry air at zero degrees centigrade and one atmosphere of pressure. As described in Chapter 14, in the late 1940s Americans received, on the average, 3 roentgens from medical and dental procedures by the time they were 30 years old.
4. FBI records, Subject George Wells Beadle, 116-HQ-6619, LA 116-468. Interview with L. Pauling on May 13, 1947. Obtained through Freedom of Information and Privacy Acts.
5. FBI records, Subject George Wells Beadle, 116-HQ-6619, SF 116-415. Interview, identity of person interviewed censored. Obtained through Freedom of Information and Privacy Acts.
6. FBI records, Subject George Wells Beadle, 77-HQ-82076-12, file AEC, no. 11348, interview with GWB on Feb. 14, 1958. Obtained through Freedom of Information and Privacy Acts.
7. Documents, box 22/5, Beadle collection, CIT; E.B. Lewis. 1954. The theory and application of a new method of detecting chromosomal rearrangements in *Drosophila melanogaster. Am. Nat.* 88: 225–239.
8. Redmond Barnett, July 22, 1997.
9. David Beadle, interview, June 27, 1997.
10. GWB to W.A. Higinbotham, Sept 20, 1948. Box 20/5, Beadle collection, CIT.
11. Ibid.
12. J. Wang. *American science in an age of anxiety*. University of North Carolina Press, Chapel Hill, 1999, p. 131.
13. Wang, "American science," p. 187.
14. GWB to C.T. Forster, July 7, 1949, box 8/13; L. Stadler to GWB, October 25, 1949. Box 7, Beadle collection, CIT.
15. Wang, "American science," p. 51.
16. T. Hager. *Force of nature: The life of Linus Pauling*. Simon and Schuster, New York, 1995, pp. 351 et seq.
17. *American Men of Science*. Tenth edition, 1960.
18. D. Beadle, interview, June 27, 1997.
19. H. Borsook and J.W. Dubnoff. 1940. Creatine formation in liver and kidney. *J. Biol. Chem.* 134: 635–639; H. Borsook and J.W. Dubnoff. 1941. The formation of glycocyamine in animal tissues. *J. Biol. Chem.* 138: 389–403; H. Borsook and J.W. Dubnoff. 1947. On the role of oxidation in the methylation of guanidoacetic acid. *J. Biol. Chem.* 171: 363–375.
20. GWB to L.A. DuBridge, Memorandum, September 2, 1950. Box 26/28, biology division collection, CIT.
21. GWB to J. Dubnoff, Memorandum, May 29, 1950. Box 26/28, biology division collection, CIT.
22. GWB to DuBridge, September 2, 1950.
23. DuBridge to E.C. Watson, September 28, 1950, and GWB to E.C. Watson, September 27, 1950. Box 26/28, biology division collection, CIT.
24. DuBridge, Letter to Alumni, January 1952, file 2, academic freedom and tenure collection, CIT.
25. E.C. Watson to G. Fling, June 9, 1951. Box 28/28, biology division collection, CIT.

26. Elga Wasserman, personal communication.
27. Ibid. and S.R. Sapirie to Mr. and Mrs. David Bonner, July 10, 1951. Box 1/21, Beadle collection, CIT.
28. A. Hollaender to GWB, February 27, 1951. Box 23/6, Beadle collection, CIT.
29. GWB to Shields Warren, March 2, 1951, box 22/5 and GWB to Hollaender, March 2, 1951, box 23/6, Beadle collection, CIT.
30. D. Bonner to GWB, July 5, 1951. Box 1/21, Beadle collection, CIT.
31. Warren to GWB, July 13, 1951. Box 1/21, Beadle collection, CIT.
32. Hollaender to GWB, July 3, 1951. Box 1/21, Beadle collection, CIT.
33. GWB to Warren, Aug. 5, 1951. Box 1/21, Beadle collection, CIT.
34. GWB. 1954. Security and science. In *Engineering and science,* California Institute of Technology, November 15–18.
35. Warren Weaver to GWB, April 1, 1952., Box 1/22, Beadle collection, CIT.
36. Hollaender to GWB, November 14, 1951. Box 1/21, Beadle collection, CIT.
37. GWB to Warren, May 13, 1952. Box 1/22, Beadle collection, CIT.
38. Warren to GWB, May 26, 1952. Box 1/22, Beadle collection, CIT.
39. H.H. Plough to GWB, April 2, 1952, handwritten on AEC stationery. Box 22/5, Beadle collection, CIT.
40. GWB to Dean E.C. Watson, May 31, 1947. Box 1/3, chemistry division collection, CIT.
41. Appel, *Shaping biology,* p. 92.
42. GWB to Dean Lacey et al., Memorandum, May 21, 1952. Box 1/4, chemistry division collection, CIT.
43. J. Bonner, A. Sturtevant, and A. von Harriveld, Committee on the Employment of Graduate Students, April 30, 1952, folder 1/4, chemistry division collection, CIT.
44. Appel, *Shaping biology,* pp. 96–98.
45. R. Dulbecco. 1966. The plaque technique and the development of quantitative animal virology. In *Phage and the origins of molecular biology* (ed. J. Cairns, G.S. Stent, and J.D. Watson), pp. 287–291. Cold Spring Harbor Laboratory of Quantitative Biology, Cold Spring Harbor, New York; R. Dulbecco, personal comunication.
46. Max Delbrück to GWB, August 11, 1950. Box 3/4, Beadle collection, CIT.
47. Delbrück to Renato Dulbecco, November 11, 1948. Box 6/22, Delbrück collection, CIT.
48. R. Dulbecco. 1952. Production of plaques in monolayer tissue cultures by single particles of an animal virus. *Proc. Natl. Acad. Sci.* 38: 747–752.
49. D. Beadle, interview, June 27, 1997.
50. FBI records, Subject George Wells Beadle, 116-HQ-6619, LA 116,468, August 31, 1956. Obtained through Freedom of Information and Privacy Acts.
51. Superior Court of the State of California in the County of Los Angeles, book 2589, p. 191, document of August 10, 1953.
52. Edward Lewis, personal communication.
53. R. Metzenberg, personal communication; FBI report, Jan 14, 1952.
54. Delbrück to GWB, Aug. 14, 1951. Box 3/4, Beadle collection, CIT.
55. GWB to Delbrück. Aug. 22, 1951. Box 2/10, Delbrück collection, CIT.
56. GWB. 1952–1953. Up Doonerak...An Arctic Adventure. The Wilderness Society, Washington, D.C., 1952–1953, pp. 1–7. Box 36/12, Beadle collection, CIT.
57. Redmond Barnett, interview, July 22, 1997.
58. D. Beadle, interview, June 27, 1997.

59. Perhaps both stories are true. George and Muriel Beadle's book, *The language of life* (Doubleday, 1966), is dedicated "To Virginia and Vernon and in Memory of Jess."
60. Muriel Barnett Beadle. Speech at the 50th reunion of her Pomona College class, April 19, 1986.
61. Barbara Wright Kalckar to GWB, January 27, 1953. Box 4/27, Beadle collection, CIT. Kalckar had been a student of Beadle's at Stanford.
62. Superior Court of the State of California in the County of Los Angeles, book 2589, p. 191, document of August 10, 1953.
63. M. Beadle. *These ruins are inhabited.* Doubleday, New York, 1961.
64. Barnett, interview.
65. D. Beadle, interview, June 27, 1997.
66. Box 65/32, biology division collection, CIT.
67. J.E. Bogen. Roger Wolcott Sperry. 1999. *Biographical memoirs.* 71: 3–19. National Academy of Sciences.
68. J.D. Watson. *The double helix.* Atheneum, New York, 1968, Chapters 2 and 3.
69. A.D. Hershey and M. Chase. 1952. Independent functions of viral protein and nucleic acid in growth of bacteriophage. *J. Gen. Physiol.* 36: 39–56.
70. R. Olby. *The path to the double helix.* University of Washington Press, Seattle, 1974, Chapter 17.
71. Watson, *Double helix*; Olby, *Path to the double helix*, Chapter 21; H.F. Judson. *The eighth day of creation: Makers of the revolution in biology,* expanded edition. Cold Spring Harbor Laboratory Press, Cold Spring Harbor, New York, 1996, Chapters 1–3; B. Maddox. *Rosalind Franklin: The dark lady of DNA.* Harper Collins Publishers, New York, 2002; V.K. McElheny. *Watson and DNA: Making a scientific revolution.* Perseus Publishing, Cambridge, Massachusetts, 2003.
72. L. Pauling and R.B. Corey. 1953. A proposed structure for the nucleic acids. *Proc. Natl. Acad. Sci.* 39: 84–97.
73. E. Chargaff. 1950. Chemical specificity of nucleic acids and mechanism of their enzymatic degradation. *Experientia* 6: 201–209.
74. J.D. Watson and F.H.C. Crick. 1953. Molecular structure of nucleic acids. A structure for deoxyribose nucleic acid. *Nature* 171: 737–738.
75. GWB to DuBridge, December 8, 1953. Box 3/4, Beadle collection, CIT.
76. GWB. 1974. Biochemical genetics: Recollections. *Annu. Rev. Biochem.* 43: 1–13.
77. Delbrück to GWB, June 2, 1954. Box 8/2. Beadle collection, CIT.
78. J.D. Watson. 1954. The structure of tobacco mosaic virus. 1. X-ray evidence of a helical arrangement of sub-units around the longitudinal axis. *Biochim. Biophys. Acta* 13: 10; A. Rich and J.D. Watson. 1954. Physical studies on ribonucleic acid. *Nature* 173: 995–996; A. Rich and J.D. Watson. 1954. Some relations between DNA and RNA. *Proc. Natl. Acad. Sci.* 40: 759–764.
79. Edward Lewis, personal communication.
80. GWB to James D. Watson, June 26, 1955. Box 8/2, Beadle collection, CIT.
81. GWB to James D. Watson, June 14, 1954. Box 8.2, Beadle collection, CIT.
82. GWB to Delbrück, June 14, 1954. Box 8.2, Beadle collection, CIT.
83. GWB to J.D. Watson, June 26, 1955.
84. Appel, *Shaping biology,* pp. 82 and 136.
85. R.L. Sinsheimer, interview by Shelley Irwin, May 30 and 31, 1990 and March 26, 1991, Pasadena, California. Oral History, CIT.

86. Hager, *Force of nature*, pp. 440 et seq.
87. GWB, "Security and science."
88. Hager, *Force of nature*, pp. 461 et seq.
89. GWB. 1955. Portrait of a scientist: A tribute to Linus Pauling. *Engineering and Science*, California Institute of Technology, April, pp. 11–14.
90. GWB. 1955. Review of Erwin N. Griswold's book, *The Fifth Amendment today. Science* 121: 549.
91. Dael Wolfle to Senator Thomas C. Hennings, Jr. June 16, 1955. Box 6, L-2-3, government relations file, Borras collection, AAAS.
92. Borsook to James Watt, December 15, 1954. Box 73/2, biology division collection, CIT.
93. National Science Board policy statement, May 21, 1954. Box 10/2, Beadle collection, CIT.
94. Borsook to Watt. December 14, 1954. Box 73/2, biology division collection, CIT.
95. A Statement of the Circumstances connected with the Termination and Reinstatement of Two Public Health Service Grants at the California Institute of Technology. Undated. Box 73/11, biology division collection, CIT.
96. Ibid.
97. H. Borsook, interview by Mary Terrell, April 15, 1978, Pasadena, California. Oral History, CIT.
98. Confidential: A Statement Concerning Certain Aspects of the U.S. Public Health Service Policy. Undated. Boxes 73/2 and 73/11, biology division collection, CIT.
99. GWB to Leonard Scheele and C.N. Van Slyke, December 7, 1954. Box 73/11, biology division collection, CIT.
100. Voting slips dated December 13, 1954. Box 73/2 biology division collection, CIT.
101. L. Bogorad, personal communication.
102. L. Scheele to DuBridge, undated telegram; Scheele to G.W. Beadle, December 23, 1954; box 73/11, biology division, CIT.
103. GWB to Alan T. Waterman, February 25, 1952. Box 53/3, biology division collection, CIT.

CHAPTER 14

1. GWB. Radiation genetics. Supplement to Nuclear Science Abstracts, 1949.
2. A.H. Sturtevant. 1954. The social implications of the genetics of man. *Science* 120: 405-407. A speech given June 22, 1954 in Pullman, Washington. Box 7/19, Beadle collection, CIT.
3. Press release published in the *Bulletin of the Atomic Scientists*, 10 (1954):164.
4. H.J. Muller. 1927. Artificial transmutation of the gene. *Science* 66: 84–87.
5. The quote is from an unsigned, undated newspaper clipping from an unidentifiable newspaper. Box 7/19, Beadle collection, CIT.
6. R. Metzenberg, personal communication; GWB. Long Range Effects of Atomic Radiation. A talk given to the Medical Research Association of California, May 16, 1955, from personal archives of David D. Perkins.
7. AAAS. 1954. Strengthening the basis of national security. *Science* 120: 957–959.
8. Editorial. Scientists and Security. *New York Herald Tribune*, December 12, 1954. Box

10/1, Beadle collection, CIT.
9. Warren Weaver to Earl Ubell, December 13, 1954. Box 10/1, Beadle collection, CIT.
10. E.A. Koester, President, American Association of Petroleum Geologists. Statement, March 28, 1955. Attached to agenda for June 1956 AAAS board meeting, AAAS.
11. National Academy of Sciences. 1956. Loyalty and research. *Science* 123: 660–662.
12. AAAS Board of Directors, Preliminary Agenda for the meeting of June 11 and 12, 1955, AAAS.
13. Montague Cobb to Dael Wolfle, May 31, 1955, AAAS.
14. AAAS Board of Directors meeting minutes, June 11 and 12 1955, AAAS; AAAS Board of Directors. 1955. Advantages of an Atlanta meeting. *Science* 121: 7A.
15. GWB. From Spanish Tile to Modern Stone: Some Recollections. Speech given at the dedication of the Beckman Auditorium at Cal Tech on February 2, 1964, CHG.
16. A New Biology Building, May 9, 1957. Box 1/3, DuBridge collection, CIT.
17. GWB. 1957. Uniqueness of man. *Science* 125: 9–11.
18. National Academy of Sciences news release, April 8 1955, AAAS; Detlev Bronk, November 20, 1955, remarks at the first meeting of the Genetics Panel of the NAS/NRC study on biological effects of atomic radiation. Committee on Biological Effects of Ionizing Radiation (BEIR), Genetics Committee, general. NAS
19. GWB to D. Bronk, December 2, 1959, NRC BEIR Genetics Committee 1958–1959, NAS.
20. J.R. St. Clair (archivist). An undated note quoting Bronk in NRC Archives, BEIR Appointments file, 1955.
21. H.J. Muller. 1955. Genetic damage produced by radiation (published Kimber Award lecture) *Science* 121: 837–840, erratum in Volume 122, p. 759.
22. *Washington Post*, September 17, 1955.
23. E.A. Carlson. *Genes, radiation, and society: The life and work of H.J. Muller*. Cornell University Press, Ithaca, New York, 1981, pp. 357–364.
24. GWB. 1955. H.J. Muller and the Geneva Conference. *Science* 122: 813.
25. News story. 1955. *Science* 122: 822.
26. Carlson, *Genes, radiation, and society*, pp. 364–366.
27. H.J. Muller to Bronk, October 1955, BEIR Genetics Committee, Appointments, 1955; minutes of meeting, November 20–22, 1955, BEIR Genetics Committee, November, 1955. NAS.
28. Genetics Committee (BEIR) meeting transcript, February 5 and 6, 1956. NRC BEIR genetics file, NAS.
29. William J. Schull. *Report of activities in Japan: July 9, 1956–September 26, 1956*. Page 7. Atomic Bomb Casualty Commission, NAS; Atomic Bomb Casualty Commission Records Group, Genetics: 1956–1957, NAS.
30. Meeting transcript of February 5 and 6, 1956. BEIR Genetics Committee, NAS.
31. Ibid., p. 77.
32. Ibid.
33. J.F. Crow. 1995. Quarreling geneticists and a diplomat. *Genetics* 140: 421–426.
34. Committee on Genetic Effects of Ionizing Radiation. 1956. The biological effects of ionizing radiation; A report to the public from the National Academy of Sciences and National Research Council. *Science* 123: 1157–1164.
35. G. DuShane. 1956. Radiation and public knowledge. *Science* 124: 101.
36. Nuclear weapons tests. 1956. *Science* 124: 925–926.

37. W. Weaver to Bronk, June 13, 1956, ADM:ORG:NAS BEIR, genetics effects file, 1956, NRC.
38. Weaver to Bronk, June 26, 1956, NRC BEIR genetics general file, 1956, NRC.
39. Weaver to GWB, July 24, 1956, NRC BEIR genetics general file, 1956, NRC.
40. E.B. Lewis, personal communication.
41. GWB to Dael Wolfle, June 25, 1956. Box 12/5, Beadle collection, CIT.
42. Lewis, personal communication.
43. GWB to R. Beadle, February 14, 1956 and March 22, 1956, RB; GWB to Chauncey E. Beadle, March 22, 1956.
44. C.E. Beadle to R. Beadle and GWB, May 10 1956. RB.
45. GWB to R. Beadle, October 12, 1956. RB.
46. Laughlin–Pullman Report, NRC BEIR general genetics file, 1955, NRC.
47. *New York Times*, July 21, 1957.
48. GWB to Bronk, September 11, 1957, BEIR genetics committee ADM: ORG, NAS.
49. Minutes of December 1 and 2, 1956 meeting of Genetics Committee. BEIR, Genetics Committee, NAS.
50. Muller, "Artificial transmutation."
51. W.F. Libby. 1956. Radioactive fallout and radioactive strontium. *Science* 123: 657–660; Edward Lewis, personal communication.
52. E.B. Lewis. 1957. Leukemia and ionizing radiation. *Science* 125: 965–972.
53. Lewis, personal communication.
54. E.B. Lewis. 1959. Thyroid radiation doses from fallout. *Proc. Natl. Acad. Sci.* 45: 894–897.
55. Lewis, personal communication.
56. R. Metzenberg, personal communication.
57. T. Hager. *Force of nature: The life of Linus Pauling*. Simon and Schuster, New York, 1995, pp. 465 and 670; E.B. Lewis, personal communication.
58. BEIR Genetics Committee General, 1956–1959. NAS.
59. M. Demerec to GWB, August 1, 1957, NRC BEIR, Genetics Committee, general, Statement on Needed Research, 1956–1959, NRC.
60. GWB to Bronk, September 11, 1957, BEIR, Genetics Committee, NAS.
61. GWB to Adrian Srb, June 13, 1957; A. Srb to GWB, June 27, 1957, courtesy of Jo Srb.
62. Genetics Committee to D. Bronk. December 4, 1959, BEIR Genetics Committee; Statement on Needed Research Drafts, December 1959, BEIR Genetics Committee, NAS.
63. Sewall Wright. 1960. On the appraisal of genetic effects of radiation in man. In *The biological effects of radiation, Summary Reports, 1960*, National Academy of Sciences–National Research Council, Washington, D.C., pp. 18–24.
64. H.J. Muller to C.I. Campbell, May 27, 1959, BEIR Genetics Committee: Appendix. NAS.
65. Genetic Summary Reports and Drafts. BEIR, Chairmen of BEIR Committees folder, 1959, NAS.
66. Committee to D. Bronk. December 4, 1959, ADM:ORG:NAS, Statement on Needed Research: Drafts, BEIR, Chairmen of BEIR Committees folder, Genetic Summary Reports and Drafts, 1959, NRC.
67. H.L. Andrews to Bronk. December 21, 1959. BEIR Genetics Committee folder on Statement on Needed Research, Appendix: Comment, 1956–1959, NAS.

68. GWB to Andrews. December 26, 1959, BEIR Genetics Committee folder on Statement on Needed Research, Appendix: Comment, 1956–1959, NAS.
69. GWB to Bronk. Sept. 28 1959. BEIR Genetics Committee, general ADM:ORG 1956–1959, NAS.
70. GWB to Bronk. December 2, 1959. BEIR Genetics Committee, general folder, 1958–1959, NAS.
71. *The biological effects of radiation. Summary Reports.* National Academy of Sciences-National Research Council. Washington, D.C., 1960.
72. Z. Jaworowski. 1999. Radiation risk and ethics. *Phys. Today* 52: 24–29; Z. Jaworowski. 2000. Radiation risk and ethics: Health hazards, prevention costs, and radiophobia. *Phys. Today* 53: 11–15 and 89–90.
73. Boxes 10.1 and 10.2, biology division collection, CIT.
74. R.L. Sinsheimer. 1957. First steps toward a genetic chemistry. *Science* 125: 1123–1128.
75. R.L. Sinsheimer. 1959. A single-stranded DNA from bacteriophage ϕX174. *J. Mol. Biol.* 1: 43–53; W. Fiers and R.L. Sinsheimer. 1962. The structure of the DNA of bacteriophage ϕX174. III. Ultracentrifugal evidence for a ring structure. *J. Mol. Biol.* 5: 424–434.
76. R.L. Sinsheimer. 1966. ϕx: Multum in parvo. In *Phage and the origins of molecular biology* (ed. J. Cairns, G.S. Stent, and J.D. Watson), Cold Spring Harbor Laboratory of Quantitative Biology, Cold Spring Harbor, New York, pp. 258–264.
77. M. Meselson and F.W. Stahl. 1958. The replication of DNA in *Escherichia coli. Proc. Natl. Acad. Sci.* 44: 671–682.
78. F.L. Holmes. *Meselson, Stahl, and the replication of DNA: A history of "the most beautiful experiment in biology."* Yale University Press. New Haven, 2001.
79. E.B. Lewis. Thomas Hunt Morgan and his Legacy. Electronic Nobel Museum, 1995, www.nobel.se/medicine/laureates/1933/
80. Ray Owen, personal communication.
81. E.B. Lewis, personal communication.
82. GWB. Lecture entitled Your DNA and Mine in the Lively Arts Forum series, February 28, 1960. Tape recording in the Cal Tech archive.
83. G. and M. Beadle. *The language of life.* Doubleday, Garden City, New York, 1966.
84. R. Metzenberg, personal communication.

CHAPTER 15

1. Muriel Beadle. *These ruins are inhabited.* Doubleday, New York, 1961, pp. 9–32 ; GWB. How to be an Eastman Visiting Professor. *The American Oxonian.* Bloomington, ID Alumni Association of American Rhodes Scholars, 1959 pp.124–128.
2. M. Beadle. *These ruins are inhabited*, pp. 14–17.
3. Ibid., p. 27.
4. GWB, "How to be an Eastman Visiting Professor."
5. Ibid.
6. M. Beadle. *These ruins are inhabited*, pp. 79–81.
7. Ibid., pp.321-322.
8. Joshua Lederberg diary entry of October 26, 1958, personal communication.
9. M. Beadle. *These ruins are inhabited*, pp. 166–167.
10. Chauncey E. Beadle, Interview. *Lincoln Evening Journal*, October 30, 1958.

11. Telegrams dated October 31, 1958. CIT, DuBridge collection Box 1.3.
12. Adrian Srb to GWB, undated but likely early November, 1958, obtained from Jo Srb.
13. GWB to Srb Nov. 9 1958, obtained from Jo Srb.
14. Cal Tech's magazine *Engineering and Science*, November 1958.
15. GWB to Max Delbrück Nov. 14, 1958. CIT, Delbrück collection. Box 2.10
16. S.W. Golomb. 1982. Max Delbrück, An appreciation. *Am. Schol.* 51: 358–359
17. M. Beadle. *These ruins are inhabited*, pp. 167–169.
18. *New York Times*, October 29, 1958, p. 10.
19. L. Fleishman. *Boris Pasternak: The poet and his politics.* Harvard University Press, 1990, pp. 272–300.
20. J.L. Schecter and V.V. Luchkov, eds. *Krushchev remembers: The glasnost tapes.* Boston Little, Brown, 1990, pp. 195–196.
21. Esther Lederberg, interview, October 19, 1997.
22. David Beadle, interview, June 27, 1997.
23. M. Beadle. *These ruins are inhabited*, p. 181.
24. *Les Prix Nobel. The Nobel Prize for Physiology or Medicine.* Stockholm, pp. 32–33.
25. Irène Ephrussi Barluet in Paris, interview, November 11, 1997.
26. *L'Express.* November 6, 1958.
27. B. Ephrussi to T. Sonneborn October 6, 1958 but more likely November 6,1958, soon after the announcement that Beadle was to receive the Prize.
28. Sonneborn to Ephrussi, November 10, 1958.
29. Ephrussi to Sonneborn, November 20, 1958.
30. *L'Express.* November 6, 1958.
31. Cited by Barluet in interview, November 11, 1997.
32. GWB to Ephrussi December 30, probably 1972; obtained from Anne Ephrussi.
33. R.M. Burian, J. Gayon, and D. Zallen. 1988. The singular fate of genetics in the history of French biology, 1900–1940. *J. Hist. Biol.* 21: 357–402.
34. Ibid. and Piotr Slonimski, interview, November 12, 1997.
35. Slonimski, interview, November 12, 1997.
36. S. Luria to Delbrück, June 28, 1976. CIT, Delbrück, collection, box 14.42.
37. M. Beadle. *These ruins are inhabited*, p. 246.
38. Ibid., pp. 332–335.

CHAPTER 16

1. Max Delbrück to GWB October 15, 1958. Delbrück collection, box 3.4, CIT archive.
2. GWB to Delbrück October 25, 1958. Delbrück collection, box 3.4, CIT archive; GWB to Delbrück, January 11, 1959. Delbrück collection, box 3.4 CIT.
3. Delbrück collection, box 3.4, CIT.
4. GWB to DuBridge, March 26, 1959. DuBridge collection, box 1.3, CIT.
5. GWB to DuBridge, April 5, 1959. Same source.
6. LGWB to DuBridge, July 7, 1960. DuBridge collection, box 1.3, CIT.
7. DuBridge to GWB, July 8, 1960. DuBridge collection, box 1.3, CIT.
8. *Los Angeles Times*, January 9, 1959, p. 12.
9. Related by M. Beadle. *Where has all the ivy gone: A memoir of university life.* Doubleday, New York, 1972, p. 2, and by GWB at a luncheon that preceded his inau-

guration as University of Chicago chancellor, May 4, 1961.
10. Education article in *Time* magazine, January 13, 1961.
11. Search committee records in the Levi Collection in the University of Chicago Department of Special Collections. Folder pertaining to the GWB search in Levi papers in CHG special collections.
12. M. Beadle, *Where has all the ivy gone,* pp. 4–6.
13. University of Chicago publication dated spring 1961; Beadle collection, CHG.
14. Levi to GWB dated January 6, 1961, CHG.
15. GWB to Levi dated January 10, 1961; ibid.
16. GWB to DuBridge dated December 7, 1960. DuBridge collection, box 1.3 CIT.
17. GWB to DuBridge dated January 4, 1961; ibid.
18. G. Lloyd and E.L. Ryerson to DuBridge, January 8, 1961. DuBridge collection, box 2.1 CIT.
19. GWB to DuBridge, January 6, 1961 DuBridge collection, box 2.11 CIT.
20. GWB to DuBridge, January 9, 1961. DuBridge collection, box 2.11 CIT.
21. GWB to DuBridge, January 9, 1961. DuBridge collection, box 2.11, CIT.
22. Ray Owen to DuBridge, January 9, 1961. DuBridge collection, box 2.11, CIT.
23. DuBridge to National Institutes of Health, January, 16, 1961. DuBridge collection, box 2.11.
24. The book "What Makes Beadle Run" was by Kent Clark and the songs and lyrics by Elliot Davis. The performance was held at Culbertson Hall on March 21, 1961; the tape and script of "What Makes Beadle Run" were provided by Professor Ray Owen of Cal Tech.
25. Related by GWB at the inauguration luncheon, May 4, 1961, and retold with various embellishments on numerous occasions.
26. University of Chicago Publication dated Spring 1961. Beadle collection CHG, special collections.
27. A comprehensive pamphlet by Muriel Beadle summarized the Hyde Park–Kenwood urban renewal years from their beginnings through to 1967, CHG, special collections.
28. W. Weaver. 1961. *Private Universities and the new unity of learning.* Weaver collection, APS, Philadelphia, 1961.
29. GWB. Comments at the civic center dinner May 3, 1961. Beadle collection, box 1.3, CHG, special collections.
30. The Midway, a double drive flanked by a grassy open space, served as a thoroughfare for the 1892 World's Columbian Exposition. The short stretch referred to as the Plaisance was developed as the exposition's amusement zone; today, the Midway separates the largely undergraduate quadrangle from several of the professional schools.
31. Yale, Princeton, and Pennsylvania Universities house the others.
32. GWB. Inaugural speech May 4, 1961. CHG, box 1.4.
33. GWB. January 10, 1962. Thoughts of a new president; Trustee-faculty dinner speech, Beadle papers, box 1.9, CHG, special collections.
34. R.D. Owen, interview. October 16, 1996.
35. Muriel Beadle to R. Beadle, May 3, 1962.
36. Board of trustees meeting minutes of April 12 and June 14, 1962, CHG, special collections.
37. Unidentified participant during interview with Mrs. K. Levi and others at Quadrangle Club, September 18, 1998.

38. M. Beadle. *Where has all the ivy gone,* pp. 102–103.
39. Note in the Search Committee file in the Levi Collection in the CHG department of special collections; confirmed by K. Levi in an interview at Quadrangle Club on September 18, 1998.
40. D. Gale Johnson, interview, October 12, 2000.
41. L. Bogorad, interview, October 10, 1998.
42. E.H. Levi papers at the CHG special collections remain unavailable as of the end of 2002.
43. Comment by Mrs. Kate Levi, during interview at Quadrangle Club, September 18, 1998.
44. E.H. Levi. Profiles in the *University of Chicago* magazine, November 1967; M. Beadle, *Where has all the ivy gone?*
45. Janet Rowley, interview, August 14, 1996.
46. M. Beadle, *Where has all the ivy gone,* pp. 226–227.
47. An unidentified participant in an interview at the Quadrangle Club on September 18, 1998.
48. M. Beadle, *Where has all the ivy gone*, pp. 65–67; Redmond Barnett, interview, July 22, 1997.

CHAPTER 17

1. Trustee Committee meeting minutes, February 8, 1962, CHG, special collections.
2. GWB. 1962. Science and religion. *Christian Century* pp. 71–72.
3. David Beadle, interview, July 27, 1997.
4. GWB to N. Horowitz, October 7, 1962, in the Horowitz collection, CIT.
5. GWB's State of the University address to the Faculty Senate on November 5, 1963. Beadle collection, box 4.7, CHG, special collections.
6. D. Gale Johnson, interview, October 12, 2000.
7. Redmond Barnett, interview, July 22, 1997.
8. A pamphlet by Muriel Beadle summarizing the Hyde Park–Kenwood urban renewal years from their beginnings through 1967, CHG, special collections.
9. M. Beadle. *Where has all the ivy gone: A memoir of university life.* Doubleday, New York, 1972. p. 63.
10. M. Beadle to R. Beadle, May 3, 1962.
11. D. Beadle, interview.
12. M. Beadle to R. Beadle, August 8, 1963.
13. G. and M. Beadle. *The language of life: An introduction to the science of genetics.* Doubleday, New York, 1966.
14. M. Beadle to R. Beadle, May 23, 1965.
15. GWB to R. Beadle, August 8, 1963.
16. Minutes of trustee meetings, September 8, 1960 and April 13, 1961. CHG, special collections.
17. G. Mitman. *The state of nature. Ecology, community, and American social thought, 1900–1950.* University of Chicago Press, 1992.
18. Ibid., pp. 29–30; p. 101.
19. Ibid., pp. 105–109.

20. Robert Yuretz, interview, August 15, 1986; Eugene Goldwasser, interview, May 19, 1998; Hewson Swift, interview, May 19, 1998.
21. Ibid., as well as others.
22. D. Gale Johnson, interview, October 12, 2000.
23. Bernard Strauss, interview, August 14, 1996; Stuart Rice, interview, September 18, 1998; Lloyd Kozloff, interview, June 26, 1998; Robert Haselkorn, interview, September 19, 1998.
24. GWB's State of the University address delivered to the University Senate on November 10, 1964.
25. E.A. Evans to GWB, October 15, 1964, CHG, special collections.
26. GWB to Evans, October 21, 1964. Administration files, box 13.5, CHG, special collections.
27. L.A. Jacobson to GWB dated June 1, 1964. Beadle collection, box 13.7, CHG, special collections.
28. Ibid.
29. W.A. Wick to GWB and Levi dated July 16, 1964; ibid.
30. Memo from GWB to H.S. Bennett dated July 17, 1964; ibid.
31. University of Chicago Office of Public Relations announcement of June 22, 1965.
32. Richard C. Lewontin, interview, October 20, 1998.
33. Lewontin to GWB and Levi dated June 23, 1966. Presidential papers, box 13.7, CHG, special collections.
34. GWB remarks to trustees regarding the campaign, October 20, 1965, and State of the University address, November 9, 1965. CHG, Beadle collection, box 7.1.
35. M. Beadle, *Where has all the ivy gone*, p. 240.
36. Ibid.
37. GWB. State of the University address, November 7, 1967. CHG, Beadle collection, box 7.1.
38. Ibid.
39. For a recounting of the student uprising, see M. Beadle, *Where has all the ivy gone*, pp. 265–281.
40. *San Francisco Chronicle*, May 14, 1966 and *Time* magazine, May 20, 1966, p. 72.
41. Beadle presidential papers, box 2.4, CHG.
42. M. Beadle, *Where has all the ivy gone*, p. 279.
43. Ibid., p. 280.
44. GWB. Speech at groundbreaking for Regenstein Library, October 23, 1967. CHG, presidential papers, box 6.8.
45. M. Beadle, *Where has all the ivy gone*, p. 356.
46. Ibid., pp. 125–140; pp. 194–195; pp. 213–215.
47. *Chicago Daily News*, June 28, 1967.
48. *Chicago Tribune*, June 28, 1967.
49. From the director of the Regenstein Library (name is unreadable but could be retrieved) to GWB, July 8, 1968. Presidential files, box 1.4, CHG, special collections.
50. Letter from John M. Shlien, June 29, 1967.
51. M. Beadle, *Where has all the ivy gone*, p. 356.
52. D. Beadle, interview, June 27, 1997.
53. *The Chicago Maroon*, June 30, 1967.
54. GWB speech at trustee-faculty dinner, January 10, 1968, CHG, Beadle collection, box 7.5.

55. GWB to James D. Watson, June 12, 1968 (obtained from Watson).
56. *Los Angeles Times,* July 5, 1967, in box 1.4 of the president's papers, CHG, special collections.
57. Profiles, *The University of Chicago Magazine*, November, 1967, pp. 30–32.
58. GWB to Delbrück, November 11, 1968.
59. Undated note from Levi to GWB. Box 31.5 of the administrative papers, CHG, special collections.

CHAPTER 18

1. GWB to Max Delbrück, July 11, 1969. Box 2.10, Delbrück collection, CIT.
2. M. Beadle. *A child's mind: How children learn during the critical years from birth to age five.* Doubleday, New York, 1970.
3. D. Gale Johnson, personal communication.
4. GWB to Delbrück, July 11, 1969.
5. Ibid.
6. Press release, Chicago Horticultural Society, May 9, 1968.
7. GWB to Delbrück, July 11, 1969.
8. GWB. 1968. President's comments. *Garden talk*, Chicago Horticultural Society, August/September, pp. 2–3, CHG.
9. GWB. 1970. The botanic garden's corn field. *Garden Talk* 11: 12–13, CHG.
10. GWB to Delbrück, July 11, 1969.
11. Muriel Beadle to Curt Stern, May 17, 1970, Curt Stern papers, MS collection 5, APS.
12. GWB to Barbara McClintock, January 22, 1972 and B. McClintock to GWB, February 14, 1972, McClintock papers, APS.
13. K.O. Pope, M.E.D. Pohl, J.G. Jones, D.L. Lentz, C. von Nagy, F.J. Vega, and I.R. Quitmyer. 2001. Origin and environmental setting of ancient agriculture in the lowlands of Mesoamerica. *Science* 292: 1370–1373.
14. GWB. 1980. The ancestry of corn. *Sci. Am.* 242: 112–119.
15. Kuntze, 1904, quoted by W.C. Galinat. 1971. The origin of maize, *Annu. Rev. Genet.* 5: 447–478.
16. Pairing efficiency depended somewhat on the extent of structural homology between the corn chromosomes and those of the teosinte strain used.
17. GWB. 1932. Studies of *Euchlaena* and its hybrids with *Zea* I. Chromosome behavior in *Euchlaena mexicana* and its hybrids with *Zea mays. Zeitschrift fur Induktive Abstammungs und Vererbungslehre*, LXII: 291–304 and plates I–IV.
18. R.A. Emerson and GWB. 1932. Studies of *Euchlaena* and its hybrids with *Zea* II. Crossing over between the chromosomes of *Euchlaena* and *Zea. Zeitschrift fur Induktive Abstammungs und Vererbungslehre* LXVII: 306–315; GWB. 1932. The relation of crossing over to chromosome association in Zea-Euchlaena hybrids. *Genetics* 17: 481–501.
19. GWB. 1939. Teosinte and the origin of maize. *J. Hered.* 30: 245–247.
20. P.C. Mangelsdorf and R.G. Reeves. 1938. The origin of maize. *Proc. Natl. Acad. Sci.* 24: 303–312.
21. E.J. Kahn, Jr. 1984. The Staffs of Life I: The Golden Thread. *The New Yorker,* June 18, pp. 46–88; H.H. Iltis. 2000. Homeotic sexual translocations and the origin of maize

(*Zea mays,* Poaceae): A new look at an old problem. *Econ. Botany* 54: 7–42.
22. E. Anderson. *Plants, man and life.* Little Brown, New York, 1952. Reissued by University of California Press, Berkeley, 1967.
23. GWB to R.A. Emerson, February 18, 1939. Box 20, department of plant breeding records, COR.
24. R.A. Emerson to GWB, March 25, 1939. Box 3/18, Beadle collection, CIT.
25. GWB. 1978. The origin of maize. In *Genes, cells, and behavior* (ed. N.H. Horowitz and E. Hutchings, Jr.), pp. 81–87. W.H. Freeman, San Francisco.
26. Robert Haselkorn, personal communication; H.H. Iltis, "Homeotic sexual translocations."
27. P.C. Mangelsdorf. 1947. The origins and evolution of maize. *Adv. Genet.* 1: 161–207; P.C. Mangelsdorf. 1958. Reconstructing the ancestor of corn. *Proc. Am. Philos. Soc.* 102: 454–463.
28. P.C. Mangelsdorf. 1959. Reconstructing the ancestor of corn. *Annu. Rep. Smithsonian Inst. 1959.*
29. W.C. Galinat. 1971. The origin of maize. *Annu. Rev. Genet.* 5: 447–478.
30. GWB, "Origin of maize."
31. H.G. Wilkes. *Teosinte: The closest relative of maize.* The Bussey Institute, Harvard University, Cambridge, Massachusetts, 1967.
32. G.W. Beadle. 1975. Biochemical genetics: Reflections. In *Three lectures, January 15–17, 1975.* The Edna H. Drane Visiting Lectureship, University of Southern California School of Medicine, CHG.
33. GWB, "Origin of maize."
34. Mangelsdorf, "Reconstructing the ancestor of corn," 1958; Mangelsdorf, "Reconstructing the ancestor of corn," 1959.
35. GWB. 1972. The mystery of maize. *Field Museum Nat. History Bull.* 43: 2–11, CHG; GWB. *The mystery of maize—An interdisciplinary saga.* Donald Forsha Jones lecture, September 14, 1972 at the Connecticut Agricultural Experiment Station, unpublished, CIT.
36. GWB to H.H. Iltis, January 26, 1970, courtesy of H.H. Iltis.
37. H.H. Iltis, personal communication.
38. E.J. Kahn, Jr., "The staffs of life," 1984.
39. Hugo Iltis. *Life of Mendel.* English translation, W.W. Norton, New York, 1932 (from German original, 1924).
40. H.H. Iltis. *The maize mystique—A reappraisal of the origin of corn.* First issued informally and circulated in January 1971, and later as a contribution from the University of Wisconsin Herbarium, no. 5. April 1985.
41. H.H. Iltis to GWB, January 21, 1970, courtesy of H.H. Iltis.
42. E.J. Kahn, Jr., "The staffs of life," 1984.
43. W.C. Galinat, "Origin of maize," 1971; GWB, *The mystery of maize*, 1972.
44 H.H. Iltis. Homeotic sexual translocations.
45. Muriel Beadle to R. Beadle, November 15, 1970, courtesy of RB.
46. GWB. 1973. Teosinte mutation hunt 1972. *Garden Talk* Feb/March 10–12, CHG.
47. GWB, "Ancestry of corn."
48. Galinat, "Origin of maize," 1971.
49. GWB to B. McClintock, February 17, 1972, McClintock collection, APS.
50. GWB, "Origin of maize."

51. N.C. Comfort. *The tangled field: Barbara McClintock's search for the patterns of genetic control.* Harvard University Press, 2001, Chapter 9.
52. B. McClintock to GWB, February 14, 1972, McClintock collection, APS.
53. GWB to McClintock, January 22, 1972, McClintock collection, APS.
54. GWB to McClintock, February 26, 1972, McClintock collection, APS.
55. McClintock to GWB, February 14, 1972.
56. GWB to McClintock, February 17, 1972.
57. GWB. Comments on Harvard Study Meeting on Archeological Maize, June 14, 1972. File 1/2, Beadle collection, Iowa State University archives; H.H. Iltis, personal communication.
58. P.C. Mangelsdorf. 1973. Field Museum of Natural History Bulletin. 44: 16; P.C. Mangelsdorf. *Corn: Its origin, evolution and improvement.* Belknap Press of Harvard University Press, Cambridge, Massachusetts, 1974.
59. GWB. 1975. Of maize and men. *Quart. Rev. Biol.* 50: 67–69.
60. GWB, *Mystery of maize,* 1972; GWB. 1975. Biochemical genetics: Reflections. In *Three lectures, January 15–17, 1975.* The Edna H. Drane Visiting Lectureship, University of Southern California, School of Medicine, CIT; GWB. The Origin of Maize, a talk at the Symposium on Genes, Cells, And Behavior; A View of Biology Fifty Years Later, CIT, 50th Anniversary Symposium, November 1–3, 1978, biology division, CIT.
61. Delbrück to GWB. January 12, 1977. Box 2/10 Delbrück collection, CIT.
62. GWB to Delbrück, January 20, 1977. Box 2/10 Delbrück collection, CIT.
63. M. Beadle. *The cat: History, biology, and behavior.* Simon and Schuster, New York, 1977.
64. GWB to R. Beadle, May 15, 1976; C.R. Worrall to GWB, July 30, 1977; GWB to Mr. Hirsch, at the University of Nebraska, August 5, 1977; GWB to R. Beadle, August 5, 1977. RB.
65. GWB to Robert Sinsheimer, February 10, 1977. Box 93/4, biology division collection, CIT.
66. GWB to Max and Manny Delbrück, December 4, 1977. Box 2/10, Delbrück collection, CIT.
67. M. Beadle to R. Beadle, August 24, 1977. RB.
68. GWB interview, Dr. Bernice Eiduson, June 6, 1978. CIT, Beadle collection, box 36.16.
69. Redmond Barnett, interview, May 20, 2001.
70. M. Beadle. *A child's mind.*
71. M. Beadle. *A nice neat operation.* Doubleday, New York, 1975.
72. M. Beadle. *The cat: History, biology, and behavior.*
73. GWB to Sinsheimer, February 10, 1977. CIT, biology division collection, box 93/4.
74. GWB to R. Beadle, May 30, 1978. RB.
75. GWB to Norman Horowitz, November 7, 1978. Folder 93.4, biology division collection, CIT.
76. Recollection of Paul Berg who also was a member of the Welch Foundation board at the time.
77. GWB to Horowitz, November 6, 1980. Folder 4.18, Beadle collection, CIT.
78. GWB to R. Beadle, November 11, 1980. RB.
79. GWB to Horowitz, November 6, 1980.
80. H.H. Iltis. 1972. Shepherds leading sheep to slaughter: The extinction of species and the destruction of ecosystems. *Am. Biol. Teacher* 34: 201–205 and 221; H.G. Wilkes.

1972. Maize and its wild relatives. *Science* 177: 1071–1077; GWB to H.H. Iltis and other memoranda of November 12, 1973, in H.H. Iltis, courtesy of H.H. Iltis; GWB, "Ancestry of corn."

81. GWB, "Ancestry of corn."
82. GWB. 1981. Origin of corn: Pollen evidence. *Science* 213: 890–892.
83. Letters between GWB and Hugh H. Iltis from January 24, 1980 through January 20, 1982. Personal files of Iltis.
84. M. Beadle to Norman and Pearl Horowitz, May, 20 1980. RB.
85. M. Beadle. 1980. "Indomitable spirit battles back after stroke." *Chicago Sunday Sun-Times,* December 21, 1980.
86. M. Beadle to "family," November 20, 1980, RB.
87. P.C. Mangelsdorf to GWB, July 30, 1981. Box 5/23, Beadle collection, CIT.
88. P.C. Mangelsdorf to GWB, August 19, 1981. Box 5/23, Beadle collection, CIT.
89. P.C. Mangelsdorf, Jr., personal communication.
90. McClintock to GWB, February 14, 1972.
91. GWB to McClintock, February 17, 1972.
92. J. Doebley. 1992. Mapping the genes that made maize. *Trends Genet.* 8: 302–307; J. Doebley. 2001. George Beadle's other hypothesis: One-gene, one-trait. *Genetics* 158: 487–493.
93. Ibid.; J. Doebley and A. Stec. 1991. Genetic analysis of the morphological differences between maize and teosinte. *Genetics* 129: 285–295.
94. J. Doebley, personal communication.
95. GWB, "Mystery of maize"; GWB, "Ancestry of corn."
96. H.H. Iltis. 1983. From teosinte to maize: The catastrophic sexual transmutation. *Science* 222: 886–894.
97. E.M. Meyerowitz. 1994. De-evolution and re-evolution of maize. *Curr. Biol.* 4: 127–130; E.A. Kellogg. 1997. Plant evolution: The dominance of maize. *Curr. Biol.* 7: R411–R413.
98. J. Doebley, A. Stec, and L. Hubbard. 1997. The evolution of apical dominance in maize. *Nature* 386: 485–488.
99. J. Doebley and R.-L. Wang. 1997. Genetics and the evolution of plant form: An example from maize. *Cold Spring Harbor Symp. Quant. Biol.* 62: 361–367.
100. J. Dorweiler, A. Stec, J. Kermicle, and J. Doebley. 1993. Teosinte glume architecture 1: A genetic locus controlling a key step in maize evolution. *Science* 262: 233–235.
101. S. White and J. Doebley. 1998. Of genes and genomes and the origin of maize. *Trends Genet.* 14: 327–332.
102. Doebley, "Mapping the genes,"; Meyerowitz, "De-evolution and re-evolution."
103. N. Lauter and J. Doebley. 2002. Genetic variation for phenotypically invariant traits detected in teosinte: Implications for the evolution of novel forms. *Genetics* 160: 333–342.
104. J. Doebley, personal communication.

EPILOGUE

1. GWB. 1945. Biochemical genetics. *Chem. Rev.* 37: 15–96; GWB. Genetics and modern biology. Jayne Lectures for 1962. American Philosophical Society, Philadelphia, 1963;

GWB. 1966. Biochemical genetics: Some recollections. In *Phage and the origins of molecular biology* (ed. J. Cairns, G.S. Stent, and J.D. Watson), Cold Spring Harbor Laboratory of Quantitative Biology, Cold Spring Harbor, New York; GWB. 1974. Biochemical genetics: Recollections. *Annu. Rev. Biochem.* 43: 1–13; GWB. 1975. Biochemical genetics: Reflections. In *Three lectures. January 15–17, 1975.* The Edna H. Drane Visiting Lectureship, University of Southern California, School of Medicine, CHG.

2. P. Berg and M. Singer. *Dealing with genes: The language of heredity.* University Science Books, Mill Valley, California, 1992.
3. T.S. Kuhn. *The structure of scientific revolutions.* University of Chicago Press, 1962.
4. J.S. Fruton. *A skeptical biochemist.* Harvard University Press, 1992.
5. GWB, "Biochemical genetics."
6. G. and M. Beadle. *The language of life.* Doubleday, New York, 1966.
7. M. Beadle to family, November 13, 1983. Box 4/18, Beadle collection, CIT.
8. M. Beadle to family, November 20, 1980. RB.
9. M. Beadle to Norman and Pearl Horowitz, June 15, 1981. RB.
10. M. Beadle to family, October 6, 1981. RB.
11. GWB to R. Beadle, November 2, 1981. RB.
12. GWB to David Perkins, November 20, 1981. Courtesy of David Perkins.
13. GWB to B. McClintock, November 20, 1981. McClintock files, APS.
14. Redmond Barnett, interview, July 22, 1997; David Savada, interview, spring 1997.
15. M. Beadle to Edith and Chauncey Harris, November 16, 1982, files of Edith Y. Harris.
16. M. Beadle to family and friends, July 13, 1984. RB.
17. William Mobley, a neurologist familiar with Alzheimer's behavior, likened these intermittent periods of awareness and amnesia to the behavior of a loose-fitting light bulb that flickers on and off with unpredictable periodicity.
18. M. Beadle to R. Beadle, March, 1986. RB.
19. M. Beadle to family and friends. June 10, 1985. RB.
20. Quoted by Joseph J. Ellis in *American sphinx.* New York, Vintage Books, 1998, p. 278.
21. D. Beadle, interview, August 13, 1997.
22. R. Barnett, email interview, May 20, 2001.
23. R. Barnett, interview, July 22, 1997.
24. GWB. Will filed for probate, July 21, 1989, Superior Court of California, County of Los Angeles.
25. Recording of the memorial service kindly provided by Ray Owen and deposited in the CIT archives.
26. J.D. Watson. Keynote Speech at the Dedication of the George W. Beadle Center on September 22, 1995. Proceedings published in Nebraska, fall 1995.

Index

L

M